Advances in Intelligent Systems and Computing

Volume 766

The series "Advances in Intelligent Systems and Computing" contains publications on theory, applications, and design methods of Intelligent Systems and Intelligent Computing. Virtually all disciplines such as engineering, natural sciences, computer and information science, ICT, economics, business, e-commerce, environment, healthcare, life science are covered. The list of topics spans all the areas of modern intelligent systems and computing such as: computational intelligence, soft computing including neural networks, fuzzy systems, evolutionary computing and the fusion of these paradigms, social intelligence, ambient intelligence, computational neuroscience, artificial life, virtual worlds and society, cognitive science and systems, Perception and Vision, DNA and immune based systems, self-organizing and adaptive systems, e-Learning and teaching, human-centered and human-centric computing, recommender systems, intelligent control, robotics and mechatronics including human-machine teaming, knowledge-based paradigms, learning paradigms, machine ethics, intelligent data analysis, knowledge management, intelligent agents, intelligent decision making and support, intelligent network security, trust management, interactive entertainment, Web intelligence and multimedia.

The publications within "Advances in Intelligent Systems and Computing" are primarily proceedings of important conferences, symposia and congresses. They cover significant recent developments in the field, both of a foundational and applicable character. An important characteristic feature of the series is the short publication time and world-wide distribution. This permits a rapid and broad dissemination of research results.

**** Indexing: The books of this series are submitted to ISI Proceedings, EI-Compendex, DBLP, SCOPUS, Google Scholar and Springerlink ****

More information about this series at http://www.springer.com/series/11156

Atilla Elçi · Pankaj Kumar Sa ·
Chirag N. Modi · Gustavo Olague ·
Manmath N. Sahoo · Sambit Bakshi
Editors

Smart Computing Paradigms: New Progresses and Challenges

Proceedings of ICACNI 2018, Volume 1

Editors
Atilla Elçi
Engineering Faculty
Aksaray University
Sağlık, Aksaray, Turkey

Chirag N. Modi
National Institute of Technology
Goa, India

Manmath N. Sahoo
Department of Computer Science
and Engineering
National Institute of Technology Rourkela
Rourkela, Odisha, India

Pankaj Kumar Sa
Department of Computer Science
and Engineering
National Institute of Technology Rourkela
Rourkela, Odisha, India

Gustavo Olague
CICESE
Ensenada, Baja California, Mexico

Sambit Bakshi
Department of Computer Science
and Engineering
National Institute of Technology Rourkela
Rourkela, Odisha, India

ISSN 2194-5357 ISSN 2194-5365 (electronic)
Advances in Intelligent Systems and Computing
ISBN 978-981-13-9682-3 ISBN 978-981-13-9683-0 (eBook)
https://doi.org/10.1007/978-981-13-9683-0

This Springer imprint is published by the registered company Springer Nature Singapore Pte Ltd.
The registered company address is: 152 Beach Road, #21-01/04 Gateway East, Singapore 189721, Singapore

Preface

It is a sheer pleasure announcing receipt of an overwhelming response from academicians and researchers of reputed institutes and organizations of the country and abroad for joining us in the 6th International Conference on Advanced Computing, Networking, and Informatics (ICACNI 2018), which makes us feel that our endeavor is successful. The conference organized jointly by the Department of Computer Science and Engineering, National Institute of Technology Silchar, and Centre for Computer Vision and Pattern Recognition, National Institute of Technology Rourkela, during June 04–06, 2018, certainly achieves a landmark toward bringing researchers, academicians, and practitioners in the same platform. We have received more than 400 research articles and very stringently have selected through peer review the best articles for presentation and publication. We, unfortunately, could not accommodate many promising works as we strove to ensure the highest quality and adhere to the recommendations of the expert reviewers. We are thankful to have the advice of dedicated academicians and experts from industry and the eminent academicians involved in providing technical comments and quality evaluation for organizing the conference with a planned schedule. We thank all people participating and submitting their works and having continued interest in our conference for the sixth year. The articles presented in the two volumes of the proceedings discuss the cutting-edge technologies and recent advances in the domain of the conference. We conclude with our heartiest thanks to everyone associated with the conference and seeking their support to organize the 7th ICACNI 2019 at the Indian Institute of Information Technology Kalyani, India, during December 20–21, 2019.

Silchar, India
November 2019

Atilla Elçi
Pankaj Kumar Sa
Chirag N. Modi
Gustavo Olague
Manmath N. Sahoo
Sambit Bakshi

Contents

About the Editors

Atilla Elçi is an Emeritus Full Professor and Chairman of the Department of Electrical and Electronics Engineering at Aksaray University, Turkey (August 2012–September 2017). He previously served as a Full Professor and Chairman of Computer and Educational Technology at Süleyman Demirel University, Isparta, Turkey (May 2011–June 2012). Dr. Elçi graduated with a degree in Electrical Engineering from METU, Turkey and completed his M.Sc. and Ph.D. in Computer Science at Purdue University, USA. He has 71 papers, 11 books, 10 conference publications and 4 talks to his credit. His research areas include semantic web technology; information security/assurance; semantic robotics; programming systems and languages; and semantics in education.

Pankaj Kumar Sa received his Ph.D. degree in Computer Science in 2010. He is currently serving as an Associate Professor in the Department of Computer Science and Engineering, National Institute of Technology Rourkela, India. His research interests include computer vision, biometrics, visual surveillance, and robotic perception. He has co-authored a number of research articles in various journals, conferences, and book chapters and has co-investigated research and development projects funded by SERB, DRDOPXE, DeitY, and ISRO. He has received several prestigious awards and honors for his contributions in academics and research.

Chirag N. Modi is an Assistant Professor, Department of Computer Science and Engineering, NIT Goa, Farmagudi. He received his B.E. (Computer Engineering) from BVM Engineering College, Gujarat, India and completed his M.Tech. and Ph.D. in Computer Engineering at SVNIT Surat, Gujarat. His research areas include cloud computing security; network security, cryptography and information security; and secured IoTs, data mining, and privacy preserving data mining. He has published several journal papers and book chapters.

Gustavo Olague received his Ph.D. in Computer Vision, Graphics and Robotics from the INPG (Institut Polytechnique de Grenoble) and INRIA (Institut National de Recherche en Informatique et Automatique) in France. He is currently a

Professor at the Department of Computer Science at CICESE (Centro de Investigación Científica y de Educación Superior de Ensenada) in Mexico, and the Director of its EvoVisión Research Team. He is also an Adjoint Professor of Engineering at UACH (Universidad Autonóma de Chihuahua). He has authored over 100 conference proceedings papers and journal articles, coedited two special issues in Pattern Recognition Letters and Evolutionary Computation, and served as co-chair of the Real-World Applications track at the main international evolutionary computing conference, GECCO (ACM SIGEVO Genetic and Evolutionary Computation Conference). Prof. Olague has received numerous distinctions, among them the Talbert Abrams Award presented by the American Society for Photogrammetry and Remote Sensing (ASPRS); Best Paper awards at major conferences such as GECCO, EvoIASP (European Workshop on Evolutionary Computation in Image Analysis, Signal Processing and Pattern Recognition) and EvoHOT (European Workshop on Evolutionary Hardware Optimization); and twice the Bronze Medal at the Humies (GECCO award for Human-Competitive results produced by genetic and evolutionary computation). His main research interests are in evolutionary computing and computer vision.

Manmath N. Sahoo is an Assistant Professor at the Department of Computer Science and Engineering, National Institute of Technology Rourkela, India. His research interest areas are in fault tolerant systems, operating systems, distributed computing, and networking. He is a member of the IEEE, Computer Society of India and the Institutions of Engineers, India. He has published several papers in national and international journals.

Sambit Bakshi is currently with Centre for Computer Vision and Pattern Recognition of National Institute of Technology Rourkela, India. He also serves as an Assistant Professor in the Department of Computer Science & Engineering of the institute. He earned his Ph.D. degree in Computer Science & Engineering in 2015. His areas of interest include surveillance and biometric authentication. He currently serves as associate editor of International Journal of Biometrics (2013 –), IEEE Access (2016 –), Innovations in Systems and Software Engineering: A NASA Journal (2016 –), Expert Systems (2017 –), and PLOS One (2017 –). He has served/is serving as guest editor for reputed journals like Multimedia Tools and Applications, IEEE Access, Innovations in Systems and Software Engineering: A NASA Journal, Computers and Electrical Engineering, IET Biometrics, and ACM/Springer MONET. He is serving as the vice-chair for IEEE Computational Intelligence Society Technical Committee for Intelligent Systems Applications for the year 2019. He received the prestigious Innovative Student Projects Award 2011 from Indian National Academy of Engineering (INAE) for his master's thesis. He has more than 50 publications in reputed journals, reports, and conferences.

Advanced Image Processing Methodologies–I

Estimation of Tyre Pressure from the Characteristics of the Wheel: An Image Processing Approach

V. B. Vineeth Reddy, H. Ananda Rao, A. Yeshwanth, Pravin Bhaskar Ramteke and Shashidhar G. Koolagudi

Abstract Improper tyre pressure is a safety issue that falls prey to ignorance of users. But a drop in tyre pressure can result in the reduction of mileage, tyre life, vehicle safety and performance. In this paper, an approach is proposed to measure the tyre pressure from the image of the wheel. The tyre pressure is classified into under pressure and normal pressure using load index, tyre type, tyre position and ratio of compressed and uncompressed tyre radius. The efficiency of the feature is evaluated using three classifiers namely Random Forest, AdaBoost and Artificial Neural Networks. It is observed that the ratio of radii plays a major role in classifying the tyres. The proposed system can be used to obtain a rough idea on whether the tyre should be refilled or not.

Keywords Image processing · Random forest · AdaBoost · Hough gradient · Neural networks

1 Introduction

Day by day, increase in automobiles and their technologies have led to innovative ideas in order to satisfy human needs. Roads being one of the most important modes of transportation, and vehicles, an integral part of it, require proper maintenance. One

---

V. B. Vineeth Reddy (✉) · H. Ananda Rao · A. Yeshwanth · P. B. Ramteke
S. G. Koolagudi
National Institute of Technology Karnataka, Surathkal 575025, India
e-mail: vineethvatti@gmail.com

H. Ananda Rao
e-mail: anandarao.h8@gmail.com

A. Yeshwanth
e-mail: akurathiyeshwanth@gmail.com

P. B. Ramteke
e-mail: ramteke0001@gmail.com

S. G. Koolagudi
e-mail: koolagudi@nitk.edu.in

A. Elçi et al. (eds.), *Smart Computing Paradigms: New Progresses and Challenges*, Advances in Intelligent Systems and Computing 766,
https://doi.org/10.1007/978-981-13-9683-0_1

such criterion in maintaining vehicles is inflating tyres. Tyres usually deflate when vehicles are run through on non-uniform roads, change in temperature or prolonged usage.

Loss in tyre stability adversely affects the stopping, cornering and handling of the tyre if it is over or under inflated. The tyres will start wearing unevenly if it is not inflated properly. The tyre becomes flatter than determined while in contact with the ground and cannot maintain its shape if it is under inflated, which results in lower life of tyres due to increased heat.

Over inflation causes the shape of the tyre to be deformed, this makes the tyre lose traction and also reduce its footprint on the road. An over inflated tyre cannot react as expected to potholes or debris as the tyres are more stiff. When accurately inflated, the tyres provide dependable traction, superior handling, and also a pleasant ride. Many people constantly find themselves going to the petrol bunk to fill air in their vehicles even when the tyre pressure is in the optimum range. As a solution to this problem, a system is developed in which a user can instantaneously and effortlessly find out whether the tyres need to be refilled or not. The basic idea is to take an image of the tyre to be checked and calculate the necessary details such as the compressed and uncompressed radii. The information is then matched by the machine with the data that with which it has already been trained. This matching, enables the machine to predict whether the air pressure is sufficient or needs a refill.

The paper is structured into 5 sections. Section 2 discusses about the existing methodologies of tyre pressure control. The proposed methodology is discussed in Sect. 3. Section 4 presents the results obtained in the experiments and gives an overview of the experimental. The work is concluded in Sect. 5 with some future directions.

2 Existing Work

The usage of vehicles, is ballooning each day, this may be in the form of more vehicles used, or the same old vehicles being used for a longer time. In the latter case there is a constant deterioration of the tyre, thus reducing its pressure too. There have been a lot of approaches to monitor this variations of the tyre pressure. One of the method is to constantly check the tyre pressure and temperature readings and based on these readings decide if the tyre is properly inflated or not [1]. A tyre pressure monitoring system (TPMS), an electronic system is used to see the fluctuations of the tyre pressure and provide the driver with proper warnings. Several other methods like using a wireless communication to detect the tyre pressure were also proposed [2]. The contact pressure is also analyzed to judge the air pressure of the tyre. As the correlation of these both ranges is observed to be between 0.803 and 0.997, one can be used to estimate the other [3]. It checks the tyre pressure automatically using a pressure gauge and signals the compressor to be turned on. Whenever the tyres are not properly inflated it takes the air from the atmosphere and fills it in the tyre. This enhances vehicle performance, human safety, fuel efficiency.

The methods proposed here although they provide methods to find the tyre pressure and also suggest methods to properly inflate the tyres a realtime solution isn't found yet. In this paper we discuss a solution which is cost effective and a novel idea which gives real-time information of whether the tyre is properly inflated.

3 Proposed Method

The objective here is to devise a method to predict the air pressure in the tyre, thereby reducing the waiting time at petrol stations and prolonging the lifetime of tyres. The proposed method predicts whether the tyre needs to be filled or not from the image of the tyre. The method can divided into the following sub-problems:

3.1 Pre-processing

Pre-processing is mainly done to enhance image contrast and reduce the noise of the image. The images are resized (Fig. 1a) to 640*480 to have a better control over quality. The images are then converted to gray-scale (Fig. 1b). Gaussian smoothing (Fig. 1c) is applied on the gray-scale images to reduce noise. Finally the images are thresholded using Otsu's Binarization (Fig. 1d) to separate tyre from the background.

3.2 Feature Extraction

Centre Detection Hough gradient method for circle detection is used to detect the rim of the wheel from which the centre of the tyre is detected [4]. The image depicting the detection of circles is shown in Fig. 1e.

Radii After obtaining the centre of the circle, the next aim is to extract the compressed and the uncompressed radii from the image. Figure 1f depicts the final image from which the radii are calculated.

- **Compressed Radius**: The image is scanned from bottom to top and a horizontal line is drawn at the pixel where there is a drastic change (relative difference) in the intensity of the pixel. This gives the first point of contact of the tyre with the ground. Distance between the centre and the contact point is the compressed radius of the tyre.
- **Uncompressed Radius**: The image is then scanned starting from the centre and two vertical lines are drawn at the pixels where there is an intensity difference. This change is measured by the relative difference between the pixels. This gives us the approximate horizontal end points of the tyre.

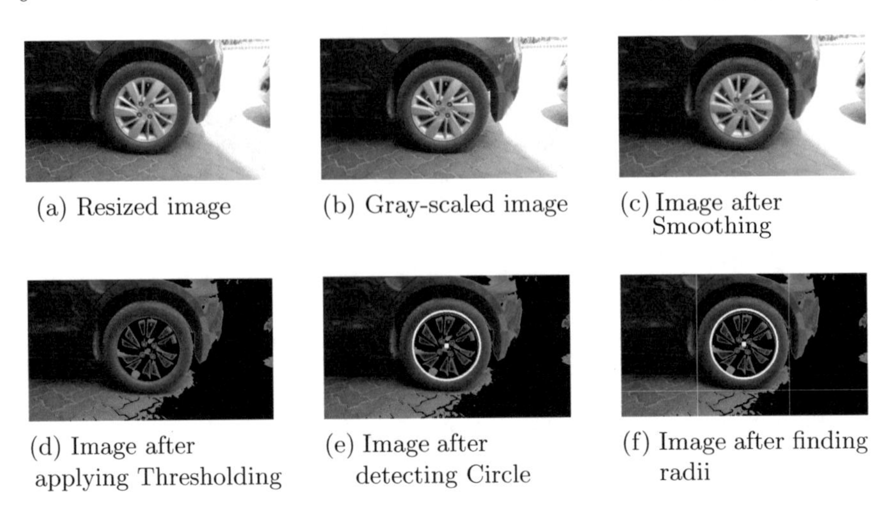

Fig. 1 Process followed to detect radii

The values of compressed and uncompressed radius may vary from image to image. Hence, to maintain consistency the ratio of the radius is taken. This ratio is then given as feature to the classifier.

3.3 Classification

The features which have an impact on the classification are load index, tyre type, tyre position and ratio. Using these features the tyres are classified into two classes, (1) Tyre pressure below required range (2) Tyre pressure in the optimum range. Both tyre type and load index is taken as an input from the user. Three different classifiers are taken into consideration, and their accuracies are compared. The different classifiers used are:

- **Random Forest**: Random forests are an ensemble learning method for classification that operate by constructing a multitude of decision trees at training time and outputting the class that is the mode of the classes [5]. For a classification problem with p features at each split, usually $\sqrt{p}$ features are used. Improved results are obtained with higher value of $n\_estimators$ (number of trees) until a certain value after which the results stabilize. Since we have 4 features $max\_features$ is set to 2 and $n\_estimators$ is set to 100.
- **AdaBoost**: AdaBoost combines relatively weak and inaccurate classifiers and creates a highly accurate classifier [6]. In this scenario Decision Trees are used as base estimators. $n\_estimators$ is set to 100 and max_depth is set to 1.
- **Neural Networks**: Artificial Neural Networks [7] are made up of multiple nodes which are fully connected and the output of each neuron (activation) is sent to

the other connected layers. And the output layer can be used to classify the input image. The architecture is composed of four input neurons, two hidden layers with ten and four neurons respectively and output layer having one neuron. ReLU is the activation function and Back-propagation [8] is the learning algorithm used in this architecture.

4 Results

A data-set of 90 images is collected. The data-set includes Tyre type, Tyre make, load, pressure, compressed radii, uncompressed radii, ratio and result fields respectively.

While classifying using Random Forest or AdaBoost classifiers only some of these features are used. Not all the features are given equal importance for classifying the data. The importance of each feature as decided by the classifier [9]. Table 1 gives the importance of each feature in both the classifiers namely Random Forest and AdaBoost. It is observed that Ratio(r) of the compressed to the uncompressed radius has the highest importance in both the classifiers. Load index and Tyre type also has decent contribution to the classification, where as Tyre type plays a meagre role.

As the data-set is small, separate sets for training and testing is not possible. To overcome this, k-fold cross-validation is used [10]. If the induction algorithm for the given data-set is stable, irrespective of the number of folds (k) the accuracy of the cross-validation estimates should be similar. But the induction algorithms used here are approximately stable.

The Random Forest classifier gives the best accuracy of 62.22% of accuracy for 2-fold cross validation while the accuracy keeps reducing as k decreases despite a few random spikes as shown in Table 2. As, k increases the Random Forest classifier tries to over fit the data thus decreasing the accuracy. The AdaBoost classifier's accuracy reduces with the decreasing k. But it is observed that this classifier performs poorly when the size of training sample is small [11]. Neural Networks consistently provide better results than the other two classifiers but sue to over fitting in this case also the accuracy improved for a while and then gets worse, which might be due to the low learning rate which prevents the classifier from having a stable convergence. Table 2 shows the results for the three different classifiers for the number of folds varying from 2 to 12. Neural forest gives the best results of around 68%.

Table 1 Feature importance

	Random Forest	AdaBoost
Tyre type	0.19	0.12
Load index	0.24	0.16
Tyre position	0.07	0.01
Ratio	0.50	0.71

Table 2 Comparing results between different classifiers and number of folds

Folds (k)	Random forest	AdaBoost	Neural networks
2	62.22	58.88	63.33
3	58.88	56.66	67.77
4	53.31	58.84	65.41
5	52.22	56.66	67.77
6	53.33	55.55	65.55
7	52.19	53.29	61.17
8	58.90	56.43	64.10
9	53.33	54.44	64.44
10	58.88	54.44	65.55
11	53.03	55.42	65.15
12	50.14	55.05	65.92
13	44.13	50.01	59.34
14	49.31	51.36	58.84
15	43.33	45.55	58.88

5 Conclusion

Three alternate classifiers for segregate tyres based on the tyre pressure have been presented here: Random Forest, AdaBoost and Neural Networks. These classifiers gave utmost importance to Ratio of radii. Other features used were Load Index, Tire Type and Tyre Position in the decreasing order of importance. Among the three classifiers Neural networks gave exceptionally good results. A accuracy of 68% was obtained using Neural networks. With this accuracy a user can get a rough idea of whether to fill the pressure in the tyre or not. Main reason for the loss in accuracy are small data-set and impact of shadows and dirt on the image. It may be mentioned here that, though the proposed method was fairly good, many more challenges are yet to be completed. These include improving the accuracy of centre detection, applying Convolutional Neural Networks to classify data, incorporating different vehicle types such as Trucks, Motor bikes and so on.

References

1. Mathai, A., Vanaja Ranjan, P.: A new approach to tyre pressure monitoring system. Int. J. Adv. Res. Electr. Electron. Instrum. Eng. (2015)
2. Mule, S., Ingle, K.S.: Review of wireless tyre pressure monitoring system for vehicle using wireless communication. Int. J. Innov. Res. Comput. Commun. Eng. (2017)
3. Minca, C.: The determination and analysis of tire contact surface geometric parameters. Review of the Air Force Academy (2015)

4. Shetty, P.: Circle detection in images. Ph.D. thesis, San Diego State University, Department of Electrical Engineering (2011)
5. Blaser, R.: Piotr Fryzlewicz random rotation ensembles. J. Mach. Learn. Res. (2015)
6. Wang, R.: AdaBoost for feature selection, classification and its relation with SVM*, a review. In: International Conference on Solid State Devices and Materials Science (2012)
7. Philipp, G., Carbonell, J.G.: Non-parametric neural networks. In: ICLR (2017)
8. Tu, J.V.: Advantages and disadvantages of using artificial neural networks versus logistic regression for predicting medical outcomes. J. Clin. Epidemiol. (1996)
9. Pedregosa, F., Varoquaux, G., Gramfort, A., Michel, V., Thirion, B., Grisel, O., Blondel, M., Prettenhofer, P., Weiss, R., Dubourg, V., Vanderplas, J., Passos, A., Cournapeau, D., Brucher, M., Perrot, M., Duchesnay, E.: Scikit-learn: machine learning in python. J. Mach. Learn. Res. (2011)
10. Vanwinckelen, G., Blockeel, H.: On estimating model accuracy with repeated cross-validation. In: BeneLearn and PMLS (2012)
11. Li, X., Wang, L., Sung, E.: Improving AdaBoost for classification on small training sample sets with active learning

A Robust and Blind Watermarking for Color Videos Using Redundant Wavelet Domain and SVD

S. Prasanth Vaidya and P. V. S. S. R. Chandra Mouli

Abstract This paper presents a color video watermarking scheme using Redundant Discrete Wavelet Transform (RDWT) and Singular Value Decomposition (SVD) with good imperceptibility and high robustness. The binary watermark is embedded into blocks of low-level sub-band of the wavelet transform. Using blind extraction process, the watermark is extracted from the watermarked color video. The proposed watermarking scheme protects the copyright information from adversaries and also helps in claiming the ownership of the video content. From the experimental results, it is found that the proposed scheme is robust to all possible attacks and has better results when compared to existing watermarking schemes.

Keywords Video watermarking · Redundant discrete wavelet transform (RDWT) · Singular value decomposition (SVD)

1 Introduction

The sudden evolution in multimedia sharing service and rapid expansion of internet has increased the distribution of multimedia data. Mostly, the video content consisting of conference videos and entertainment videos is shared and distributed with ease. To claim the ownership of pirated data and to protect intellectual property rights (IPR) of video content, digital watermarking is used. In digital watermarking, a watermark that can be in the form of text, image, audio and video is embedded into multimedia data [1–3]. The watermark can be extracted from the watermarked multimedia data

S. Prasanth Vaidya
Department of Computer Science and Engineering, Gayatri Vidya Parishad College of Engineering (A), Visakhapatnam, Andhra Pradesh, India
e-mail: vaidya269@gmail.com

P. V. S. S. R. Chandra Mouli (✉)
School of Computer Science and Engineering, Vellore Institute of Technology, Vellore, Tamil Nadu, India
e-mail: chandramouli@vit.ac.in

A. Elçi et al. (eds.), *Smart Computing Paradigms: New Progresses and Challenges*, Advances in Intelligent Systems and Computing 766,
https://doi.org/10.1007/978-981-13-9683-0_2

for authentication [4]. The significant properties of robust watermarking approach are imperceptibility, robustness and security.

At present, online video piracy has become a major perturb for video production. Using digital video watermarking scheme, video files can be protected against piracy from unauthorized users. In video watermarking scheme, original content is slightly modified by embedding watermark into selected frames adaptively, and is highly affected by attacks like frame dropping, frame swapping, frame averaging, frame skipping, interpolation and so on. In image watermarking, these attacks are not possible. A video consists of three dimensional characteristics where temporal variations should be considered while embedding the watermark. So it is very difficult to maintain high imperceptibility of watermark in video watermarking algorithms. The main issue in these schemes is whether it is appropriate to embed an identical watermark throughout or different watermarks in each frame [5].

Generally, embedding watermark in transform domain provides robustness to watermarking scheme [6]. Faragallah et al. [7] proposed DWT based SVD video watermarking scheme by embedding watermark with spatial and temporal redundancy. Agilandeeswari et al. [8] proposed a bit plane sliced, scramble color image watermark embedded in color video using hybrid transforms. Using bit plane slicing mechanism, color watermark image is divided into 24 slices. To achieve security, slices are scrambled using Arnold transform. These scrambled slices are embedded into one of the DWT mid frequency coefficients of successive contourlet transformed non-motion frames of host video. Based on shot boundary detection algorithm, the non-motion frames are identified. In [9], digital video watermarking scheme based on multi-resolution wavelet decomposition is presented. Binary watermark is embedded into wavelet coefficients of sub-bands by quantization, by using error correction codes and by repeatedly embedding same watermark in different frames of the video. This method achieved good resilience against different attacks, but still has some detection problems after compression and filtering.

From the literature, it can be noticed that most of the watermarking schemes can't resist attacks. The proposed video watermarking scheme shows high robustness against attacks by utilizing RDWT and SVD. The watermark is embedded into low frequency sub-band of the video frames. Experimental results show that the proposed color video watermarking scheme is robust against attacks. The major contributions of the proposed scheme are the using of RDWT and SVD for video processing where the watermark is embedded in different locations of the selected frames, and blind extraction.

The rest of this paper is organized as follows. An overview of the proposed video watermarking scheme is presented in Sect. 2. The experimental results are shown in Sect. 3. Finally, Sect. 4 concludes the paper.

2 Proposed Method

In this section, watermark embedding and watermark extraction processes are discussed in detail.

2.1 Watermark Embedding

The block diagram of proposed watermark embedding scheme is shown in Fig. 1. In this process, video is converted into frames and for frame selection, scene change detection is used. The selected frames are converted from RGB to YCbCr color space. Further, the luminance (Y) frames are decomposed using 2D-RDWT with L = 1 resolution level. The watermark is embedded in LL sub-band where the size of sub-band is same as frame size. The LL sub-band of size $m \times n$ is divided into size of $\alpha \times \beta$ blocks. The number of blocks is set optimal based on prior tests and is set to 4. Increasing the block size does not have any significant improvements rather increase computational complexity.

Let $M = \frac{m}{\alpha}$ and $N = \frac{n}{\beta}$ where m, n are width and height of the frame/sub-band. If $\alpha = \beta = 4$, then each block can be represented as $B_{m,n}, m \in \{1 \ldots M\}, n \in \{1 \ldots N\}$. SVD is applied to each block B_{ij} as shown in Eq. (1).

$$[U_{mn} S_{mn} V_{mn}] = SVD(B_{mn}) \tag{1}$$

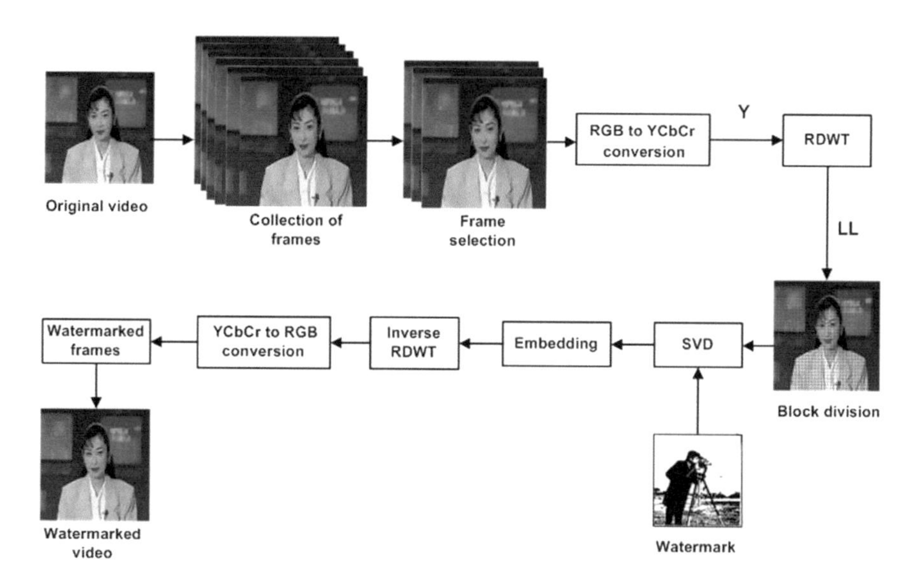

Fig. 1 Block diagram of the proposed watermark embedding process

The watermarked U component matrix is estimated by adding binary watermark bit and embedding factor θ to U component values (decomposed from original frame). Here, θ controls the strength of watermark as shown in Eqs. (2) and (3). The values of i and j vary from frame to frame and are calculated as given in algorithm. The following values are combinations of i and j [(1, 2), (1, 3), (1, 4), (2, 1), (2, 3), (2, 4), (3, 1), (3, 2), (3, 4), (4, 1), (4, 2), (4, 3)].

$$\text{If } w = 1, \begin{cases} u_i^w = sign(u_i) \times (U_{avg} + \theta) \\ u_j^w = sign(u_j) \times (U_{avg} - \theta) \end{cases} \tag{2}$$

$$\text{If } w = 0, \begin{cases} u_i^w = sign(u_i) \times (U_{avg} - \theta) \\ u_j^w = sign(u_j) \times (U_{avg} + \theta) \end{cases} \tag{3}$$

where sign(x) represent sign of x and $U_{avg} = \frac{u_i + u_j}{2}$.

After a series of experiments, the embedding process maintains fidelity and provides robustness, when $\theta = 0.25$. Inverse SVD is applied to the blocks by combining unitary matrix V and singular matrix Σ with watermarked U matrix, as shown in Eq. (2.1).

$$B_{mn}^w = U_{mn}^w \times S_{mn} \times V_{mn}^T$$

All the watermarked blocks are combined to get the watermarked low-level frequency sub-band as shown in Eq. (4) and inverse RDWT is used to get watermarked luminance image. Instead of LL, watermarked LL^w is used as shown in Eq. (5).

$$LL^w = \sum_{m=1}^{M} \sum_{n=1}^{N} B_{m,n}^w \tag{4}$$

$$Y^w = IRDWT[LL^w, LH, HL, HH] \tag{5}$$

Finally, watermarked luminance component is combined with blue-difference (Cb) and red-difference (Cr) chroma components to generate watermarked RGB frame. All the color frames are combined to form the watermarked color video.

2.2 *Watermark Extraction*

The block diagram of extraction process is given in Fig. 2. Here, the inverse process of embedding is done. In extraction process, the watermarked video is divided into frames and by using scene change detection, the watermarked frames are selected. The selected RGB frames are converted to YCbCr. Using RDWT, the Y component is decomposed for 1 level. Since the watermark is embedded in LL sub-band, the

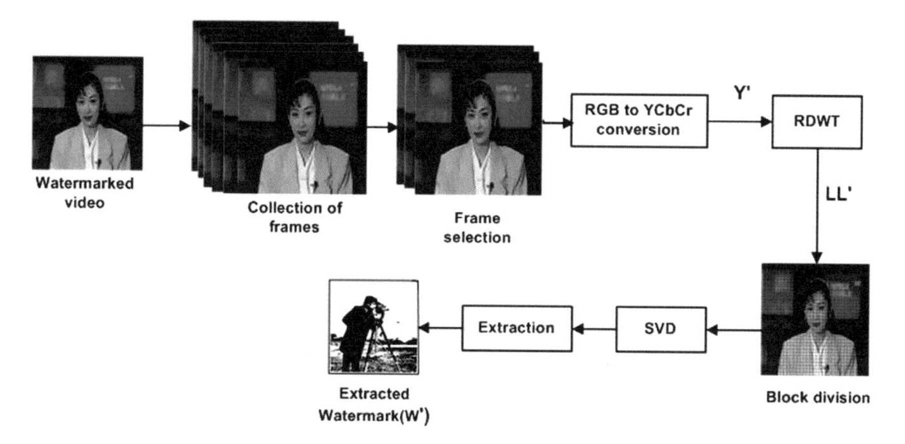

Fig. 2 Block diagram of the proposed watermark extraction process

same is considered for watermark extraction. The LL sub-band is divided into blocks of size $\alpha \times \beta$ where $\alpha = \beta = 4$. To each and every block SVD is applied. The U component matrix is evaluated to extract the watermark using Eq. (6).

$$\text{Watermark bit} = \begin{cases} 0, & \text{if}(u_i^{'} > u_j^{'}) \\ 1, & \text{If}(u_i^{'} \leq u_j^{'}) \end{cases} \tag{6}$$

Finally, the watermark is formed by combining all the watermark bits.

3 Experimental Results

In this simulation, "Akiyo, coastguard, foreman, Suzie, container, mother-daughter, hall and Rhinos" video sequences with resolution 256 × 256 and binary watermark of size 64 × 64 are utilized. The sample avi color videos with 10 s duration and 30 frames/second are considered. To estimate imperceptibility and quality of the watermarked image, peak signal to noise ratio (PSNR) is used [10]. Higher PSNR value provides better image quality and low PSNR value implies high numerical difference between the host frame and the watermarked frame. The obtained PSNR values are given in Table 1.

The robustness of proposed scheme is estimated using normalized correlation coefficient (NCC). The NCC is calculated between extracted watermark and original watermark [11]. Closer the value of NCC to 1, closer the extracted watermark to the original watermark i.e., $W = W'$ [12]. The obtained NCC values are given in Table 2. Various attacks like salt and pepper (S&P) noise, Gaussian noise, mean filter, median filter, cropping, scaling, blurring and JPEG compression were applied to analyze the robustness.

Table 1 PSNR values of samples videos with no attack for Frame 13

Video sequence	PSNR	Video sequence	PSNR
Akiyo	39.53	Container	35.62
Coastguard	37.20	Mother_Daughter	38.33
Foreman	35.28	Hall	36.00
Suzie	39.09	Rhinos	36.92

Table 2 NCC values of samples videos with different attacks for Frame 13

Video sequence	No attack	S&P	Gaussian	Mean filtering	Median filtering	Cropping	Scaling	Blurring	JPEG
Akiyo	0.9935	0.9645	0.9598	0.9644	0.9622	0.9789	0.9573	0.9789	0.9856
Coastguard	0.9785	0.9645	0.9636	0.9230	0.9608	0.9657	0.9484	0.9771	0.9684
Foreman	0.9963	0.9849	0.9924	0.9435	0.9868	0.9800	0.9296	0.9790	0.9963
Suzie	0.9980	0.9757	0.9918	0.9808	0.9924	0.9822	0.9481	0.9792	0.9980
Container	0.9796	0.9704	0.9768	0.9363	0.9709	0.9655	0.9324	0.9651	0.9796
Mother_Daughter	0.9982	0.9798	0.9795	0.9725	0.9920	0.9813	0.9356	0.9772	0.9982
Hall	0.9844	0.9734	0.9788	0.9545	0.9722	0.9710	0.9654	0.9647	0.9844
Rhinos	0.9925	0.9856	0.9763	0.9659	0.9857	0.9834	0.9558	0.9742	0.9873

Table 3 Comparison of NCC values with other methods

Attacks	[13]	[14]	[15]	Proposed method
No attack	0.9298	0.9945	0.9300	0.9980
S&P noise	–	–	0.7150	0.9823
Gaussian noise	0.9121	0.9562	0.6950	0.9897
Median filtering	0.8026	0.9904	0.6700	0.9924
Frame swapping	0.8477	0.9438	–	0.9512
Frame averaging	0.8132	0.8562	–	0.8925

The proposed scheme is compare with Xu et al. [13] blind method, Masoumi et al. [14] wavelet based method and Sahu et al. [15] block based method. Attacks considered for comparison are salt and pepper noise with density (0.001), Gaussian noise with zero mean and variance (0.001), median filtering, frame swapping and frame averaging. From the comparison results given in Table 3, it can be observed that the proposed method is better than the existing methods.

4 Conclusion

In this paper, a robust and blind video watermarking scheme for copyright protection and ownership identification is presented based on RDWT and SVD. Embedding the watermark bits by modifying only two elements of U matrix provide imperceptibility and robustness against attacks. Experimental results show that the scheme is robust against noise attacks, geometric attacks, denoising attacks and also against compression attack. Compared to existing watermarking schemes, the proposed method is more robust. The proposed watermarking scheme can be used for public watermarking applications since the extraction process is blind.

References

1. Cox, I.J., Miller, M.L., Bloom, J.A., Honsinger, C.: Digital Watermarking, vol. 1558607145. Springer, Berlin (2002)
2. Prasanth Vaidya, S., Chandra Mouli, P.V.S.S.R.: Adaptive digital watermarking for copyright protection of digital images in wavelet domain. Procedia Comput. Sci. **58**, 233–240 (2015)
3. Prasanth Vaidya, S., Chandra Mouli, P.V.S.S.R.: Adaptive, robust and blind digital watermarking using Bhattacharyya distance and bit manipulation. Multimedia Tools and Applications, pp. 1–27 (2017)
4. Cox, I.J., Kilian, J., Leighton, F.T., Shamoon, T.: Secure spread spectrum watermarking for multimedia. IEEE Trans. Image Process. **6**(12), 1673–1687 (1997)
5. Doerr, G., Dugelay, J.L.: A guide tour of video watermarking. Signal Process. Image Commun. **18**(4), 263–282 (2003)
6. Langelaar, G.C., Setyawan, I., Lagendijk, R.L.: Watermarking digital image and video data. A state-of-the-art overview. IEEE Signal Process. Mag. **17**(5) (2000)
7. Faragallah, O.S.: Efficient video watermarking based on singular value decomposition in the discrete wavelet transform domain. AEU-Int. J. Electron. Commun. **67**(3), 189–196 (2013)
8. Agilandeeswari, L., Ganesan, K.: A robust color video watermarking scheme based on hybrid embedding techniques. Multimed. Tools Appl. **75**(14), 8745–8780 (2016)
9. Preda, R.O., Vizireanu, D.N.: A robust digital watermarking scheme for video copyright protection in the wavelet domain. Measurement **43**(10), 1720–1726 (2010)
10. Wang, Z., Bovik, A.C., Sheikh, H.R., Simoncelli, E.P.: Image quality assessment: from error visibility to structural similarity. IEEE Trans. Image Process. **13**(4), 600–612 (2004)
11. Huang, H.Y., Yang, C.H., Hsu, W.H.: A video watermarking technique based on pseudo-3-D DCT and quantization index modulation. IEEE Trans. Inf. Forensics Secur. **5**(4), 625–637 (2010)
12. Campbell, J.Y., Lo, A.W.C., MacKinlay, A.C.: The econometrics of financial markets. Princeton University Press, Princeton (1997)
13. Xu, D.W.: A blind video watermarking algorithm based on 3D wavelet transform. In: 2007 International Conference on Computational Intelligence and Security, pp. 945–949. IEEE (2007)
14. Masoumi, M., Amiri, S.: A blind scene-based watermarking for video copyright protection. AEU-Int. J. Electron. Commun. **67**(6), 528–535 (2013)
15. Sahu, N., Tiwari, V., Sur, A.: Robust video watermarking resilient to temporal scalability. In: 2015 Fifth National Conference on Computer Vision, Pattern Recognition, Image Processing and Graphics (NCVPRIPG), pp. 1–4. IEEE (2015)

Optimized Object Detection Technique in Video Surveillance System Using Depth Images

Md. Shahzad Alam, T. S. Ashwin and G. Ram Mohana Reddy

Abstract In real-time surveillance and intrusion detection, it is difficult to rely only on RGB image-based videos as the accuracy of detected object is low in the low light condition and if the video surveillance area is completely dark then the object will not be detected. Hence, in this paper, we propose a method which can increase the accuracy of object detection even in low light conditions. This paper also shows how the light intensity affects the probability of object detection in RGB, depth, and infrared images. The depth information is obtained from Kinect sensor and YOLO architecture is used to detect the object in real-time. We experimented the proposed method using real-time surveillance system which gave very promising results when applied on depth images which were taken in low light conditions. Further, in real-time object detection, we cannot apply object detection technique before applying any image preprocessing. So we investigated the depth image by which the accuracy of object detection can be improved without applying any image preprocessing. Experimental results demonstrated that depth image (96%) outperforms RGB image (48%) and infrared image (54%) in extreme low light conditions.

Keywords Depth image · Object detection · Kinect · Real-time video surveillance

1 Introduction

Over the recent years, object detection, as well as object recognition has been studied extensively since it is used in various areas like manufacturing, security, surveillance, medical and robotic system with promising results. Human vision is very much

Md. Shahzad Alam (✉) · T. S. Ashwin · G. Ram Mohana Reddy
National Institute of Technology Karnataka, Surathkal, India
e-mail: shahzadalam98@gmail.com

T. S. Ashwin
e-mail: ashwindixit9@gmail.com

G. Ram Mohana Reddy
e-mail: profgrmreddy@gmail.com

A. Elçi et al. (eds.), *Smart Computing Paradigms: New Progresses and Challenges*, Advances in Intelligent Systems and Computing 766,
https://doi.org/10.1007/978-981-13-9683-0_3

accurate and allows us to perform and process the complex image and analyze it very quickly. But to perform the same in computers, we require a lot of computation to detect and identify an object in images and if we apply this on live video streaming in real-time then even using the state-of-the-art technique we will not get desired results. So to overcome this problem, we used YOLO (You Only Look Once) [1] to detect the object in the real-time. YOLO as compared to other techniques like R-CNN [2], Fast-CNN [3] and Faster R-CNN [4] is very fast since other techniques need more than one convolution network. But YOLO considers object detection as a single regression problem, straight from image pixels to bounding box coordinates and class probabilities.

All the abovementioned techniques fail to detect objects in low light condition if RGB images are taken. Further, if we take surveillance application we may not get the sufficient light for RGB image or video to work perfectly for the said object detection task. Video surveillance system uses infrared images during night and these images are monitored manually to detect any abnormal activity. This entire procedure is tedious and time-consuming if the surveillance area is too large. In order to overcome the abovementioned limitations, we need to automate the entire process for the real-time object detection.

Thus motivated us to propose an object detection system in different light intensity conditions and also we used depth as well as infrared images from Kinect sensor to improve the efficiency of real-time object detection. We have also shown the relationship between light intensity and the accuracy of object detection using YOLO architecture.

The key contributions of this paper are as follows:

- Design and development of novel framework for object detection in video surveillance system.
- Performance evaluation of object detection for various light intensities for RGB, Infrared and Depth Images.
- Dataset creation to test the object identification for surveillance system in real-time.

The rest of paper is organized as follows; related work is discussed in Sect. 2. The proposed methodology is explained in Sect. 3. Experimental Setup Results and Analysis are discussed in Sect. 4. Finally, Sect. 5 concludes the paper with future directions.

2 Related Work

Several researchers in the past have tried to overcome the problem of low accuracy of object detection [5] by using the depth images from depth sensors like Kinect, etc. Kinect is a very reliable tool in areas such as robotics, object detection, and character recognition.

Table 1 Summary of existing work

Authors	Methodology	Advantages	Limitations
Hou et al. [7]	A two-stage learning framework with property derivation using color and depth images	It effectively uses depth and RGB features which boosts the performance of object detection	It is not effective in real-time as it uses R-CNN which is slow
Cao et al. [8]	It combines the features of depth and RGB images for object detection	It improves the accuracy of both object detection and segmentation by augmenting RGB images with estimated depth images	It is still not effective in real-time as YOLO [1]
Pham et al. [9]	To detect the motion, it used the reference image as a background model and updates whenever a new static scene comes	Its detection technique is based on depth frame difference from a reference background depth image. Hence it is faster and accurate	It uses depth information only and does not consider RGB features of object
Manap et al. [6]	To improve the accuracy of object it investigated the properties and features of object and it's distance from the Kinect	It considered the object properties like shape, color while detecting the object in depth image	It fails to detect the shiny and curved object

Various works [6] related to depth image inferred that the object detection accuracy varies significantly with respect to distance between the object and the Kinect depth sensor. Hou et al. [7] used depth information to boost the performance of RGB-Depth object detection using CNN model. Table 1 shows the summary of existing works.

3 Proposed Methodology

Figure 2 shows the proposed framework for effective video surveillance where we are not only capturing RGB image based videos but also depth image based videos. It consists of two phases where in the first phase we measured the light intensity which tells how much light is present in a room and on the basis of that we are considering which type of video we have to capture. So, if the Lux measurement is less than 50 then the depth image is captured as Lux meter less than 50 denotes the low light condition. And if Lux meter is greater than 50 then RGB image is captured as Light in the room is sufficient for object detection in RGB image.

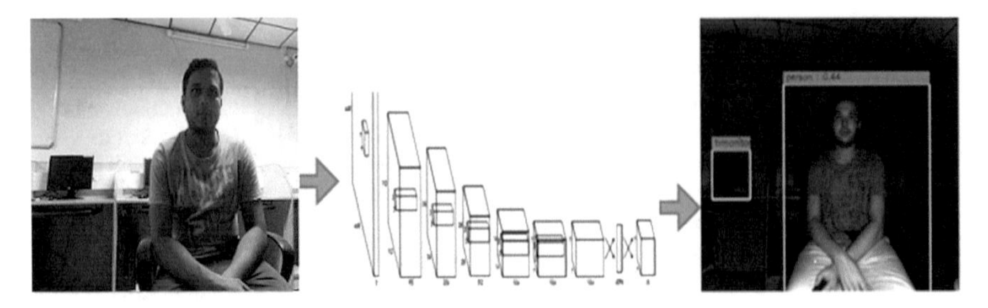

Fig. 1 The architecture of YOLO [1] which uses one convolution network in single

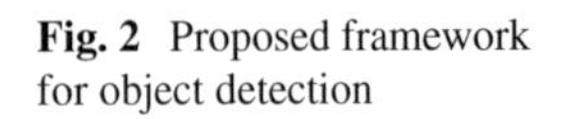

Fig. 2 Proposed framework for object detection

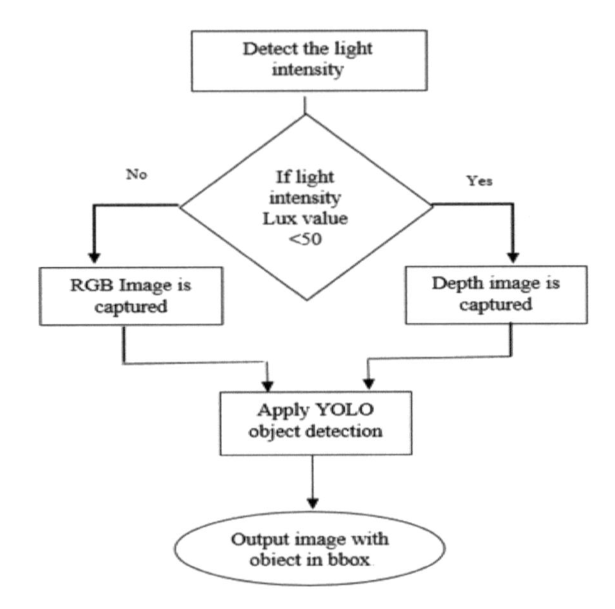

In the second phase in order to get the probability of object detection in real-time, we used YOLO architecture [1] which is state-of-the-art technique to detect the object in real-time. Figure 1 shows YOLO architecture which uses only one Convolution Neural Network. Further, we observe that this technique works very fast in this kind of scenario since this technique is simple and straightforward as shown in Fig. 1.

3.1 Training with Depth image

For training the YOLO model we have trained the model with our own dataset of our lab in which we have captured 20 depth videos of 2 min duration in various light condition. Initially, we converted the videos into image frames and then we manually annotated the object in depth images of size 448 * 448 for training.

3.2 Real-Time Object Detection

After training the model we detected the object in real-time in RGB, depth, and infrared images to find out the accuracy of object detection and also recorded the Light intensity from Lux meter for the particular image frame. Further, we plotted the graph to study the effect of light intensity with respect to the accuracy of object detection. So overall methodology for object detection is shown in Fig. 2.

4 Experimental Result and Analysis

4.1 Experimental Setup: Device Used

To extract exact amount of light intensity for every object present in the image frame we used Digital LUX meter which measures the light intensity in Lux. In our experimental setup, we have also used Kinect Depth sensor which is directly connected to the system through which we captured the depth image as well as infrared image.

4.2 Experimental Setup: Software Used

We used special software iPi Recorder 3 which records depth video and depth image in color as shown in Fig. 3. We have used colored depth image as it increases the accuracy of detecting an object and this iPi Recorder 3 captures image more clear as compared to the default software provided by the Microsoft.

4.3 Data Collection

For our experiment, we took 20 videos of 2 min duration from depth camera of Kinect using iPi software for all light conditions. Another 20 videos of 2 min duration are taken in RGB for the same light conditions and we created dataset of 4000 images of single object as well as multiple object in RGB, infrared, and depth images. All the image frames are equally distributed on the range of 0 to 180 LUX light intensities. In order to obtain video frames with different light intensities, we collected the data from both natural and artificial source (Light Emitting Diodes, Florescent Tubes, etc.). There was only one participant who is directly involved for the data collection.

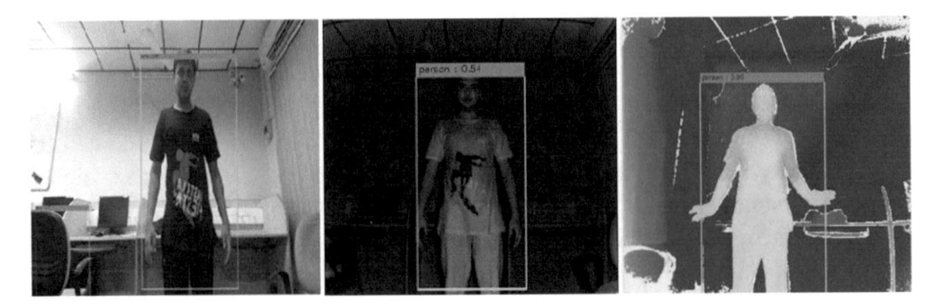

Fig. 3 YOLO object detection in RGB, infrared, depth images

Table 2 Probability of object detection

Object type	Probability		
	RGB	Infrared	Depth
Person	0.48	0.54	0.96
Chair	0.27	<0.10	<0.10

4.4 Results and Analysis

Result Analysis in Full Light Condition. Results obtained by taking the image sample from RGB, infrared, and depth images in full light conditions with Digital Lux meter reads greater than 150 Lux. YOLO object detection technique was applied on all three images. Probabilities of object recognition for different objects present in all three images are shown in Table 2. It is also observed from Table 2 that the probability of detecting a person is increasing from 0.48 (RGB image) to depth image 0.96 (depth image). This means that there is an improvement in the accuracy of object detection in full light condition. Sample image frames from RGB, infrared, and depth are shown in Fig. 3.

Result Analysis in Low Light Condition. In the low light conditions, the images taken from Kinect RGB and Depth Cameras are analyzed. RGB image frame fails to detect the object but from the depth image, we can easily detect the person even in extremely dark condition. A sample image frame in dark light condition is shown in Fig. 4 where the light intensity value of this image is 15 LUX. But For the same image frame with same light intensity LUX value is taken from depth sensor we can detect the person with a probability of 0.68 as shown in Fig. 4.

We have also tested on multiple persons using depth image and it detected everyone. Depth image detection fails only when the person is overlapped in depth image as shown in Fig. 4.

From Fig. 5, we can observe that if Lux value increases then the probability of object detection also increases with RGB image frames. This shows that if light intensity increases then the probability of object detection also increases. It is observed from graph in Fig. 5 that the depth image also varies with respect to light intensity in which the probability value decreases slightly as Lux value increases. This shows

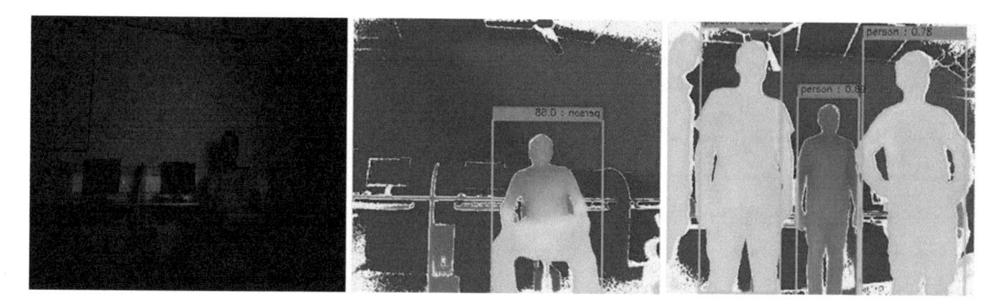

Fig. 4 Original RGB image in low light with no detection and in same scenario with depth image we can detect person

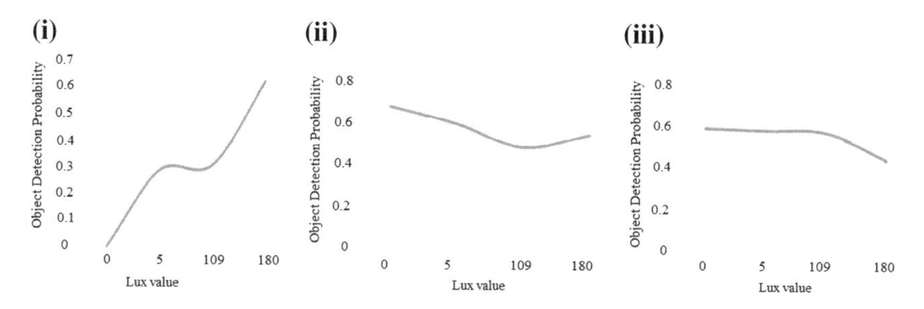

Fig. 5 (i) Light intensity versus object probability graph in RGB image where X-axis is light intensity in LUX and Y-axis is object detection probability. (ii) Light intensity versus object probability in depth image. (iii) Light intensity versus object probability in infrared image

that the depth image works better in low light condition as compared to infrared and RGB images. But there is no further decrease in object detection accuracy for LUX values more than 180 as the object detection probability get saturated. From Table 3, we can observe that the accuracy of object detection varies with respect to the LUX value in RGB, depth, and infrared images.

4.5 Limitation of Object Detection in Depth Image

After applying object detection in depth image we observed that we can detect the person easily with greater accuracy. But this technique sometimes fails to detect the objects like monitor, a chair in depth image which can be easily detected in RGB image. So this can be improved by training the model with different objects and more images. Since we are focusing on intrusion detection or surveillance application so the detection of person is sufficient.

Table 3 Accuracy of detecting person in different light intensity

LUX value	Accuracy		
	RGB	Infrared	Depth
10	0	59	85
20	0	57	83
30	0	54	83
40	18	51	71
50	31	50	57
60	64	43	62
80	69	41	65
100	72	45	61

5 Conclusion

Experimental results demonstrate the effectiveness of object detection under different light intensities. Our proposed framework overcomes this problem by using depth image using Kinect device. So, in the applications like surveillance, intrusion detection we need to use not only an RGB image but also depth image so that the object detection is made more effective in low light conditions. The proposed model can be further optimized by adding different object classes for the recognition of moving object detection in real-time video surveillance system.

Acknowledgements Authors have obtained all ethical approvals from the Institutional Ethics Committee (IEC) of National Institute of Technology Karnataka Surathkal, Mangalore, India and a written consent was also obtained from the human subject.

References

1. Redmon, J., Divvala, S., Girshick, R., Farhadi, A.: You only look once: unified, real-time object detection (2015). arXiv:1506.02640
2. Girshick, R., Donahue, J., Darrell, T., Malik, J.: Rich feature hierarchies for accurate object detection and semantic segmentation. In: Proceedings of the IEEE Conference on Computer Vision and Pattern Recognition, pp. 580–587 (2014)
3. Girshick, R.: Fast R-CNN. In: Proceedings of the IEEE International Conference on Computer Vision, pp. 1440–1448 (2015)
4. Ren, S., He, K., Girshick, R., Sun, J.: Faster R-CNN: towards real-time object detection with region proposal networks. In: Advances in Neural Information Processing Systems, pp. 91–99 (2015)
5. Southwell, B.J., Fang, G.: Human object recognition using color and depth information from an RGB-D Kinect sensor. Int. J. Adv. Robot. Syst. **10**, 171 (2013)
6. Manap, M.S.A., Sahak, R., Zabidi, A., Yassin, I., Tahir, N.M.: Object detection using depth information from Kinect sensor. In: 2015 IEEE 11th International Colloquium on Signal Processing

7. Hou, S., Wang, Z., Wu, F.: Deeply exploit depth information for object detection. In: 2016 IEEE Conference on Computer Vision and Pattern Recognition Workshops (CVPRW)
8. Cao, Y., Shen, C., Shen, H.T.: Exploiting depth from single monocular images for object detection and semantic segmentation. IEEE Trans. Image Process. **26**(2) (2017)
9. Pham, T.T.D., Nguyen, H.T., Lee, S., Won, C.S.: Moving object detection with Kinect v2. In: 2016 IEEE International Conference on Consumer Electronics-Asia (ICCE-Asia)

Evidence-Based Image Registration and Its Effect on Image Fusion

Ujwala Patil, Ramesh Ashok Tabib, Rohan Raju Dhanakshirur and Uma Mudenagudi

Abstract In this paper, we propose Evidence-based technique for image registration. In our previous work, we proposed hierarchical model for image registration using Normalized Mutual Information (NMI) as similarity metric. In few cases, we observe atypical behavior of NMI and infer NMI alone is not sufficient to optimize the transformation matrix, to address this problem in this paper we propose evidence-based image registration using Structural Similarity (SSIM) and NMI as evidences. Atypical behavior of NMI is addressed in evidence- based image registration. We also propose evidence-based framework for image fusion and show image fusion is sensitive to the registration of input observations. Multi-temporal image fusion is challenging due to the presence of high mutual information among them. To address this, we formulate an evidence-based fusion framework with weighted combination of observations, considering Confidence Factor (CF) as weights. CFs for fusion are generated using principal components and distance of registered input observations from reference as evidences. Dempster–Shafer Combination Rule (DSCR) is used to combine the evidences to generate CF. We compare the results with state-of-the-art registration techniques.

Keywords Image registration · Image fusion · Confidence factor (CF) · Dempster–Shafer combination rule (DSCR) · Evidence parameter

---

U. Patil (✉) · R. A. Tabib · R. R. Dhanakshirur · U. Mudenagudi
School of Electronics and Communication, KLE Technological University, Hubballi 580031, India
e-mail: ujwalapatil@bvb.edu

R. A. Tabib
e-mail: ramesh_t@bvb.edu

R. R. Dhanakshirur
e-mail: rohandhanakshirur@gmail.com

U. Mudenagudi
e-mail: uma@bvb.edu

A. Elçi et al. (eds.), *Smart Computing Paradigms: New Progresses and Challenges*, Advances in Intelligent Systems and Computing 766,
https://doi.org/10.1007/978-981-13-9683-0_4

1 Introduction

The main objective of the image fusion is to combine multiple images of the same seen taken at different instances, from different view points and/or by different sensors in order to provide more information and semantic interpretation. The spatial and spectral resolution of monocular imaging system limits the information of single image. Typically, image fusion problem is addressed at pixel level, region level, feature level, and decision level. Several image fusion algorithms in the spatial and transform domain are available in [6, 7, 9, 10, 12, 15]. The fused image obtained by averaging the gray level of input observations introduces blur artifacts. Pixel-based fusion algorithms are used in applications where both pixel spacing and spectral properties of source images are same. Since useful information concerned is not only in one pixel but also in features such as shape, size, edges, are used for fusion in [6, 7]. The authors in [14] show that information content of the restored image is influenced by the number of input observations. In our previous work [16], we address region-based fusion framework and observe, handling high mutual information in multi-temporal image observations is challenging. To address this, we propose evidence-based fusion framework.

The quality of the fused image is sensitive to the registration of input observations. The goal of image registration is to bring two images into alignment. We propose to find spatial mapping between images to bring them into alignment by considering registration as an optimization problem. We use hierarchical model and propose iterative solution for image registration. Different image registration algorithms are available in the literature [4, 5, 8, 11, 13, 17, 18, 20]. The authors in [22] discuss the pipeline for registration, which includes detection of features, matching of features, estimation of transform model. Feature detection includes extraction of salient features like significant regions, edges, lines, intersection points. The features matching algorithms using cross correlation, mutual information, and cross entropy are discussed in [21]. The wavelets and hierarchical pyramid-based feature matching are discussed in [21]. Motion estimation using hierarchical model and Sum of Square Difference (SSD) as similarity metric is discussed in [3] and observed improvement in the the computational efficiency. The authors in [2] present compression-based framework for image registration using similarity metric, where two images are registered properly when it is possible to maximally compress one image given the information in the other image. The next step in the registration is to estimate the transform matrix for the registration to warp the target image toward the reference image. Different methods for transformation estimation are discussed in [22]. The final estimation of transformation is used as forward transformation or backward transformation with the target image to warp it to wards the reference image. The authors in [1] discuss image registration using SSIM as similarity metric. SSIM defines homologous structures of images.

In our previous work [17], we have proposed a hierarchical model-based algorithm for registration of images, related by affine transformation using NMI as a similarity measure. We observed atypical behavior of NMI and corresponding transformation

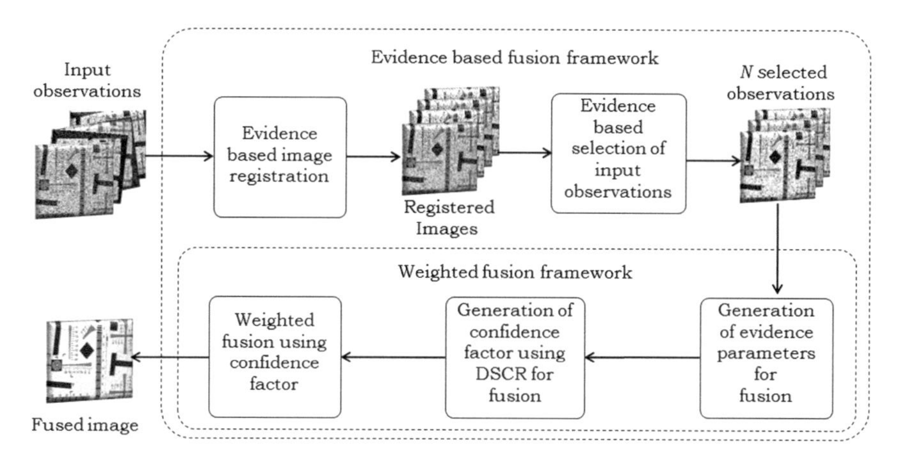

Fig. 1 Evidence-based fusion framework

matrix is considered for the percolation to the higher resolution level in the hierarchical model and quality of registration is not satisfactory. To address this problem, we propose evidence- based image registration using NMI and SSIM to register the input observation for fusion as shown in Fig. 3. To address this, we propose evidence-based framework to consider appropriate transformation matrix for registration from a set of transformation matrices generated at every level of the hierarchy using confidence factor as the selection criteria. Confidence factor for consideration of transformation matrix is generated using NMI and SSIM as evidences.

In Fig. 1, we demonstrate the proposed evidence-based fusion pipeline, with hierarchical registration, selection of input images, and weighted combination of input observations. Toward this, we make the following contributions:

- We propose an evidence-based image fusion framework.
  - We provide an evidence-based hierarchical model for image registration using confidence factor as similarity index to register selected input observations.
    We propose to generate evidence parameters for registration based on NMI and SSIM.
    We generate confidence factor by combining evidence parameters using DSCR.
  - We propose to select best input observations from a set of inputs based on confidence factor.
    Evidence parameters for selection of input observations are generated using NMI and SSIM.
  - We provide an evidence-based technique to estimate the contribution of each input pixel from selected input images for the reconstruction of pixel in the fused image by considering confidence factors as weights for input pixels.

We generate confidence factor for fusion by combining evidence parameters using DSCR.
We propose to generate evidence parameters for fusion based on Euclidean distance and PCA of input observations.

- We demonstrate the effect of registration on quality of the fused image and compare our results with different state-of-the-art methods.

In Sect. 2, we discuss the proposed evidence-based fusion framework. In Sect. 3, we discuss the results of proposed framework and conclude in Sect. 4.

2 Evidence-Based Fusion Framework

In this section, we discuss the proposed evidence-based fusion framework. The proposed pipeline for image fusion is shown in Fig. 1.

2.1 Hierarchical Image Registration

The input observations are registered using proposed registration technique. We also show the effect of different registration techniques on fusion.

In our previous work [17], we used NMI as similarity metric as it is more robust to the change in entropy compared to the mutual information. We experiment the NMI-based registration technique on real and synthetic data set. In few cases we observe, decrease in NMI with iterations and infer NMI alone is not sufficient as similarity metric, as shown in Fig. 2. Typically, error between the reference and the warped image (SSD) decreases with iterations as shown in Fig. 2c. For few cases, we observe atypical behavior in NMI and SSD as shown in Fig. 2a, b. To address this, in this paper, we propose evidence-based registration using NMI and SSIM as evidences. SSIM defines the objective criterion used to estimate registration quality between the homologous structures of images [1].

Atypical behavior of NMI and SSD in NMI-based image registration demands another parameter along with NMI to calculate the similarity metric. We choose SSIM along with NMI and propose evidence-based image registration technique as shown in Fig. 3. Unlike in NMI-based registration, Confidence Factor (CF) (δ^I) is considered as similarity metric and the transformation matrix with highest confidence factor is percolated to the next level in the hierarchical model. NMI and SSIM are combined using DSCR to generate δ^I. Change in δ^I is considered as stopping criteria.

Confidence factor (δ^I) for registration: Let $\mathcal{E}_{r1}$ and $\mathcal{E}_{r2}$ be the evidence parameters considered for evidence-based registration. Evidence parameter $\mathcal{E}_{r1}$ is generated using NMI. $\mathcal{E}_{r2}$ is generated considering the SSIM index. Better the alignment between warped and reference image, higher is the NMI and SSIM index. Evidence parame-

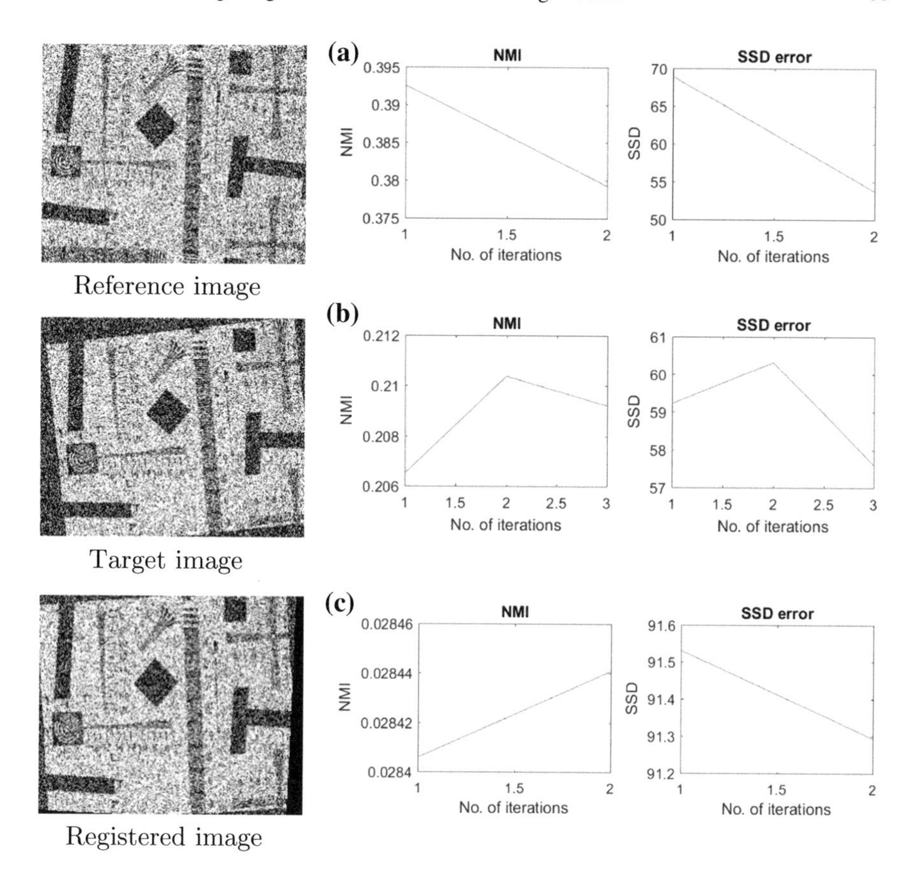

Fig. 2 Variation in NMI and SSD along the iterations at **a** level 3, **b** level 2, **c** level 1 (atypical behavior in NMI and SSD at level 2 and level 3)

ters $\mathcal{E}_{r1}$ and $\mathcal{E}_{r2}$ are used as mass of belief toward the set hypothesis $\mathcal{H}$ and are used to compute $\mathcal{E}_{r1}$ as shown [19]. The δ^I confidence factor in favor of aligning the target image and the reference image. The confidence factor at every iteration is computed and the transformation matrix with highest confidence factor is percolated to the next hierarchal level. We use change in δ^I as stopping criteria in the evidence-based image registration.

2.2 *Evidence-Based Selection of Input Observations*

Input observations selected based on the confidence factor δ^S. δ^S is generated using NMI and SSIM as evidences $\mathcal{E}_{s1}$ and $\mathcal{E}_{s2}$ respectively. Evidence parameters $\mathcal{E}_{s1}$ and

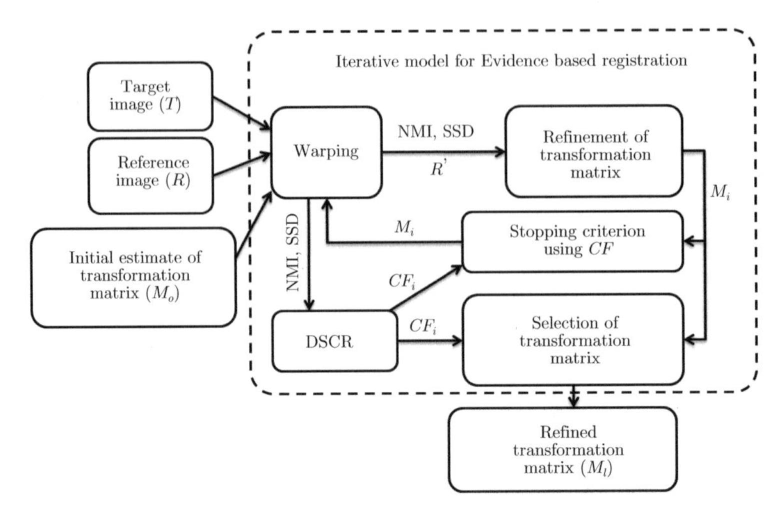

Fig. 3 Evidence-based registration: T is is warped toward R using the transformation matrix, refined using NMI and SSD at every iteration. CF, (δ^I) generated using NMI and SSIM is used as stopping criterion for the iterations and the transformation matrix with highest δ^I is percolated to the next level

$\mathcal{E}_{s2}$ are used as $\mathcal{E}_1$ and $\mathcal{E}_2$, masses in [19] to compute confidence factor δ^S. Input observations having the δ^S above set threshold are selected for fusion.

2.3 *Weighted Fusion Framework*

Evidence parameters for fusion are computed using PCA and the distance of registered input observations from the reference image. Evidence parameters are combined using DSCR to generate confidence factor and use them as weights for fusion. The pixel F in fused image is the weighted combination of all the observations with corresponding weights and is given by

$$F = \frac{\sum_{k=1}^{N} \delta_k^W \cdot I_k}{\sum_{k=1}^{N} \delta_k^W} \tag{1}$$

where δ_k^W and I_k are the weight and intensity value of the pixel of kth input observation, respectively. The δ_k^W is the confidence factor calculated by combining the evidences using DSCR.

Confidence factor δ^F for fusion: Let $\mathcal{E}_{f1}$ and $\mathcal{E}_{f2}$ be the evidence parameters. p_1, p_2, ..., p_k are the pixel from the selected input observations for the generation of corresponding pixel F in the fused image. Evidence parameter $\mathcal{E}_{f1}$ is computed using

Euclidean distance d_k between pixel in the reference image p_r and corresponding pixel p_k in the kth observation. The normalized distance α_k is given by

$$\alpha_k = \frac{d_k}{d_M}$$

where the distance is normalized using maximum distance d_M among the considered observations. The evidence parameter $\mathcal{E}_{f1}$ is given by

$$\mathcal{E}_{f1} = 1 - \alpha_k \tag{2}$$

Lesser the *euclidean distance* between corresponding input pixel and pixel in the reference image, higher is the evidence. Evidence parameter $\mathcal{E}_{f2}$ is generated considering region-based PCA [17]. Let C_k be the principal component of the pixel in kth observation. The C_k is normalized using maximum principal component C_M among the considered observations and used as evidence parameter $\mathcal{E}_{f2}$ and is given by

$$\mathcal{E}_{f2} = \frac{C_k}{C_M} \tag{3}$$

Evidence parameters $\mathcal{E}_{f1}$ and $\mathcal{E}_{f2}$ are used as masses to compute confidence factor δ^F for fusion. The confidence factor δ^W and used as weight for fusion.

3 Results and Discussions

In this section, we demonstrate the results of proposed registration algorithms on both synthetic and real data. We compare the results of evidence-based registration with different registration techniques. We show the effect of registration on fusion.

We discuss the results of proposed evidence-based registration and the results are compared with NMI-based registration [17] and different state-of-the-art techniques of registration. The registration of input observations affects the quality of the fused image as shown in Fig. 5. Figure 4 shows the number of iterations, corresponding NMI, Sum of Square Distances (SSD) at every level and iteration at which the transformation matrix is percolated for the observation Fig. 4. Unlike to the relation between SSD and NMI, SSD is decreasing with NMI as shown in Fig. 4. To address this, we use confidence factor as similarity metric generated using NMI and SSIM. Along with mutual information the need for structural similarity between reference and target image for registration motivate us to consider SSIM for generation of confidence factor. The transformation matrix at every hierarchical level with highest confidence factor is percolated to the next level. We observe decrease in SSD error using proposed evidence-based registration. We demonstrate the quality of fused image using different quality metrics and register using different registration techniques. We observe the quality of the fused image, input images registered using

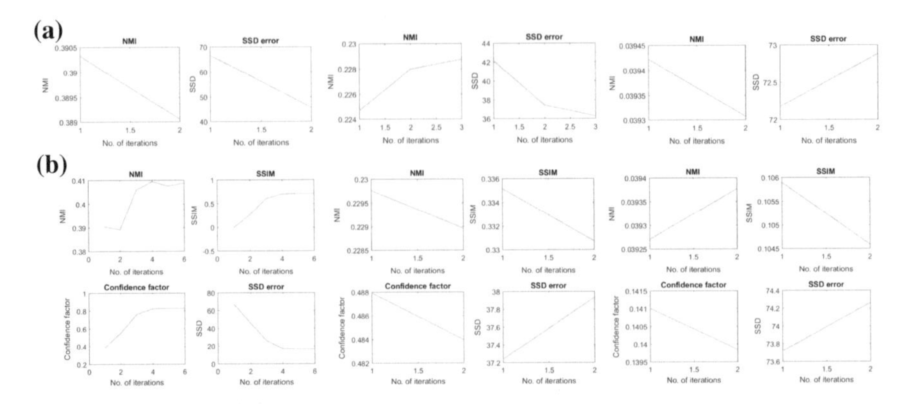

Fig. 4 NMI-based registration [17], **a** variation of NMI and SSD at level 1 using NMI-based registration, **b** plot of NMI, SSIM, confidence factor, and SSD at level 1 using proposed evidence-based registration

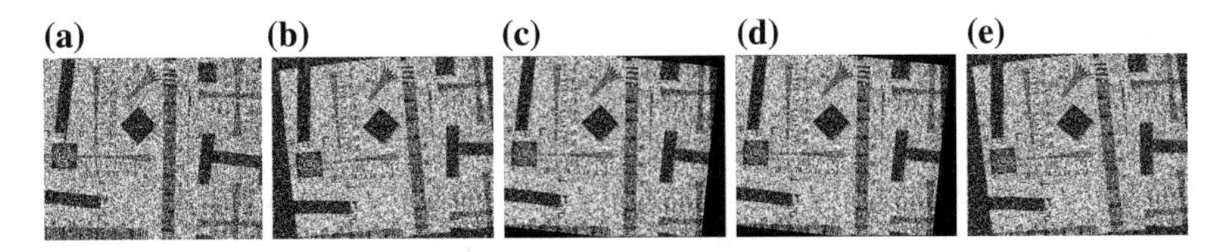

Fig. 5 Comparison of image registration for noisy dataset **a** reference image, **b** target image; Registered image using **c** NMI method [17] SSD = 109.87, **d** proposed evidence-based method SSD = 98.6, **e** RIR method [5] SSD = 127.9

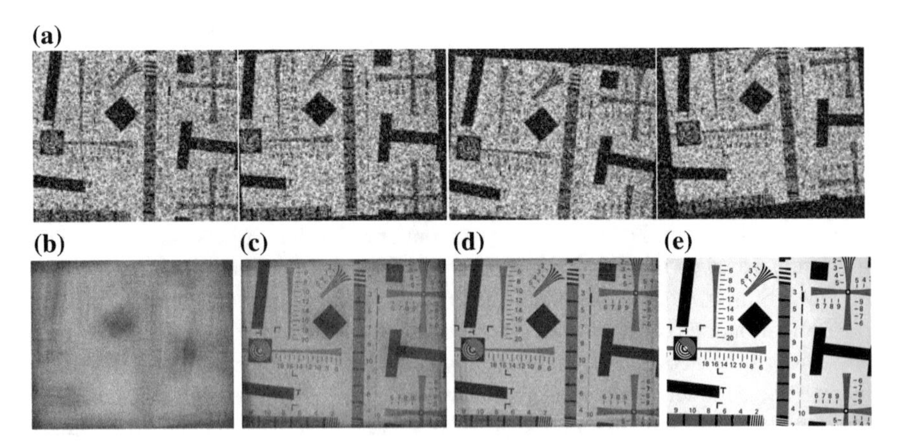

Fig. 6 Effect of registration on fusion **a** input observations; Fusion after **b** RIR [5]-based registration, **c** NMI-based registration [17], **d** evidence-based registration, **e** ground truth

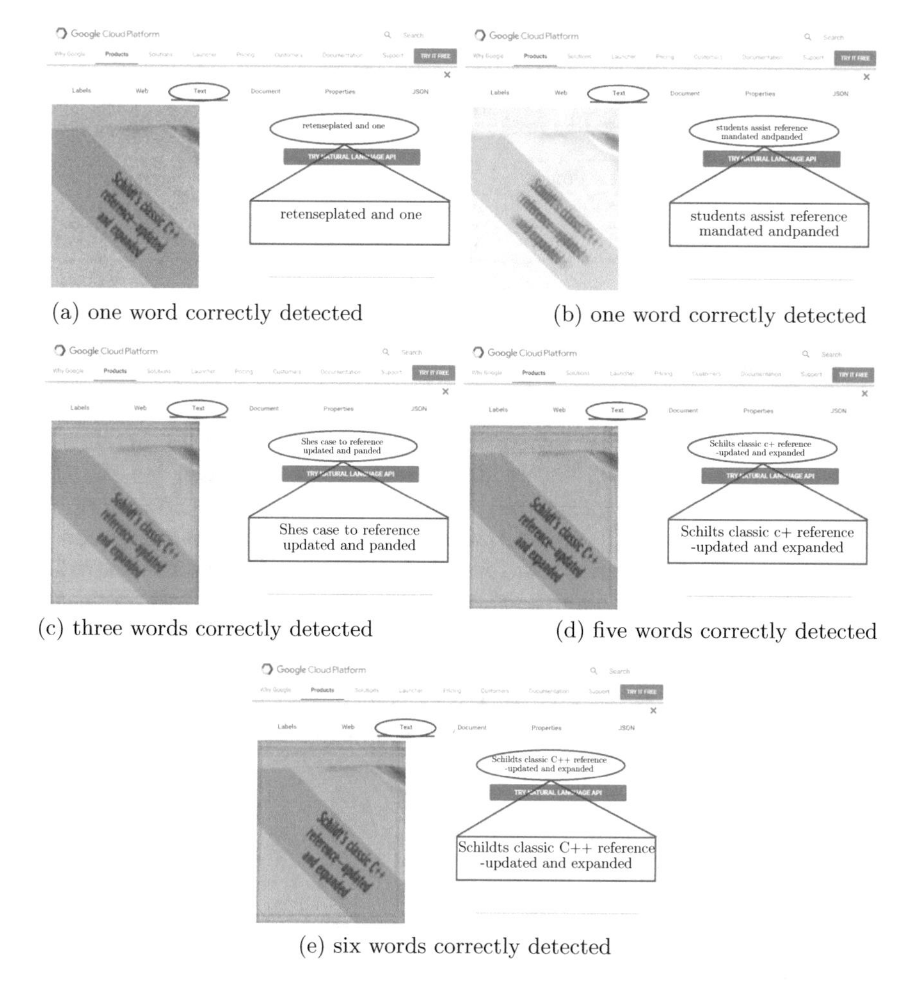

(a) one word correctly detected

(b) one word correctly detected

(c) three words correctly detected

(d) five words correctly detected

(e) six words correctly detected

Fig. 7 Effect of registration on evidence-based fusion using 40 input observations [Real dataset 2] using GoogleAPI **a** Reference image; Registered using **b** RIR [5], **c** LLT [11], **d** NMI [17] Text not detected, **e** evidence-based registration. Actual text is of seven words "**Schildt's classic C++ reference-updated and expanded**"

proposed evidence-based registration is better as shown in Table 1 for the images shown in Fig. 6. We also demonstrate the quality of the fused image using Google Vision API and is shown in Fig. 7. We observe the text detected in the fused image generated using proposed framework is similar to the actual text.

Table 1 Comparison of quality of fused images fused using proposed evidence- based fusion; input images registered using different registration techniques for images shown in Fig. 6

Quality metrics →	MI	SSIM	PSNR	RMS
Input image	0.298	0.127	8.026	101.209
Proposed evidence-based registration	**1.320**	**0.500**	**13.405**	**54.483**
NMI-based registration [17]	0.942	0.371	11.094	71.089
RIR [5]	0.133	0.120	9.304	87.360

4 Conclusions

In this paper, we proposed evidence-based image registration and show the effect of registration on image fusion. Toward this, we included image registration, selection of input observations, and image fusion as modules in the proposed framework. We proposed evidence-based hierarchical model for image registration using confidence factor generated by considering SSIM and NMI as evidences. We observed the reduction in registration error and intern improvement in the fused image. We formulated a scheme to generate fused image from weighted combination of registered input observations. Confidence factor for fusion were generated by combining evidence parameters using DSCR and were used as weights for fusion. We demonstrated the results on synthetic and real dataset.

References

1. Amintoosi, M., Fathy, M., Mozayani, N.: Precise image registration with structural similarity error measurement applied to super resolution. EURASIP J. Adv. Signal Process. **12**, 1–7 (2009)
2. Bardera, A., Feixas, M., Boada, I., Sbert, M.: Compression based image registration. IEEE Int. Symp. Inf. Theory. **6**, 436–440 (2006)
3. Bergen, J.R., Anandan, P., Hanna, K.J., Rajesh, H., Zhiyong, Bin Gu, Lin.: Hierarchical model based motion estimation. In: Proceedings of the European Conference on Computer Vision, vol. 2, pp. 164–173 (1992)
4. Bhist, S.S., Gupta, B., Rahi, P.: Image registration concepts and techniques: a review. Int. J. Eng. Res. Appl. (2014)
5. Forsberg, D.: Robust image registration for improved clinical efficiency. Ph.D. thesis, Linkoping University (2013)
6. Gayathri, N., Deepa, P.L.: Multi-focus color image fusion using NSCT and PCNN. In: 2016 International Conference on Communication Systems and Networks (ComNet), pp. 173–178 (2016)
7. Kalaivani, K., Phamila, Y.A.V.: Analysis of image fusion techniques based on quality assessment techniques. Indian J. Sci. Technol. 1–8 (2016)
8. Lakshmi, K.D., Vaithiyanathan, V.: Image registration techniques based on the scale invariant feature transform. IETE Tech. Rev. **34**(1), 22–29 (2017)
9. Li, S., Kang, X., Hu, J.: Image fusion with guided filtering. IEEE Trans. Image Process. **22**(7), 2864–2875 (2013)

10. Liu, Y., Liu, S., Wang, Z.: A general framework for image fusion based on multi-scale transform and sparse representation. Inf. Fusion **24**, 147–164 (2015)
11. Ma, J., Zhou, H., Zhao, J., Gao, Y., Jiang, J., Tian, J.: Robust feature matching for remote sensing image registration via locally linear transforming. IEEE Trans. Geosci. Remote Sens. **53**(12), 6469–6481 (2015)
12. Ma, K., Li, H., Yong, H., Wang, Z., Meng, D., Zhang, L.: Robust multi-exposure image fusion: a structural patch decomposition approach. IEEE Trans. Image Process. **26**(5), 2519–2532 (2017)
13. Mohod, N.P., Ladhake, S.A.: Polar transform in image registration. Int. J. Adv. Res. Comput. Sci. Softw. Eng. 603–606 (2013)
14. Mudenagudi, U., Banerjee, S., Kalra, P.K.: Space-time super-resolution using graph-cut optimization. IEEE Trans. Pattern Anal. Mach. Intell. **33**(5), 995–1008 (2011)
15. Naidu, V.P.S., Elias, B.: A novel image fusion technique using DCT based Laplacian pyramid. Int. J. Inven. Eng. Sci. (IJIES) ISSN 2319–9598 (2013)
16. Patil, U., Mudengudi, U.: Image fusion using hierarchical PCA. In: 2011 International Conference on Image Information Processing (ICIIP), pp. 1–6 (2011)
17. Patil, U., Mudengudi, U., Ganesh, K., Patil, R.: Image fusion framework. In: Second International Conference CNC 2011, Bangalore, India, 10–11 March 2011. Proceedings, pp. 653–657. Springer, Berlin (2011)
18. Patil, U., Patil, R., Kalyani, R., Mudenagudi, U.: Robust registration for image fusion, pp. 1–5
19. Tabib, R.A., Patil, U., Ganihar, S.A., Trivedi, N., Mudenagudi, U.: Decision fusion for robust horizon estimation using Dempster Shafer combination rule. In: 2013 Fourth National Conference on NCVPRIPG, pp. 1–4 (2013)
20. Ward, G.: Fast, robust image registration for compositing high dynamic range photographs from handled exposures. J. Graph. Tools **8**, 17–30 (2012)
21. Wolberg, G., Zokai, S.: Robust image registration using log polar transform. In: IEEE Conference on Image Processing, Canada (2000)
22. Zitova, B., Flusser, J.: Image registration methods: a survey. J. Image Vis. Comput. **21**, 977–1000 (2003)

Two Stream Convolutional Neural Networks for Anomaly Detection in Surveillance Videos

Adarsh Jamadandi, Sunidhi Kotturshettar and Uma Mudenagudi

Abstract In this paper we propose a deep learning framework to identify anomalous events in surveillance videos. Anomalous events are those which do not adhere to normal behaviour. We propose to use two discriminatively trained Convolutional Neural Networks, to capture the spatial and temporal features of videos, the classification scores obtained from the two streams are later fused to assign one final score. Since our approach is scenario-based, this eliminates the need for adopting a particular definition of anomaly. We show that the Two Stream CNNs perfectly capture the intricacies involved in modelling a video data by demonstrating the framework on airport and mall surveillance datasets respectively. We achieve a final test accuracy of 99.1% for spatial stream and 91% for temporal stream for airport scenario and an accuracy of 94.7% for spatial and 90.1% for the temporal stream for the mall scenario. Our framework can be easily implemented in real-time and is capable of detecting anomaly in each frame fed by a live surveillance system.

1 Introduction

Convolutional Neural Networks have been largely successful in modelling image data and is constantly being bettered to learn exhaustive learning representation techniques for videos. Some of the key areas where Deep Learning algorithms have found its applications include Crowded Scene Understanding, Video Classification, Action Recognition, Anomaly Detection [1–3] etc. Anomaly detection mainly refers

A. Jamadandi (✉) · S. Kotturshettar
B. V. Bhoomaraddi College of Engineering and Technology, Hubli, India
e-mail: adarsh.jam@gmail.com
URL: https://adarshmj.github.io

S. Kotturshettar
e-mail: sunidhikshettar@gmail.com

U. Mudenagudi
KLE Technological University, Hubli, India
e-mail: uma@kletech.ac.in

A. Elçi et al. (eds.), *Smart Computing Paradigms: New Progresses and Challenges*,
Advances in Intelligent Systems and Computing 766,
https://doi.org/10.1007/978-981-13-9683-0_5

to detection and modelling of unusual activities or activities that do not adhere to normalcy. Detection of anomalies in surveillance videos is a non-trivial task, because the definition of "anomaly" is subjective and it's difficult to generalise this definition to every scenario at hand. It forms a pre-requisite feature for any claimed intelligent surveillance system to detect anomalous events, with almost minimal human intervention. We propose a methodology to model anomalous events in specific scenarios, for example, consider airport surveillance videos, a person entering the airport with baggage but leaving it unattended for longer periods of time could be considered an anomaly or people engaged in an angry brawl in the airport premises can be detected as an anomalous event. Thus adopting a scenario-based anomaly detection framework eliminates the hassles of assuming a generalised definition of "anomaly". In this paper we propose a scenario-based anomaly detection framework, we model the anomalous events in surveillance videos using a Two Stream Convolutional Neural Network architecture (Two Stream CNN), the two streams trained discriminatively encode the spatial and temporal aspects of the query video, we reduce the problem of anomaly detection to a binary classification problem, The classification score given by the two separately trained CNNs are later fused by taking the average of both the scores and a final classification score is obtained. Our contributions can be summarised as follows:

1. We provide a scenario-based Two Stream CNN framework to detect anomalies in surveillance videos by,
   - training an image recognition CNN architecture to model spatial features of the video.
   - training a CNN architecture to model the temporal features of the video.
   - provides an averaging technique to fuse the classification scores of both the streams to assign a final score.
2. We provide an intelligence enabled anomaly detection framework which is capable of detecting anomalies either on pre-recorded videos or video-feed from the cameras can be directly fed into our system to get classification in real-time.
3. We demonstrate our framework on two scenarios—airport and mall surveillance videos.

2 Related Work

The standard approach for modelling video data is to use an architecture that is able to effectively capture the long-term dependencies of the video. A video is basically images that evolve dynamically with time, RNNs, particularly LSTMs have been extremely successful in capturing the long term dependencies that exist in the video-data [4]. Anomaly detection in videos is also usually tackled by using the LSTMs approach [5], the idea is to model normal behaviour and check for any deviations, the errors observed from the normal behaviour is calculated and parametrised by a regularity score, the LSTMs are used to predict the possible future behaviour by observing the video at hand, any deviations from the predictions made by the model are treated as anomaly. For example, a pedestrian walking on pedestal if modelled using the LSTM approach, the prediction usually will be that in a given later time, the pedestrian continues to walk on the pedestal, any deviation from this behaviour is calculated as an error which is later quantified by a regularity score to indicate the anomalous event.

There have been more traditional approaches that hinge on using Hand-crafted features to detect anomalies such as Cong et al. in [6] and in works proposed in [7], while such methods might prove successful in some cases, they become difficult to scale to more complex situations, because of non-ubiquitous nature of the hand-crafted features. In this paper, we have tried to exploit the idea of Two Stream CNN first introduced by Simonyan et al. [3], the rationale behind using two separate streams to model the spatial and temporal information stems from the fact that human beings tend to perceive information like shape, colour, texture and motion information through two distinct channels. This approach is extended to train two separate neural networks, one neural network is tasked to learn spatial information while the other neural network learns temporal-information, this temporal information is usually fed into, in the form of motion information. The motion information can be either optical flow or trajectory based motion-information, as discussed in [3]. The classification score obtained from both the streams are combined using different fusion techniques [8].

3 Methodology

In this section of the paper we discuss in depth the dataset chosen, the CNN architecture employed for training and testing, the scenario considered for testing and training, the various preprocessing steps involved in the training, the implementation details and evaluation of the results.

3.1 Dataset

Since the framework developed by us hinges on contextual based anomaly detection, we considered two different scenarios—airport and mall, the airport and mall surveillance datasets were acquired from Youtube. The datasets used for training and testing were challenging because most of the videos available on Youtube are either captured from very low-end devices by the bystanders present in and around the situation, which have lot of jittering and adverse lighting conditions or the videos could be a result of the surveillance systems installed at that particular location. There were totally 22 videos of airport scenario which were used for training and testing, each of varying length with events covering from passenger baggage theft, shootouts, brawls, bomb blasts etc. The mall surveillance dataset had 10 videos for training and testing with again a variety of situations.

3.2 Two Stream Architecture for Anomaly Detection

An overview of the Two Stream CNN architecture is provided in Fig. 1, a two stream architecture mainly involves training two different streams which capture complementary information. In case of video data, it is essential to learn the temporal dependencies so that the relationship between the consequent frames is established. Thus we have two streams—a spatial-stream and a temporal-stream that encode complementary information about the video. The spatial stream is basically a state-of-the-art CNN architecture which is trained for image-classification. We use the famous Inception V3 model which was used for the ImageNet Large Visual Recognition Challenge (2012). The architecture was trained for over a 1000 different classes, and reports a top-5 error percent of 3.46 [9]. Since our major motive is to deploy the solution in real-time, we decided to use a technique called Transfer-Learning described in the work [10], the idea is to use a well-trained model architecture like Inception v3, chop off the final layers and retrain them for new categories. Since the architecture already has been trained on vast amount of image data, this type of training provides us with quicker results and can be easily trained on systems that dont incorporate GPUs or have very limited training resources. The spatial stream was trained with RGB frames sampled at 30 fps from the video clips. The temporal stream was trained with the motion information across the video frames. To capture the motion information we performed the dense optical flow and saved them as grey optical images. Saving them as images, allowed us to once again train the Inception V3 architecture for new categories. Each stream gives a separate classification score for a given query

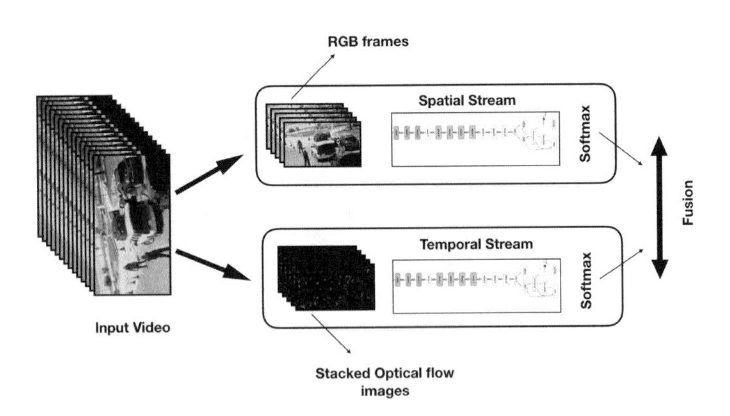

Fig. 1 An overview of the proposed two stream architecture for anomaly detection

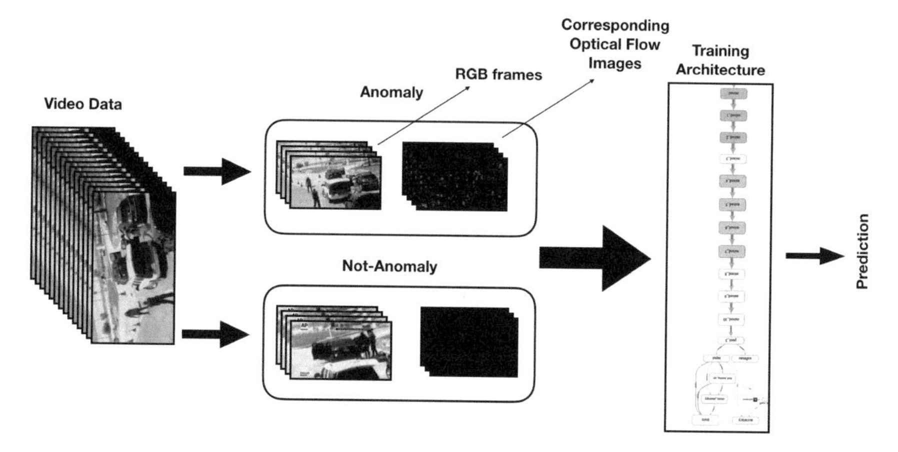

Fig. 2 The frames in the form of RGB images and their corresponding optical flow images are separately trained on a different CNN architecture to obtain two independent classification scores

frame which will later be combined using different fusion techniques [8]. In our case, we have considered simple averaging technique, wherein we fuse the classification scores from the two streams by taking a mean of those individual values to get a final classification score.

3.3 Implementation Details

The videos are trimmed to separate out the anomaly events and the preceding events which could be termed as normal behaviour. These video clips are converted to frames using the FFMPEG tool and are stored in two separate folders—anomaly and not-anomaly. The RGB frames will serve as input for the spatial stream for training. To train the temporal-stream, we consider the Farneback dense optical flow method [11], the video clips which contained the anomaly are subjected to optical flow, so that we have RGB frames and their corresponding motion information via the optical flow images. The optical flow images are saved as grey images with the white content indicating the motion and black, a lack of. The software used for training was TensorFlow, all the training and testing was done on a local machine, with configuration—8GB RAM and i5 processor with no dedicated GPU support. The implementation workflow is depicted in Fig. 2.

4 Results and Discussions

In this section we discuss the different scenarios considered for training and testing, their respective test-train accuracies. We discuss about the fusion technique used to combine the classification score of both the streams. Tables 1 and 2 shown below, summarises the train-test accuracy for the airport and mall surveillance videos. We have considered 80% of the images for main training, 10% of the images for validation and the final 10% of the images are kept for final prediction. We achieve a final test accuracy of 99.1 and 91% for airport scenario and a final testing accuracy of 94.7 and 90.1% for the mall scenario. From the tables we can infer that as the number of training images for the temporal stream has decreased, correspondingly we see a drop in the accuracy. This is compensated by the fusion technique, which helps us to arrive at a final score by combining the outputs of both the stream instead of basing our classification score on just one stream. Many different fusion strategies have been adopted to combine the scores of both the streams, literature [12] talks about different fusion techniques, in the context of action recognition, in our framework we have adopted a simple averaging technique that involves getting the final prediction score of both the streams for both the categories—anomaly and not-anomaly. We take the classification score of both the streams for both the categories and compute the mean, that is, the mean of the anomaly score from both the streams is computed, similarly the mean of the not-anomaly score is also computed, the final resultant score is assigned as a final classification score for the query frame. This method is demonstrated in Fig. 3, where we have a query frame with its corresponding RGB and Optical flow image, the Two stream CNN gives a classification score for the frame under consideration, later we compute the mean and assign a final classification score as to whether the given frame is anomalous or not-anomalous. In Fig. 3, the first image is the RGB image and the second image is its corresponding optical flow image. The classification score shown in the image, is the final score obtained after averaging the scores from both the streams.

Table 1 Scenario: airport surveillance videos

No. of training images	Train accuracy (%)	Cross-validation accuracy (%)	Final test accuracy (%)
3,44,128 (RGB)	98	98	99.1
3532 (Optical flow images)	91	89	91

Table 2 Scenario: mall surveillance videos

No. of training images	Train accuracy (%)	Cross-validation accuracy (%)	Final test accuracy (%)
61,946 (RGB)	96	93	94.7
3320 (Optical flow images)	94	87	90.1

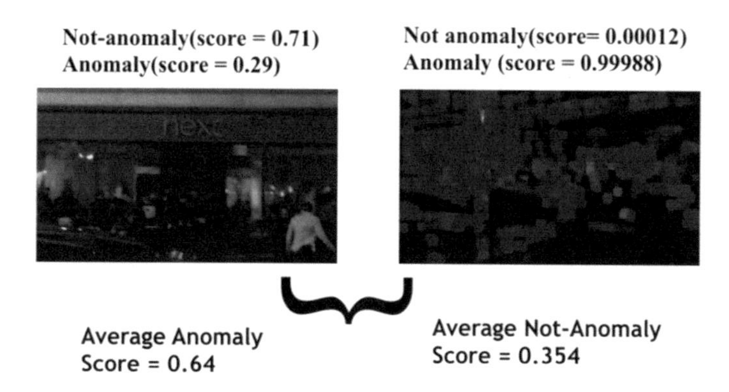

Fig. 3 The classification scores obtained for the RGB image and the optical flow image of a query frame are fused by taking averages, and a final score is assigned for the frame

5 Conclusion

In this paper, a Two Stream CNN for anomaly detection in surveillance videos is proposed. The proposed model exploits the spatial and temporal components of the video to provide effective classification of events as anomalous or not anomalous events. We have furthered this approach by creating a framework that is capable of working on live video feeds and classify each frame in real-time on the fly. This framework can further be improved by training an object classification architecture to localise anomalous events by annotating the frames, thus not only classifying a given frame as anomalous or not, but also localising where the anomaly is happening. The localising mechanism can also be based on a Reinforcement learning architecture, where the policy function helps in focusing on the relevant parts of an image.

References

1. Li, T., Chang, H., Wang, M., Ni, B., Hong, R., Yan, S.: Crowded scene analysis: a survey. arXiv:1502.01812v1 [cs.CV]. Last accessed 6 Feb 2015
2. Karpathy, A., Toderici, G., Shetty, S., Leung, T., Sukthankar, R., Fei-Fei, L.: Large-scale video classification with convolutional neural networks. In: CVPR (2014)
3. Simonyan, K., Zisserman, A.: Two-stream convolutional networks for action recognition in videos. In: Proceedings of the 27th International Conference on Neural Information Processing Systems, vol. 1 (2014)
4. Srivastava, N., Mansimov, E., Salakhudinov, R.: Unsupervised learning of video representations using LSTMs. arXiv:1502.04681v3 [cs.LG]. Last accessed 4 Jan 2016
5. Medel, J.R., Savakis, A.: Anomaly detection in video using predictive convolutional long short-term memory networks. arXiv:1612.00390; Cong, Y., Yuan, J., Liu, J.: Sparse reconstruction cost for abnormal event detection. In: CVPR, pp. 3449–3456 (2011)
6. Cong, Y., Yuan, J., Liu, J.: Sparse reconstruction cost for abnormal event detection. In: CVPR, pp. 3449–3456 (2011)

7. Mahadevan, V., Li, W., Bhalodia, V., Vasconcelos, N.: Anomaly detection in crowded scenes. In: Proceedings of IEEE Conference on Computer Vision and Pattern Recognition (2010)
8. Feichtenhofer, C., Pinz, A., Zisserman, A.: Convolutional two-stream network fusion for video action recognition. arXiv:1604.06573v2 [cs.CV]. Last accessed 26 Sep 2016
9. Szegedy, C., Vanhoucke, V., Ioffe, S., Shlens, J., Wojna, Z.: Rethinking the inception architecture for computer vision. arXiv:1512.00567 [cs.CV]
10. Donahue, J., Jia, Y., Vinyals, O., Hoffman, J., Zhang, N., Tzeng, E., Darrell, T.: DeCAF: a deep convolutional activation feature for generic visual recognition. arXiv:1310.1531v1 [cs.CV]. Last Accessed 6 Oct 2013
11. Farneback, G.: Two-frame motion estimation based on polynomial expansion. In: Proceedings of the 13th Scandinavian Conference on Image Analysis (2003); Liu, W., Anguelov, D., Erhan, D., Szegedy, C., Reed, S., Fu, C.Y., Berg, A.C.: SSD: single shot multibox detector. arXiv:1512.02325 [cs.CV]
12. Lin, K., Chen, S.-C., Chen, C.-S., Lin, D.-T., Hung, Y.-P.: Abandoned object detection via temporal consistency modelling and back-tracing verification for visual surveillance. IEEE Trans. Inf. Forensic Secur. (TIFS) (2015)

Detection and Classification of Road Signs Using HOG-SVM Method

Anant Ram Dubey, Nidhi Shukla and Divya Kumar

Abstract Artificially intelligent systems are becoming a crucial part of the future we are heading to. In this paper, we aim to produce an intelligent software system that can recognize the traffic warning signs. This paper also describes a road sign recognition (RSR) system, a step toward driverless cars. It uses the color-based localization and classification capabilities of support vector machines over the features described as histogram of oriented gradients. The proposed system is exhaustively tested on a dataset of realistic images. The obtained encouraging results validate the potential of the proposed method.

Keywords Computer vision · Color-based recognition · Color thresholding · HOG · Support vector machine

1 Introduction

With the increasing speed of vehicles, it has become prominently important to ensure the follow-up of traffic rules and traffic signs. Traffic signs are the integral part of road safety as these signs not only guide the drivers but also alert them in potentially dangerous situations. Thus, the automatic detection and recognition of road signs has become an intriguing area of computer vision. In the recent times, a lot of research is being done in making artificially intelligent objects, and the importance of safety for drivers, occupants and pedestrians has received an increasing impetus in the research community. For ensuring the road safety, there is an increasing demand for a precise software application that could capture images from the surrounding using an itinerant camera. From the captured image it should recognize the road

---

A. R. Dubey · N. Shukla · D. Kumar (✉)
Department of Computer Science and Engineering, Motilal Nehru National Institute of Technology, Allahabad 211004, India
e-mail: divyak@mnnit.ac.in

A. Elçi et al. (eds.), *Smart Computing Paradigms: New Progresses and Challenges*, Advances in Intelligent Systems and Computing 766,
https://doi.org/10.1007/978-981-13-9683-0_6

signs from natural scene. The proposed road sign recognition (RSR) system fulfills these requirements and finds its utility in being used for creating "driver-less cars". Also, the algorithm developed can be used in making an application to assist blind, visually challenged and color blind people in their normal day-to-day life. To describe the proposed system the flow of this paper goes in the following manner: first, we discuss the related works in Sect. 2, then in Sect. 3 we present the RSR localization and classification system. Finally, the experimental details, results and limitations are detailed in Sects. 4, 5 and 6, respectively.

2 Related Works

There is a lot of research work that is being done in the area of road sign detection. On the basis of various models, scientists and researchers have worked on various techniques. Among the techniques based on RGB model, Benallal et al. [1] used the changes in illumination on road signs and captured the nature of RGB components with the change. For color segmentation, the difference of any two components was taken. Not just RGB, but models like HSV, HSI, YIQ, YCbCr and CIExyz have also been used for the research. Using the HSV model, Ching-Hao Lai [2] proposed color quantization. Owing to the fact that YCbCr color space is independent of variable color illumination, Jitendra et al. [3] used it for color segmentation. Many research works demonstrate the use of HSI model as well. This color space is similar to the human eye color perception. Color thresholding is yet another technique that identifies pixels of the image as object pixels and background pixels. The pixels closer to the required color are picked up as object pixels, rest as background pixels. Shape is another feature that can be used for road sign detection and recognition. Barnes and Zelinsky [4] detected circular road signs using the variation of Hough transform.

There are other methods for road sign detection that involve sliding window. These methods make use of histogram of oriented gradients [5], support vector machines, multi-scale shape filter or classical Viola-Jones-like detector [6]. The algorithm in [7] proposes to detect traffic signs using a rough-to-fine sliding window scheme wherein rough detection of ROIs is done. For feature extraction, recent methods such as the methods discussed in [8, 9] use HOG features. Making use of the CIELAB and YCbCr color models, Creusen et al. [8] used the HOG algorithm and incorporated color information. In New Zealand, two variant formulations of HOG features were used by Overett et al. [9] for the detection of speed signs. Apart from the above, the Adaboost algorithm [10] is used to solve the supervised pattern recognition problem by combining (weak) classifiers.

However, it is a common glitch in most of the proposed methods that their performance is really slow, rendering them incapable for application in real-time problems. There are some methods like [11–15] where many seconds are required to work on a single frame. Some commercial methods have also been introduced in the market, including Volkswagen Media Services [16] and Mobileye [17]. However, these commercial systems are able to recognize a handful of traffic signs only; consider the example of the system developed by Mobileye [17]. This system can only detect speed limit signs and no-overtaking signs.

3 Traffic Sign Detection Module

The traffic signs that we see around have well-defined shape and color. They have a high contrast that separates them from the natural surroundings. By making use of these characteristics, we extract the traffic sign candidates from a natural scene image. These extracted candidates may contain lots of false positives. To filter them out, we have used SVM classifier that has been trained using HOG features. The detection module used in our project makes use of color-based recognition and extracts HOG features of the road sign candidate extracted. The details of the module are given in the following sections.

3.1 Road Sign Candidate Extraction

This stage involves the use of color-based recognition for extracting out those sections of the image that may qualify as a traffic sign candidate. The colors used in road signs in different countries are mostly primary colors. Here the traffic signs are segmented according to the colors, and a region of interest is produced which is used in the further process.

The color model used in the process is HSV, since it is closer to human eye perception. Also, the hue value is illumination-invariant. Carrying out segmentation operation is easier in HSV color space as compared to RGB in which the three-dimensional and highly correlated coordinates make it even more difficult. After the image is converted into HSV, we find out the background and foreground pixels using color thresholding. Each color has its own HSV values. Authors in [18] suggested that in a HSV model, the normalized hue values for red color must be greater than 0.95 or lesser than 0.05. Similarly, the ranges for saturation and value, used in the experiments, are (0.5, 1] and (0.1, 1] respectively. The values enlisted here for red can be used while color thresholding. If a pixel's value meets the threshold color criteria, it would be selected as object pixel or foreground pixel. Object pixel is represented as 1, while background pixel is represented by 0. From this binary image, the eight-connectivity of the selected pixels is found out and the image is labeled accordingly. Also, since smaller objects like red flowers and so on could also be selected, we consider only those clusters which are between 200 and 5000 pixels. This labeled road sign candidate is then extracted and represented in YCbCr model, as shown in Fig. 1.

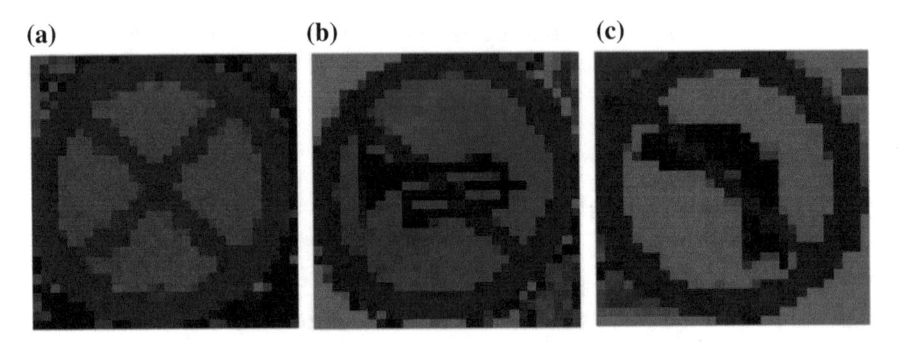

Fig. 1 The extracted images of sign boards in YCbCr model: **a** No Parking **b** No Horn **c** No Turn Left

3.2 Traffic Sign Detection Using HOG

The extracted image reduces the size of the search space but at the same time the proposed traffic sign candidates may also contain a lot of false positives. To remove these false positives, we classify these extracted regions using SVM classifier which has been trained using the HOG features of various signs. Edges and their orientations have a crucial role to play in defining the important features that help recognize an object in any natural scene. HOG is the shape descriptor based on gradient direction that is used in face detection, human detection, traffic sign detection and so on. While extracting features, the image gets divided into several overlapping blocks which comprise non-overlapping cells.

The gradient orientation and magnitude of each pixel is calculated after which a histogram of these orientations is formed for all the cells. Normalization is done to improve the result. All these histograms are later on concatenated to form the final HOG descriptor. The various steps for implementation of feature extraction by HOG algorithm are gradient calculation, orientation binning, descriptor block and block normalization. In the detection module, we ascertained that the extracted object belongs to the category of traffic signs. The classification module further helps to put them in their respective sub-class, for example, whether the extracted road sign is a speed limit sign or a stop sign. We have used support vector machines as the classifier and all the images are resized to 30×30 since all the inputs to SVM must be of the same size as shown in Fig. 2.

4 Implementation Details

The detection and classification algorithm was tested on images which comprise the following different road signs: Stop Signs, No parking Signs, Give Way Signs, No Horn zone, No turn left/right and Speed Limit Signs. There were 720 images that we

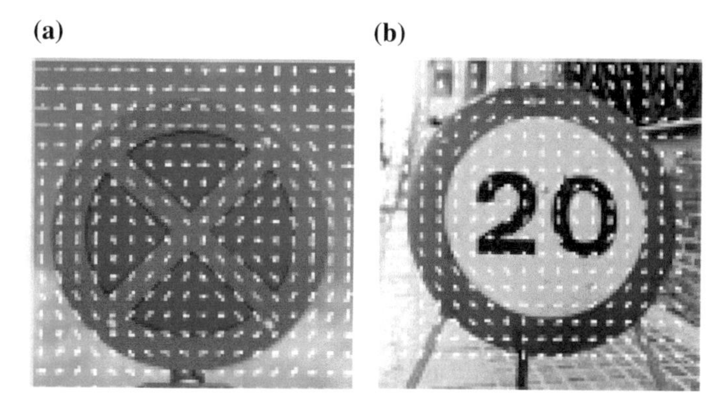

Fig. 2 Extracted HOG features of **a** No Parking Road sign **b** Speed Limit Road sign

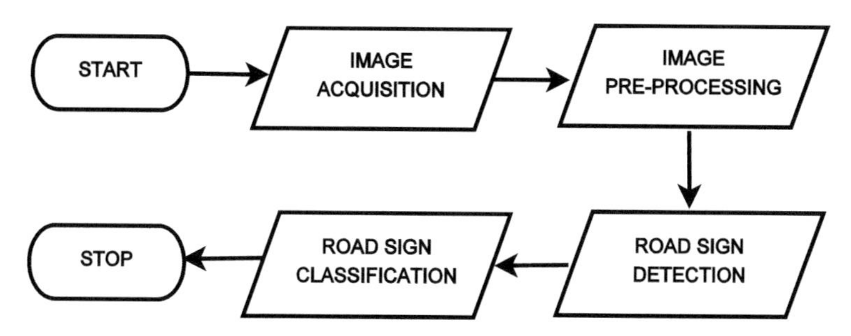

Fig. 3 Workflow adopted wherein the acquired image containing the traffic sign is preprocessed to remove noise and convert to HSV model. SVM classifier, trained using HOG features, is used to classify the images into their respective classes

used in the dataset. The traffic signs being classified are Stop Sign, No Parking Sign, Give Way Sign, No Entry Sign, No Horn Sign, No Turn Left/Right Sign and Speed Limit (20 kmph) Sign. Each of them has a representation of 90 images in the dataset. For training 600 signs have been used, while for testing 120 were used. The images used for training were sized as 150 × 150 pixels. We used MATLAB 2015a for the implementation and analysis of the proposed methodology. Applications available in MATLAB like Classification Learner App were used for training SVM classifier. For extraction of HOG features, "extractHOGFeature" function was used. Figure 3 shows the workflow of the experimental setup.

5 Results and Conclusion

The algorithm was tried on 120 test images belonging to each category of traffic signs outlined in previous section. The images were taken in varying daylight. To avoid

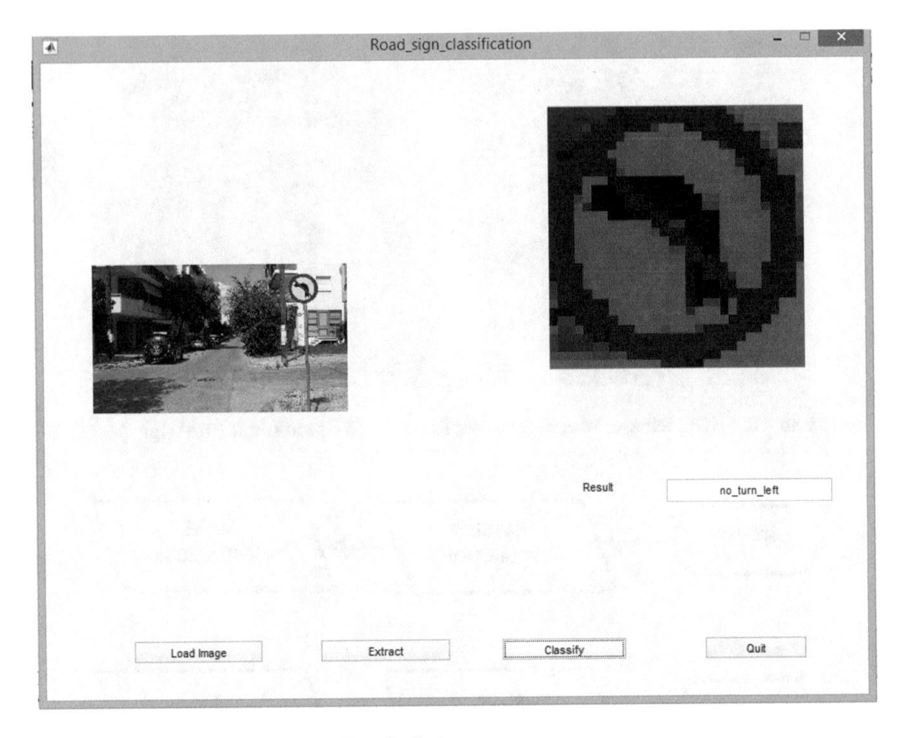

Fig. 4 Result obtained for Do Not Turn Left sign

over fitting at any stage of project, we also kept a set of images which contained blurred traffic signs and low contrast signs. The images in which the traffic signs were obstructed by some object were also considered in out experiment. Hence, the validity of the results can be justified on real-time images also. Screenshots of sample results obtained from the program are shown in Figs. 4 and 5. The summary of results is presented in Table 1. From the results it can be claimed that the proposed RSR system exhibits a considerable precision and it can be used as a driving-assistance system. Further, RSR can also be deployed for automated surveillance of road traffic devices which is desirable to design smart car control systems, assisting visually challenged people and GPS-based navigation system.

6 Limitations

The suggested approach does not work efficiently for extremely low-contrast images. The images clicked in late evening hours and the images clicked with viewing angle also face troubles in recognition. It is also observed that the chances for detection of road signs located far away become considerably less.

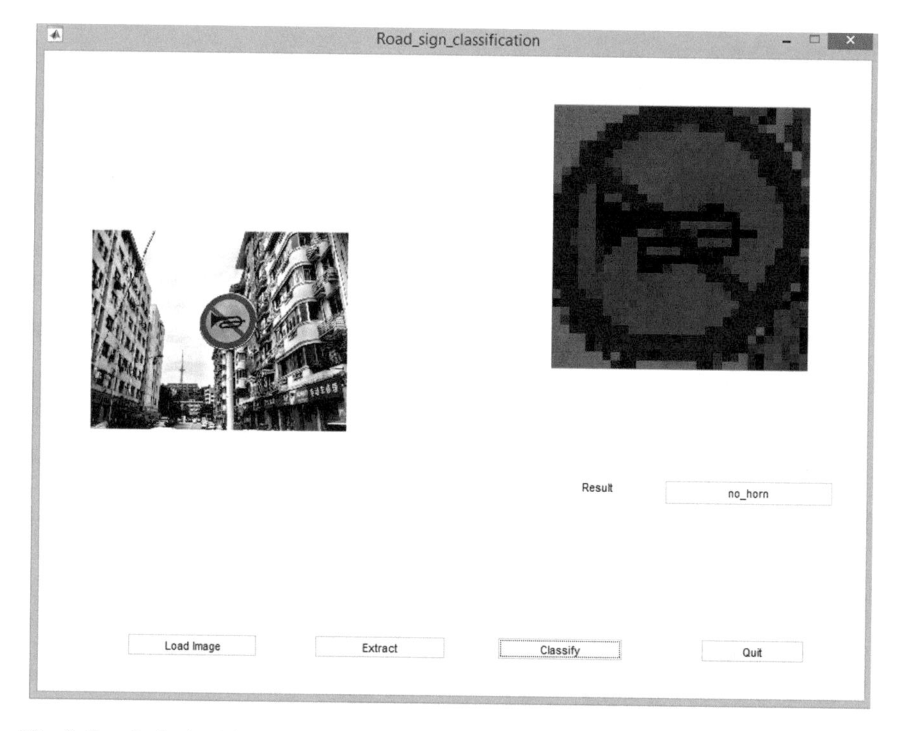

Fig. 5 Result obtained for No Horn sign

Table 1 The results obtained on test data. Each sign type is tested through 15 different images

Sign type	Sign localized	Sign identified	Percentage accuracy	Confusing sign
No Parking	12	10	66.67	Stop, Speed Limit
Speed Limit (20)	15	15	100	Speed Limit 40, Give Way
Give Way	13	13	86.67	
Stop	13	13	86.67	
No Entry	15	13	86.67	No Turn Right
No Horn	11	10	66.6	No Parking
No Right Turn	15	15	100	
No Left Turn	13	12	80	No Parking

References

1. Benallal, M., Meunier, J.: Real-time color segmentation of road signs. In: Proceedings of the Electrical and Computer Engineering Conference, vol. 3, pp. 1823–1826. Canada (2003)
2. Lai, C.H., Yu, C.-C.: An efficient real-time traffic sign recognition system for intelligent vehicles with smart phones. In: International Conference on Technologies and Applications of Artificial Intelligence, pp. 195–202 (2010)
3. Chourasia, J.N., Bajaj, P.: Centroid based detection algorithm for hybrid traffic sign recognition system. In: Third International Conference on Emerging Trends in Engineering and Technology, pp. 96–100 (2010)
4. Barnes, N., Zelinsky, A.: Real-time radial symmetry for speed sign detection. In: Proceedings of the Intelligent Vehicles Symposium, pp. 566–571. Italy (2004)
5. Dalal, N., Triggs, B.: Histograms of oriented gradients for human detection. In: 2005 IEEE Computer Society Conference on Computer Vision and Pattern Recognition, CVPR 2005, vol. 1, pp. 886–893 (2005)
6. Jones, M.J., Viola, P.: Robust real-time object detection. In: Workshop on Statistical and Computational Theories of Vision, vol. 266 (2001)
7. Wang, G., Ren, G., Wu, Z., Zhao, Y., Jiang, L.: A robust, coarse-to-fine traffic sign detection method. In: The 2013 International Joint Conference on Neural Networks (IJCNN), pp. 1–5. IEEE (2013)
8. Creusen, I., Wijnhoven, R., Herbschleb, E.: Color exploitation in HOG-based traffic sign detection. In: 2010 17th IEEE International Conference on Image Processing (ICIP), pp. 2669–2672. IEEE (2010)
9. Overett, G: Large scale sign detection using HOG feature variants. In: Intelligent Vehicles Symposium (IV), 2011 IEEE, pp. 326–331. IEEE (2011)
10. Freund, Y., Schapire, R.E.: A Decision Theoretic Generalization Online Learning and an Application to Boosting. J. Comput. Syst. Sci. **55**(1), 119–139 (1997)
11. Vitrià, Jordi, Baró, Xavier: Traffic Sign Detection on Greyscale Image. Congrés Català d'Intelligència Artificial, Spain (2004)
12. H. Pazhoumand-Dar and M. Yaghobi, "DTBSVMs: A New Approach for Road Sign Recognition", in Proc. ICCICSN, pp. 314–319, Jul. 2010
13. Maldonado-Bascón, S., Lafuente-Arroyo, S., Gil-Jimenez, P., Gomez-Moreno, H., Lopez-Ferreras, F.: Road-sign detection and recognition based on support vector machines. IEEE Trans. Intell. Transp. Syst. **8**(2), 264–278 (2007)
14. Reiterer, A.: Automated traffic sign detection for modern driver assistance systems. In: Proceedings of the Facing the Challenges Building the Capacity, pp. 11–16. Australia (2010)
15. Khan, J.F., Bhuiyan, S.M.A., Adhami, R.R.: Image segmentation and shape analysis for road-sign detection. IEEE Trans. Intell. Transp. Syst. **12**(1), 83–96 (2011)
16. Volkswagen Media Services: Phaeton debuts with new design and new technologies. https://www.volkswagen-media-services.com/. Last accessed 15 April 2018
17. Mobileye: Traffic Sign Detection. http://mobileye.com/technology/applications/traffic-sign-detection/. Last accessed 15 April 2018
18. Liu, H.X., Ran, B.: Vision-based stop sign detection and recognition system. J. Transp. Res. **1748**, 161–166 (2001)

Segmentation of Calcified Plaques in Intravascular Ultrasound Images

Tara Chand Ulli and Deep Gupta

Abstract Intravascular ultrasound (IVUS) imaging is mostly used in the diagnosis and treatment of coronary artery diseases, especially in atherosclerosis, because it becomes very difficult to identify in the calcified regions manually. The IVUS images allow to visualize the inner portion of the coronary artery with enhanced resolution and also to acquire the cross-sectional images of arteries. Therefore, this paper presents a computational framework to identify the calcified region in IVUS images. In this paper, spatial fuzzy C-means approach is used to extract the exact boundary of the calcified plaque region in the IVUS images along with the wavelet transform decomposition. This clustering approach is capable of incorporating additional spatial information obtained from the neighboring pixels and also overcoming the limitations of noise and artifacts in IVUS coronary images. Several experiments have been performed on the different IVUS data and their experimental results are analyzed in terms of both quantitative and qualitative manner. The results revealed that the spatial fuzzy C-means provides better segmentation accuracy by extracting the calcified region as compared with other approaches.

Keywords IVUS · Coronary artery · Spatial fuzzy C-means · Calcified plaque

1 Introduction

Atherosclerosis is a common cardiovascular disease having a disastrous effect on the human life all around the world [1, 2]. It is a disease in which an artery wall gets thickened due to accumulation of dead white blood cells, following the low-density lipoproteins forming a hard inelastic tissue known as plaque. Plaque formed in the artery creates a blockage of the oxygen-rich blood flowing into the heart, causing

---

T. C. Ulli (✉) · D. Gupta
Department of Electronics and Communication Engineering, Visvesvaraya National Institute of Technology, Nagpur 440010, India
e-mail: ullitarachand@gmail.com

D. Gupta
e-mail: deepgupta@ece.vnit.ac.in

A. Elçi et al. (eds.), *Smart Computing Paradigms: New Progresses and Challenges*,
Advances in Intelligent Systems and Computing 766,
https://doi.org/10.1007/978-981-13-9683-0_7

serious health issues [3, 4]. It is obligatory for a cardiologist to identify the exact location of the calcified plaque and its volume before performing the percutaneous coronary interventional (PCI) procedure [2, 5–7]. There are various ways of imaging coronary artery, which include optical coherence tomography (OCT), computed tomography (CT), magnetic resonance imaging (MRI), and so on. All these medical imaging modalities play a very prominent role in the diagnosis and treatment of cardiovascular diseases [8]. Among all the imaging modalities, IVUS is the most commonly used imaging modality, to diagnose cardiovascular diseases, and it is comparatively less expensive, require less time and capable of producing real-time data [2, 9, 10].

In recent days, the normal procedure of calcium volume estimation is carried out by processing the image taken from an IVUS scanner, and applying various region extraction methods to find the calcium volume [11–13]. The most commonly used segmentation methods are k-means [14], fuzzy C-means (FCM) [15–17], active contour model [18, 19], level set approach [20, 21], and so on. IVUS scanner generates videos consisting of large number of frames (around 2040 approximately); therefore, it becomes very tedious and time-consuming, if the analysis is done manually. However, it may produce a lot of errors that affect the diagnosis efficiency [2, 22]. Therefore, there is a need to estimate the calcium volume by developing the most efficient computational approaches. IVUS images are larger in size, so they are required to be downsample, using different multi-resolution techniques in order to reduce the processing time [23]. Multi-resolution techniques are of two types: adaptive and non-adaptive [2]. Adaptive techniques make use of the factors like edge information and texture intensity values as landmarks [24]. There are various adaptive techniques, namely seam carving, context-aware resizing, segment-based and wrapping-based [25]. Non-adaptive multi-resolution techniques do not make use of image features; instead, they directly manipulate the pixels intensities. Non-adaptive multi-resolution techniques are very easy to perform and are most likely to be preferred [23]. The most popularly known non-adaptive multi-resolution methods are bilinear [26], bicubic [27], discrete wavelet [28], Lanczos [29] and Gaussian pyramid [2, 30]. In the case of non-adaptive techniques, only low-frequency components are stored, thereby leading a path toward blurring and artifacts. The high-frequency components must be present in the image in order to retain the visual quality of an image. Adaptive multi-resolution techniques are capable of preserving the high-frequency components, however, these methods are mathematically complex and require more time compared to non-adaptive multi-resolution techniques [23, 25]. Therefore, in the presented paper, non-adaptive techniques combined with suitable bias correction are used to identify the calcified plaque regions in the IVUS images.

2 Methodology

This section explains the various methodologies that are used to identify the calcified regions. The process to identify the calcified plaque is shown in Fig. 1, as similarly discussed in [2], except the segmentation technique, that is, spatial fuzzy

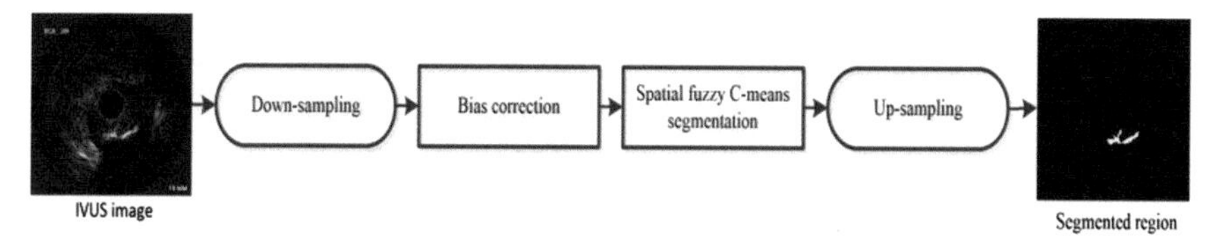

Fig. 1 Block diagram of the overall system flow

C-means [16]. The first step is to process the image with any one non-adaptive technique followed by bias correction. After the bias correction is done, spatial fuzzy C-means clustering approach is utilized on the IVUS image to extract the calcified plaque region with the multi-resolution property of discrete wavelet transformation technique.

2.1 Discrete Wavelet Transform

The implementation of discrete wavelet transform (DWT) is carried out by a simple filtering operation with well-specified filter coefficients [25, 28]. The forward discrete wavelet transform computed as the input signal (x) is first filtered by a low-pass filter ($\hat{g}$), and the same input (x) is filtered by a high-pass filter ($\hat{j}$), separately. In the outputs of both the filters, alternate output samples are dropped out to produce (y_L) and (y_H) as low-pass and high-pass outputs, respectively, as shown in Fig. 2. The mathematical expressions are given as follows:

$$y_L(n) = \sum_{i=0}^{\tau_L - 1} \hat{g}(\mathrm{i}) \times \mathrm{x}(2\mathrm{n} - \mathrm{i}) \tag{1}$$

$$y_H(n) = \sum_{i=0}^{\tau_H - 1} \hat{j}(\mathrm{i}) \times \mathrm{x}(2\mathrm{n} - \mathrm{i}) \tag{2}$$

where τ_L and τ_H refer to the lengths of low-pass and high-pass filters, respectively.

In the inverse process of the DWT, both the low-pass and high-pass outputs of the forward system are taken as inputs for the reverse system (synthesis). Both the outputs of the forward system y_L and y_H are taken separately and up-sampled simply by inserting zeros and filtered by the low-pass filter (g) and a high-pass filter (j),

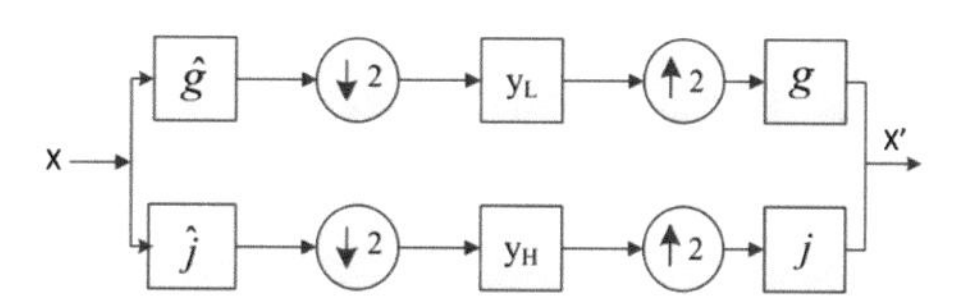

Fig. 2 DWT system analysis followed by synthesis

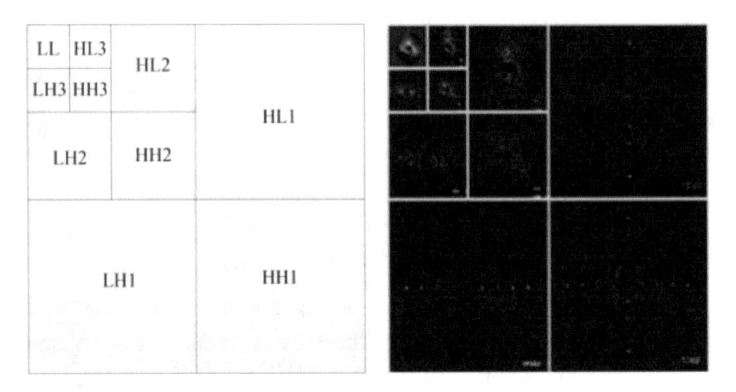

Fig. 3 Third-level DWT decomposition of an IVUS image

respectively. The respective outputs after filtering are summed up to produce x' as shown in Fig. 2. Third-level decomposition of IVUS images is shown in Fig. 3.

2.2 *Spatial Fuzzy C-Means Clustering*

Fuzzy clustering is an adaptive thresholding technique in which the cluster centers and the cluster elements are estimated in an iterative manner by minimizing the cost function [31]. FCM is a standard clustering algorithm that is being extensively implemented in the medical applications. For a data set given by X ($X = x_j$, j = 1, 2……n), and given c (number of clusters) the mathematical expression of the cost function in conventional FCM is expressed as:

$$J_{FCM} = \sum_{i=1}^{c}\sum_{j=1}^{N} \mu_{ij}^{m} d_{ij}^{2} \quad (3)$$

$$d_{ij}^{2} = \left\| x_j - v_i \right\|^{2},\ 1 \le \mathrm{i} \le \mathrm{c} \text{ and } 1 \le \mathrm{j} \le \mathrm{n} \quad (4)$$

where m is fuzziness controlling parameter that is taken >1. The membership functions obtained by minimization of the cost function should satisfy the following constraints:

$$\sum_{i=1}^{c} \mu_{ij} = 1;\ 0 \le \mu_{ij} \le 1;\ \sum_{j=1}^{n} \mu_{ij} > 0 \quad (5)$$

The cluster centers and the membership functions updated iteratively are expressed as:

$$\mu_{ij} = \frac{1}{\sum_{l=1}^{c}\left(\frac{d_{ij}^2}{d_{lj}^2}\right)^{\frac{1}{m-1}}} \tag{6}$$

$$v_i = \frac{\sum_{j=1}^{N}\mu_{ij}^m x_j}{\sum_{j=1}^{N}\mu_{ij}^m} \tag{7}$$

The conventional FCM approach does not provide any spatial information. Moreover, the medical images are also vulnerable to noise and artifacts that decrease the performance of the segmentation techniques. Therefore, there is a need to incorporate the spatial information of the pixels that will provide the better results and segmentation accuracy also.

As mentioned in [31], spatial information is included in the conventional fuzzy membership functions directly such that the performance of the segmentation method is enhanced. The membership function after including the spatial information is expressed as given in Eq. 8.

$$\mu_{ij}' = \frac{\mu_{ij}^p h_{ij}^q}{\sum_{l=1}^{c}\mu_{lj}^p h_{lj}^q} \tag{8}$$

$$h_{ij} = \sum_{k \varepsilon N_n}\mu_{jk} \tag{9}$$

where p and q are contribution parameters and h_{ij} is a variable that incorporates the spatial information. N_n denotes a window centered on the image pixel j.

3 Implementation Steps

In the presented paper, discrete wavelet transform and spatial fuzzy C-means (SpFCM) approach are utilized. The salient implementation steps are given as:

Step 1 *First, start with the IVUS images acquired from the different patient data set.*

Step 2 *Downsample the image using the multi-resolution technique (discrete wavelet transform).*

Step 3 *Do the bias correction because high-frequency components may be lost during the downsampling process.*

Step 4 *Apply spatial fuzzy C-means segmentation approach as mentioned in Sect. 2.2.*

Step 5 *Choose the number of clusters (c) and minimize the cost function J_{FCM} by updating the membership function μ_{ij} and the cluster center v_i.*

Step 6 *Extra spatial information h_{ij} is also included in the conventional membership function, making the segmentation process more immune to image noise and instrumental disturbances.*

Table 1 Performance parameters and their respective mathematical equations

Parameters used	Mathematical expression
Jaccard index (JCI) [2]	$J_{XY} = \frac{\lvert X \cap Y \rvert}{\lvert X \cup Y \rvert}$
Dice similarity coefficient (DSC) [2]	$D_{XY} = \frac{2\lvert X \cap Y \rvert}{\lvert X \rvert + \lvert Y \rvert}$
Sensitivity (SN) [32]	$S_{XY} = \frac{TP}{TP+FN}$
Specificity (SP) [32]	$sp_{XY} = \frac{TN}{TN+FP}$
Accuracy (AC) [32]	$ACC_{XY} = \frac{TP+TN}{TP+TN+FP+FN}$

Step 7 *Apply some morphological operations to extract the identified region and to generate the binary image to visualize the calcified plaque region properly*

4 Performance Measures

There are various performance evaluation parameters available in the literature; the most commonly used performance measures are true positive (TP), true negative (TN), false positive (FP) and false negative (FN). Based on all these four parameters, different measures such as Jaccard index, dice similarity coefficient, sensitivity, specificity and accuracy are used to evaluate the performance of the presented approach and the other existing approaches. Their mathematical expressions are given in Table 1.

X is a binary image region segmented using computational approach and Y is the reference region segmented manually by an expert.

5 Results and Discussions

In this paper, the calcified plaque region is identified by spatial fuzzy C-means approach using the multi-resolution property of wavelet decomposition. To evaluate the performance of SpFCM, a large number of patient IVUS data are taken from [12], out of which three are considered to present the performance of the segmentation approach. Moreover, the segmentation performance of the presented approach is also compared with the other existing multi-resolution approaches and without multi-resolution technique, as given in the following:

Method 1 Bilinear multi-resolution technique as described in [26] with the size of the source image (192 × 192).

Method 2 Bicubic multi-resolution algorithm as described in [27] with the size of the source image (192 × 192).

Method 3 Lanczos multi-resolution algorithm as described in [28] with the size of the source image (192 × 192).

Method 4 Gaussian pyramid multi-resolution algorithm as described in [29] with the size of the source image (192 × 192).

Method 5 Wavelet multi-resolution algorithm as described in [30] with the size of the source image (192 × 192).

Method 6 Without multi-resolution technique using spatial fuzzy C-means approach with the size of the source image (384 × 384)

To assess the performance of the above-mentioned segmented approaches, three different frames of IVUS data are taken and shown in Figs. 4, 5 and 6, respectively. Their corresponding segmented results are shown in Figs. 4, 5 and 6 in column 2–7, respectively. In addition to visual segmented results, quantitative measures are also computed as mentioned above. The comparative performance of all the above-mentioned approaches is shown in Table 2. From Table 2, it is observed that the wavelet-based approach provides better segmentation results with the higher averaged accuracy of 92.60%, when the resolution is reduced to half. The presented approach produces better segmentation accuracy as compared with the other multi-

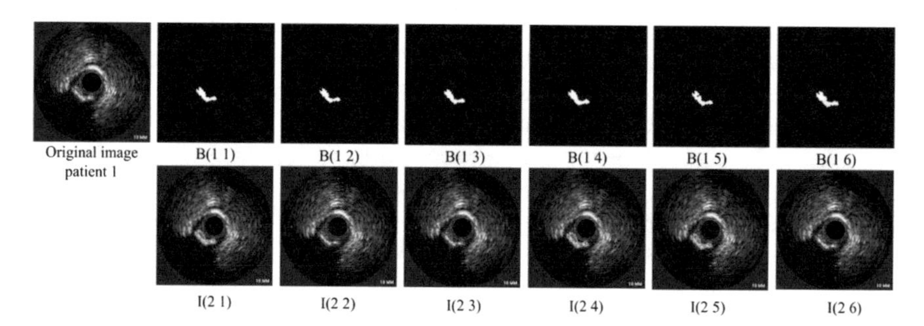

Fig. 4 Images B(1 1)–B(1 5) are binary images after spatial fuzzy C-means segmentation using bilinear, bicubic, Lanczos, Gaussian pyramid and wavelet multi-resolution methods, respectively, and B(1 6) is without any multi-resolution. I(2 1)–I(2 6) are their respective images after boundary detection

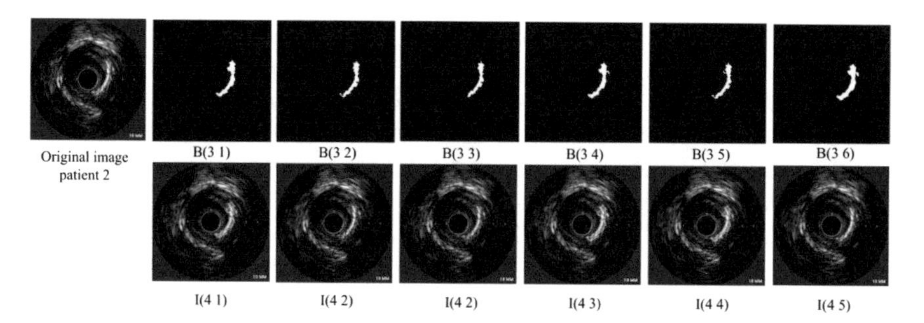

Fig. 5 Images B(3 1)–B(3 5) are binary images after spatial fuzzy C-means segmentation using bilinear, bicubic, Lanczos, Gaussian pyramid and wavelet multi-resolution methods, respectively, and B(3 6) is without any multi-resolution. I(4 1)–I(4 6) are their respective images after boundary detection

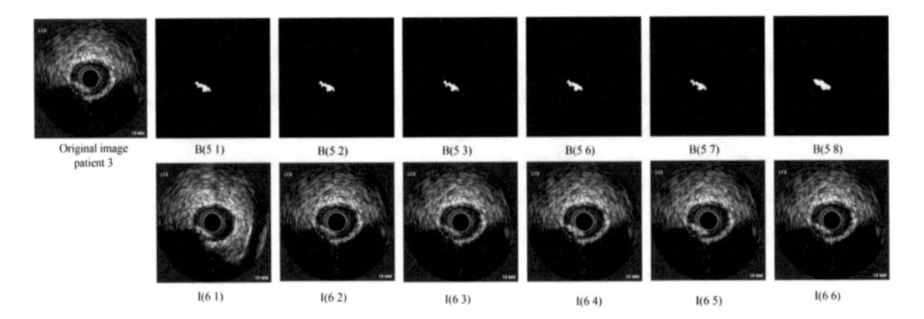

Fig. 6 Images B(5 1)–B(5 5) are binary images after spatial fuzzy C-means segmentation using bilinear, bicubic, Lanczos, Gaussian pyramid and wavelet multi-resolution methods, respectively, and B(5 6) is without any multi-resolution. I(6 1)–I(6 6) are their respective images after boundary detection

resolution methods, such as bilinear, bicubic, Lanczos and Gaussian pyramid, which provide averaged accuracy of 91.30, 91.69, 92.17 and 92.01%, respectively. Furthermore, without multi-resolution approach, SpFCM-based technique provides highest segmentation accuracy approx. 93.33% among all the approaches; however, the processing time is too large and wavelet-based technique requires less processing time and also produces approximately similar results along with all the quantitative measures. In addition to accuracy, other performance measures, such as JCI, DSC, SN and SP, are also computed for each individual approach separately. From their results shown in Table 2, a similar observation is provided such that the spatial fuzzy C-means approach with wavelet decomposition provides better segmentation results with higher accuracy. The JCI, DSC, SN and SP values are close to 6 as compared with the other approaches.

6 Conclusions

In this paper, spatial fuzzy C-means (SpFCM) is used as a clustering approach to extract the calcified plaque region in the IVUS image. The position and area of the classified plaque inside the blood vessel are identified properly, and its boundary is detected on the image which is commonly used in the treatment of patients suffering from atherosclerosis. Before segmenting the images, its resolution is reduced using a set of multi-resolution methods, so that the processing time required for segmentation is reduced as compared with the other methods. Along with the information provided by conventional FCM in clustering algorithm, spatial information is also included improving the segmentation performance resulting in the correct identification and estimation of boundaries and increasing the immunity toward image noise. From the experimental results, it is concluded that without multi-resolution, the SpFCM approach provides better results, but it takes too much time. Moreover,

Table 2 Quantitative results obtained by spatial fuzzy C-means segmentation for various multi-resolution methods

No.	Methods	Performance measures				
		Jaccard index	Dice similarity coefficient	Sensitivity	Specificity	Accuracy (%)
1	Method 1	0.6182 ± 0.064	0.7388 ± 0.049	0.6542 ± 0.108	0.9701 ± 0.0288	92.4177 ± 1.4630
	Method 2	0.6001 ± 0.1091	0.726 ± 0.0831	0.6324 ± 0.1266	0.9719 ± 0.0135	92.6879 ± 1.8692
	Method 3	0.5875 ± 0.113	0.741 ± 0.0866	0.669 ± 0.1280	0.9732 ± 0.0129	92.5476 ± 1.9566
	Method 4	0.6254 ± 0.0451	0.7391 ± 0.0351	0.6572 ± 0.1035	0.9704 ± 0.0215	92.7521 ± 1.0787
	Method 5	0.6453 ± 0.0823	0.741 ± 0.0653	0.6727 ± 0.090	0.9721 ± 0.0096	93.3869 ± 1.5861
	Method 6	0.6577 ± 0.1055	0.780 ± 0.0215	0.7371 ± 0.0112	0.9747 ± 0.0112	93.6807 ± 0.9985
2	Method 1	0.5461 ± 0.0997	0.7006 ± 0.0286	0.618 ± 0.1549	0.9068 ± 0.0286	90.6758 ± 3.6047
	Method 2	0.5199 ± 0.0239	0.6745 ± 0.0247	0.5799 ± 0.1689	0.9207 ± 0.0247	90.8344 ± 4.2267
	Method 3	0.6276 ± 0.0884	0.7676 ± 0.0698	0.8094 ± 0.1304	0.9298 ± 0.0698	91.1486 ± 4.3606
	Method 4	0.590 ± 0.0805	0.7387 ± 0.0832	0.8167 ± 0.1257	0.8947 ± 0.0832	89.2664 ± 5.4676
	Method 5	0.6292 ± 0.0913	0.7583 ± 0.0895	0.8179 ± 0.1295	0.9495 ± 0.0895	91.4850 ± 5.9025
	Method 6	0.6542 ± 0.0753	0.7865 ± 0.0258	0.8221 ± 0.1265	0.9633 ± 0.0465	92.6521 ± 4.5067
3	Method 1	0.5974 ± 0.0640	0.7341 ± 0.0497	0.6418 ± 0.108	0.906 ± 0.0288	90.8085 ± 1.4630
	Method 2	0.6015 ± 0.1091	0.7388 ± 0.0831	0.6467 ± 0.1266	0.9327 ± 0.0135	91.5625 ± 1.8692
	Method 3	0.6197 ± ±0.113	0.7516 ± 0.0866	0.6827 ± 0.128	0.9578 ± 0.0129	92.8365 ± 1.9566
	Method 4	0.6237 ± 0.0451	0.7371 ± 0.0351	0.6799 ± 0.1035	0.9496 ± 0.0215	91.8053 ± 1.0787
	Method 5	0.6417 ± 0.0823	0.746 ± 0.0653	0.6945 ± 0.0909	0.9641 ± 0.0096	92.9334 ± 1.5861
	Method 6	0.7368 ± 0.0428	0.8465 ± 0.0743	0.8745 ± 0.0149	0.9652 ± 0.0047	93.6589 ± 1.4789

the SpFCM approach with discrete wavelet transform also provides approximate results as compared with Method 6, along with the better processing time.

References

1. Ulusoy, F.R., Yolcu, M., İpek, E., Korkmaz, A.F., Gurler, M.Y., Gulbaran, M.: Coronary artery disease risk factors, coronary artery calcification and coronary bypass surgery. J. Clin. Diagn. Res. **9** (2015)
2. Banchhor, S.K., Araki, T., Londhe, N.D., Ikeda, N., Radeva, P., Elbaz, A., Saba, L., Nicolaides, A., Shafique, S., Laird, J.R.: Five multiresolution-based calcium volume measurement techniques from coronary IVUS videos: A comparative approach. Comput. Methods Programs Biomed. **134**, 237–258 (2016)
3. Banchhor, S.K., Londhe, N.D., Saba, L., Radeva, P., Laird, J.R., Suri, J.S.: Relationship between automated coronary calcium volumes and a set of manual coronary lumen volume, vessel volume and atheroma volume in Japanese diabetic cohort. J. Clin. Diagn. Res. **11** (2017)
4. Kubo, T., Imanishi, T., Takarada, S., Kuroi, A., Ueno, S., Yamano, T., Tanimoto, T., Matsuo, Y., Masho, T., Kitabata, H.: Assessment of culprit lesion morphology in acute myocardial infarction: ability of optical coherence tomography compared with intravascular ultrasound and coronary angioscopy. J. Am. Coll. Cardiol. **50**, 933–939 (2007)
5. Araki, T., Nakamura, M., Utsunomiya, M., Sugi, K.: Visualization of coronary plaque in arterial remodeling using a new 40-MHz intravascular ultrasound imaging system. Catheter Cardiovasc. Interv. **81**, 471–480 (2013)
6. Araki, T., Ikeda, N., Dey, N., Acharjee, S., Molinari, F., Saba, L., Godia, E.C., Nicolaides, A., Suri, J.S.: Shape-based approach for coronary calcium lesion volume measurement on intravascular ultrasound imaging and its association with carotid intima-media thickness. J. Ultrasound Med. **34**, 469–482 (2015)
7. Mintz, G.S., Popma, J.J., Pichard, A.D., Kent, K.M., Satler, L.F., Chuang, Y.C., Ditrano, C.J., Leon, M.B.: Patterns of calcification in coronary artery disease: a statistical analysis of intravascular ultrasound and coronary angiography in 1155 lesions. Circulation **91**, 1959–1965 (1995)
8. Coutts, S.B., Modi, J., Patel, S.K., Demchuk, A.M., Goyal, M., Hill, M.D.: CT/CT angiography and MRI findings predict recurrent stroke after transient ischemic attack and minor stroke: results of the prospective CATCH study. Stroke **43**, 1013–1017 (2012)
9. Schoenhagen, P., Nissen, S.: Understanding coronary artery disease: tomographic imaging with intravascular ultrasound. Heart **88**, 91–96 (2002)
10. Katouzian, A., Angelini, E., Sturm, B., Konofagou, E., Carlier, S.G., Laine, A.F.: Applications of multiscale overcomplete wavelet-based representations in Intravascular Ultrasound (IVUS) images. Ultrasound Imaging, pp. 313–336. Springer, Berlin (2012)
11. Araki, T., Banchhor, S.K., Londhe, N.D., Ikeda, N., Radeva, P., Shukla, D., Saba, L., Balestrieri, A., Nicolaides, A., Shafique, S.: Reliable and accurate calcium volume measurement in coronary artery using intravascular ultrasound videos. J. Med. Syst. **40**, 51 (2016)
12. Gao, Z., Guo, W., Liu, X., Huang, W., Zhang, H., Tan, N., Hau, W.K., Zhang, Y.-T., Liu, H.: Automated detection framework of the calcified plaque with acoustic shadowing in IVUS images. PloS one **9** (2014)
13. Cardinal, M.H., Meunier, J., Soulez, G., Maurice, R.L., Therasse, E., Cloutier, G.: Intravascular ultrasound image segmentation: a three-dimensional fast-marching method based on gray level distributions. IEEE Trans. Med. Imaging **25**, 590–601 (2006)
14. Santos Filho, E., Saijo, Y., Tanaka, A., Yoshizawa, M.: Detection and quantification of calcifications in intravascular ultrasound images by automatic thresholding. Ultrasound Med. Biol. **34**, 160–165 (2008)
15. Suganya, R., Shanthi, R.: Fuzzy c-means algorithm-a review. Int. J. Sci. Res. Publ. **2**, 1–3 (2012)

16. Chuang, K.-S., Tzeng, H.-L., Chen, S., Wu, J., Chen, T.-J.: Fuzzy c-means clustering with spatial information for image segmentation. Comput. Med. Imaging Graph. **30**, 9–15 (2006)
17. Yang, M.-S., Tsai, H.-S.: A Gaussian kernel-based fuzzy c-means algorithm with a spatial bias correction. Pattern Recogn. Lett. **29**, 1713–1725 (2008)
18. Yuan, Y., He, C.: Adaptive active contours without edges. Math. Comput. Model. **55**, 1705–1721 (2012)
19. Chan, T.F., Vese, L.A.: Active contours without edges. IEEE Trans. Image Process. **10**, 266–277 (2001)
20. Paragios, N., Deriche, R.: Geodesic active contours and level sets for the detection and tracking of moving objects. IEEE Trans. Pattern Anal. Mach. Intell. **22**, 266–280 (2000)
21. Li, B., Chui, C., Ong, S.H., Chang, S.: Integrating FCM and level sets for liver tumor segmentation. In: Lim, C., Goh, J.H. (eds.) Proceedings: 13th International Conference on Biomedical Engineering, vol. 23, pp. 202–205. Springer, Heidelberg (2009)
22. Puertas, E., Escalera, S., Pujol, O.: Generalized multi-scale stacked sequential learning for multi-class classification. Pattern Anal. Appl. **18**, 247–261 (2015)
23. Mahajan, S.H., Harpale, V.K.: Adaptive and non-adaptive image interpolation techniques. In: International Conference on Computing Communication Control and Automation, pp. 772–775 (2015)
24. Harb, S.M.E., Isa, N.A.M., Salamah, S.: New adaptive interpolation scheme for image upscaling. Multimed. Tools Appl. **75**, 7293–7325 (2016)
25. Acharya, T., Tsai, P.-S.: Computational foundations of image interpolation algorithms. Ubiquity **2007**, 1–17 (2017)
26. Press, W.H.: Numerical Recipes in Pascal: the Art of Scientific Computing. Cambridge University Press, Cambridge (1989)
27. De Boor, C., De Boor, C., Mathématicien, E.-U., De Boor, C., De Boor, C.: A practical guide to splines. Springer, New York (1978)
28. Mallat, S.G.: A theory for multiresolution signal decomposition: the wavelet representation. IEEE Trans. Pattern Anal. Mach. Intell. **11**, 674–693 (1989)
29. Turkowski, K.: Filters for common resampling tasks. Graphics gems, pp. 147–165. Academic Press Professional, Inc. (1990)
30. Adelson, E.H., Burt, P.J.: Image data compression with the Laplacian pyramid. University of Maryland, Computer Science (1980)
31. Li, B.N., Chui, C.K., Chang, S., Ong, S.H.: Integrating spatial fuzzy clustering with level set methods for automated medical image segmentation. Comput. Biol. Med. **41**, 1–10 (2011)
32. Gupta, D., Anand, R.S.: A hybrid edge-based segmentation approach for ultrasound medical images. Biomed. Signal Process. Control **31**, 116–126 (2017)

Copy–Move Attack Detection from Digital Images: An Image Forensic Approach

Badal Soni, Pradip K. Das, Dalton Meitei Thounaojam and Debalina Biswas

Abstract Due to the new development of image handling tool or software, copy–move attack is increasingly becoming a common practice and on the other hand, the detection of such type of attack from digital images has become the challenging and active research area. This paper presents the recent block and keypoints-based Copy–Move Forgery Detection (CMFD) techniques. In this paper, we cover the critical discussions of different blocks and keypoints-based CMFD techniques with their pros and cons. The paper also describes the different publicly available databases and performance evaluation measures. Some unsolved research issues in the field of copy–move forgery detection is identified and present in this paper.

Keywords Copy–move forgery · SIFT · SURF · Image forensics · Image processing · Image security · Image matching · Image feature · Image stenography

---

Please note that the LNCS Editorial assumes that all authors have used the western naming convention, with given names preceding surnames. This determines the structure of the names in the running heads and the author index.

B. Soni (✉) · P. K. Das · D. M. Thounaojam · D. Biswas
National Institute of Technology Silchar, Silchar, Assam, India
e-mail: soni.badal88@gmail.com
URL: http://cs.nits.ac.in/badal/, http://www.nits.ac.in/

P. K. Das
e-mail: pkdas@iitg.ernet.in
URL: http://www.iitg.ernet.in/pkdas/, http://www.nits.ac.in/

D. M. Thounaojam
e-mail: dalton.meitei@gmail.com
URL: http://cs.nits.ac.in/dalton/, http://www.nits.ac.in/

B. Soni · P. K. Das · D. M. Thounaojam · D. Biswas
Indian Institute of Technology Guwahati, Guwahati, Assam, India

A. Elçi et al. (eds.), *Smart Computing Paradigms: New Progresses and Challenges*,
Advances in Intelligent Systems and Computing 766,
https://doi.org/10.1007/978-981-13-9683-0_8

1 Introduction

Due to the advancement of many high-quality tools, softwares, and techniques in the field of multimedia and image processing, the modifications and manipulation of digital contents of the image has become easy and a common practice; this process is called the digital image tampering. Digital image tampering detection techniques can be classified into two ways: Active [1, 2] and Passive [3, 4].

Passive approach can be defined as a process that takes the feature of the image into consideration without taking any other new information. Image CMFD is one popular passive approach. In this technique, some portion of the image is copied and pasted in some other location of the same image to hide or duplicate some component of the original image. To make the forgery detection more complex, some operations like rotation, scaling, noise, filtering are performed either in the whole image or only the copied region of the image.

In literature, there is a lot of copy–move forgery detection approaches have been proposed in the past decade. In paper [5], a detailed survey on different Copy–Move Forgery Detection (CMFD) techniques have been given. This paper presents the descriptions of performance evaluation measures and different standard datasets used in CMFD techniques. It is observed that this paper critically discuss and highlight the pros and cons of each of the techniques covered in paper. It is found that currently, keypoints-based copy–move forgery detection techniques have emerged due to its outstanding performance against the geometric transformations. However, still, there is a scope for further improvement.

In this paper, we present the general phases of copy–move forgery detection and descriptions of various copy–move forgery detection techniques. This paper also enlists the different databases used for forgery detection and different evaluation parameters.

The paper is organized as follows: Sect. 2 described the different phases of CMFD approach. Block-based and keypoints-based approaches are described in Sects. 3, and 4 respectively. Sections 5 and 6 present the databases and performance evaluation description, respectively. Research issues are given in Sect. 7. Finally, the paper is concluded in Sect. 8.

2 Steps for Image CMFD Process

There are four basic steps, which is followed by every image copy–move forgery detection technique. The block diagram of general copy–move forgery detection process is shown in Fig. 1.

- Image Preprocessing: To remove noise present in the image.
- Image Features Extraction: Local as well as global features extraction.
- Features Matching: Extracted features are matched using difference features matching techniques.

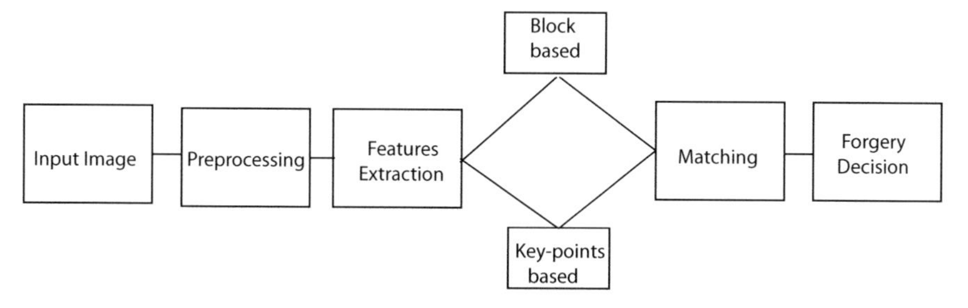

Fig. 1 Block diagram of general copy–move forgery detection process

- Post-Processing Opertaion: It is an optional step. Some techniques are required to do post-processing to remove the outliers.

3 Block-Based CMFD Approaches

In [6], present a new copy–move forgery detection technique, which is based on Discrete Radial Harmonic Fourier Moments (DRHFMs). This technique analyses the aspect of RHFMs and creating DRHFMs for forgery detection. For reducing the processing time, in this paper, the image is divided into 9×9-sized overlapping circular block rather than 16×16 sized. It is observed that this technique required less processing time than traditional block-based method.

Discrete Cosine Transform (DCT) and the package clustering-based method is proposed in [7]. In this method, initially, the image is divided into small blocks and DCT is applied into each block to extract the block features. To improve the detection results, package clustering algorithm is used instead of lexicographic sorting. It is observed that this method is able to detect forged region from images in which the addition of white Gaussian noise, Gaussian blurring, and mixed operations are applied.

In [8], an efficient approach for copy–move forgery detection is given. To reduce the computing time, in this paper, an improved expanding block clustering approach is utilized. It is observed that the false positive rate of this method is less due to the use of block variance.

Recently, in [9], an block phase correlation approach for CMFD system has proposed.

In [10], a blur invariant CMFD system is proposed using Stationary Wavelet Transform (SWT) and Singular Value Decomposition (SVD). SWT is shift invariant, and contribute to find the match blocks. In this method, each block is represented by SVD features. It is observed from experimental results that the detection accuracy of this technique is high in comparison to existing block-based methods.

In [11], Local Binary Pattern Histogram Fourier Feature (LBP-HF) based block approach is proposed. In this paper, initially, image is divided into fixed size overlapping blocks and LBP-HF features are extracted from each block and matches for forgery detection.

4 Keypoints-Based CMFD Approaches

The main limitations of block-based CMFD techniques are their high computational cost and poor affine transformation invariance. Most of the block-based techniques failed in forgery detection when the copied portion of the image undergoes some operation like scaling, rotation, etc. To overcome this problem, another type of approach known as keypoint-based approach has been widely used by the researchers. The keypoints-based method is based on identification and selection of high-entropy image regions, which takes into account the features which are extracted from the complete image. The advantage of key point-based method over block-based method is that because of its low post-processing operation it has low computational complexity. Keypoints-based method mainly uses the scale and rotation-invariant feature detector and descriptors algorithms. In this paper, we surveyed and categorized different keypoints-based copy–move forgery detection techniques based on their feature extraction procedure.

In [12], a method based on Maximally Stable Extremal Regions (MSER) and SIFT is proposed for CMFD. To improve the detection results finally, this method is also utilized Particle Swarm Optimization based clustering.

In [13], a hybrid features-based novel copy–move forgery detection technique is proposed. In this method, SIFT and KAZE are combined to extract more keypoints. For detecting multiple forgery in this paper, a n-best matched features an improved matching algorithm is used which can find the n-best matched features. It is observed from experimental results that the method is able to detect copied regions even after the geometric transformation is applied.

In order to overcome the shortcomings of SIFT feature extraction techniques like locating copy–move forgery in flat regions, Guo et. al in [14] proposed a method using DAISY descriptor to improve the result. Yu et. al in [15] discussed a method using Multi-support Region Order based Gradient Histogram(MROGH) and Hue Histogram descriptor.

In [16], a noble method has been proposed which deals with reflection-based attacks and combination of it. This method uses SIFT-based CFMD (as it provides better performance in rotation and scaling) and symmetry matching techniques.

In [17], a robust copy–move forgery detection have been proposed, which detect geometrical transformation and deformation in tampered regions as it is based on Affine SIFT (ASIFT). It is totally an affine invariant descriptor. The proposed method is robust against some post-processing operations like blurring, Gaussian noise addition.

In [18], an efficient CMFD approch has propsed by the authors for multiple forgeries detection using SIFT and DBSCAN.

Table 1 Description of the image CMFD datasets

Sl.no.	Database	Type/Version	Image size	No. of images
1	CASIA [19]	CASIA v1.0	374×256	1725
		CASIA v2.0	240×160 to 900×600	12323
2	CoMoFoD [20]		512×512	10400
			3000×2000	3120
3	MICC [21]	MICC-F220	722×480	220
		MICC-F2000	2048×1536	2000
		MICC-F8 multi	2048 × 1536	8
		MICC-F600	800×533 to 3888×2592	600

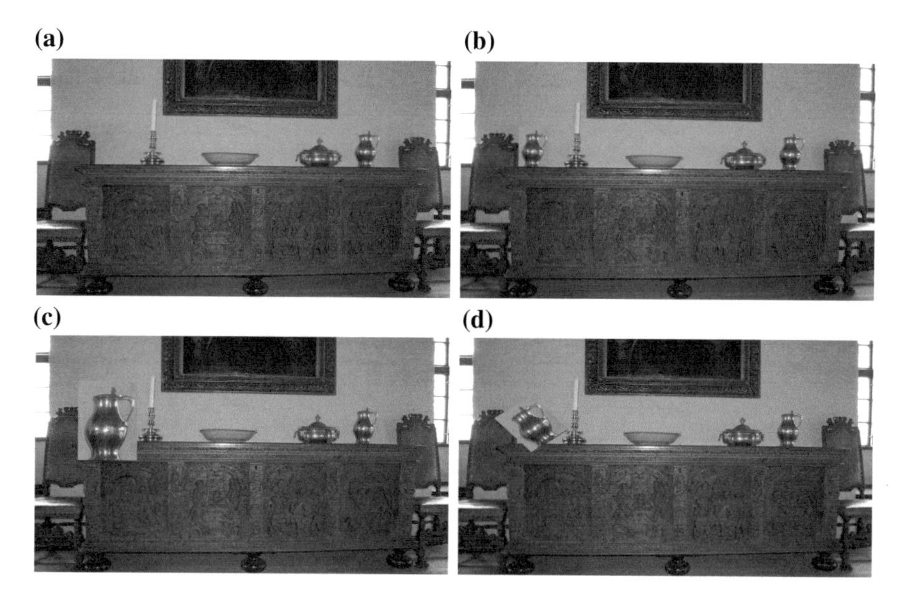

Fig. 2 Example of CMFD dataset (**a**) Original image from MICC-F2000 (**b**) Tampered image without attack (**c**) Tampered image under scale attack (**d**) Tampered image under rotation attack

5 Database Description

Generally, datasets used in copy–move forgery detection system are CASIA, CoMoFoD, and MICC. These datasets are publicly available for academic and research purpose. Table 1 describes these dataset and the example of the MICC-F2000 dataset images is given in Fig. 2.

6 Performance Evaluation Measures

Performance of forgery attack approaches are determined by calculating False Negative Rate F_N, some images have been classified as original but they were actually forged and False Positive Rate F_P, some images have falsely detected as forged.

$$F_N = \frac{|forged\ image\ detected\ as\ original|}{|forged\ images|} \tag{1}$$

$$F_P = \frac{|original\ image\ detected\ as\ forged|}{|original\ images|} \tag{2}$$

At the pixel level, precision and recall are calculated. Precision is the probability that a detected forgery is a true forgery. Recall is the probability that a forged image is detected. Precision and Recall are described as

$$Precision = \frac{T_P}{T_P + F_P} \tag{3}$$

$$Recall = \frac{T_P}{T_P + F_N} \tag{4}$$

Another performance measure used in copy–move attack detection is *F1 score*, which combined the value of Precision and Recall measures.

$$F1 = 2.\frac{Precision \times Recall}{Precision + Recall} \tag{5}$$

7 Research Issues

From the survey, it is found that mostly papers address the single forgery detection scenario, and very few papers address the multiple forgeries detection techniques but the accuracy limitation are still there. Therefore, there is special attention and focus is required to propose the system for detection of multiple forgery. Therefore, from this study, several research issues are identified and some of them are given as follows:

- Find an effective local keypoint features for fast and accurate forgery detection in flat and highly similar regions so that CMFD system can be used in real-time applications.
- Explore in video manipulation detection. In videos, existing CMFD techniques can be extended to search for duplicated regions in multiple image frames.
- Using of soft computing techniques for accurate copy–move forgery detection system.

8 Conclusions and Future Scope

Copy–move forgery detection in digital images is becoming a popular research area. Generally, the attacker tries to manipulate the original content of the image by performing the copy–move attack. To detect this type of forgery, many techniques have been proposed in past decade. It is observed from the literature that keypoint-based techniques are exceptionally good in terms of computational time and detection accuracy. Keypoints-based techniques are also robust against different geometric transformations like scaling, rotation and post-processing operations like JPEG compression, Gaussian noise, where block-based methods are not effective against these operations. Although some of the block-based methods are effective in locating the forgery region precisely, keypoints-based methods are effective in all other terms. Therefore, keypoints-based methods are given more preference than block-based methods in CMFD. It is also observed that in keypoints-based CMFD, still, many challenges are there such as robustness in flat or uniform image regions and complexity. Therefore, still, improvements are required for proper and accurate forgery detection in digital images.

References

1. Luo, X.-Y., Wang, D.-S., Wang, P., Liu, F.-L.: A review on blind detection for image steganography. Signal Process. **88**(9), 2138–2157 (2008)
2. Huo, Y., He, H., Chen, F.: A semi-fragile image watermarking algorithm with two-stage detection. Multimed. Tools Appl. **72**(1), 123–149 (2014)
3. Piva, A.: An overview on image forensics. ISRN Signal Process. **2013** 2013
4. Birajdar, G.K., Mankar, V.H.: Digital image forgery detection using passive techniques: a survey. Digit. Investig. **10**(3), 226–245 (2013)
5. Soni, B., Das, P.K., Thounaojam, D.M.: CMFD: a detailed review of block based and key feature based techniques in image copy-move forgery detection. IET Image Process. **12**(11), 167–178 (2018)
6. Zhong, J., Gan, Y., Young, J., Huang, L., Lin, P.: A new block-based method for copy move forgery detection under image geometric transforms. Multimed. Tools Appl. **76**(13), 14887–14903 (2017)
7. Wang, H., Wang, H.-X., Sun, X.-M., Qian, Q.: A passive authentication scheme for copy-move forgery based on package clustering algorithm. Multimed. Tools Appl. **76**(10), 12627–12644 (2017)
8. Lin, C.S., Chen, C.C., Chang, Y.C.: An efficiency enhanced cluster expanding block algorithm for copy-move forgery detection. In: International Conference on Intelligent Networking and Collaborative Systems, pp. 228–231 (2015)
9. Soni, B., Das, P.K., Thounaojam, D.M.: An efficient block phase correlation approach for cmfd system. In: Pattnaik, P.K., Rautaray, S.S., Das, H., Nayak, J., (eds.) Progress in Computing, Analytics and Networking, pp. 41–49. Springer, Singapore (2018)
10. Dixit, R., Naskar, R., Mishra, S.: Blur-invariant copy-move forgery detection technique with improved detection accuracy utilising swt-svd. IET Image Process. **11**(5), 301–309 (2017)
11. Soni, B., Das, P.K., Thounaojam, D.M.: Copy-move tampering detection based on local binary pattern histogram fourier feature. In: International Conference on Computer and Communication Technology, ICCCT-2017, pp. 78–83. ACM, New York (2017)

12. Sekhar, R., Shaji, R.S.: An investigation on the use of mser and sift for image forgery detection (2017)
13. Yang, F., Li, J., Wei, L., Weng, J.: Copy-move forgery detection based on hybrid features. Eng. Appl. Artif. Intell. **59**, 73–83 (2017)
14. Guo, J.-M., Liu, Y.-F., Zong-Jhe, W.: Duplication forgery detection using improved daisy descriptor. Expert Syst. Appl. **40**(2), 707–714 (2013)
15. Liyang, Y., Han, Q., Niu, X.: Feature point-based copy-move forgery detection: covering the non-textured areas. Multimed. Tools Appl. **75**(2), 1159–1176 (2016)
16. Warif, N.B.A., Wahab, A.W.A., Idris, M.Y.I., Salleh, R., Othman, F.: Sift-symmetry: a robust detection method for copy-move forgery with reflection attack. J. Vis. Commun. Image Represent. **46**, 219–232 (2017)
17. Shahroudnejad, A., Rahmati, M.: Copy-move forgery detection in digital images using affine-sift. In: 2016 2nd International Conference of Signal Processing and Intelligent Systems (ICSPIS), pp. 1–5 (2016)
18. Soni, B., Das, P.K., Thounaojam, D.M.: multicmfd: fast and efficient system for multiple copy-move forgeries detection in image. In: Proceedings of the 2018 International Conference on Image and Graphics Processing, ICIGP 2018, pp. 53–58. ACM, New York (2018)
19. Dong, J., Wang, W., Tan, T.: Casia image tampering detection evaluation database. In: 2013 IEEE China Summit and International Conference on Signal and Information Processing (ChinaSIP), pp. 422–426. IEEE (2013)
20. Tralic, D., Zupancic, I., Grgic, S., Grgic, M.: CoMoFoD new database for copy-move forgery detection. In: ELMAR, 2013 55th International Symposium, pp. 49–54. IEEE (2013)
21. Amerini, I., Ballan, L., Caldelli, R., Del Bimbo, A., Serra, G.: A sift-based forensic method for copy-move attack detection and transformation recovery. IEEE Trans. Inform. Forensics Secur. **6**(3), 1099–1110 (2011)
22. Soni, B., Das, P.K., Thounaojam, D.M.: Blur invariant block based copy-move forgery detection technique using fwht features. In: International Conference on Watermarking and Image Processing, ICWIP 2017, pp. 22–26, ACM, New York (2017)
23. Yang, B., Sun, X., Guo, H., Xia, Z., Chen, X.: A copy-move forgery detection method based on cmfd-sift. Multimed. Tools Appl. **77**(1), 837–855 (2018)

A Comparative Study of Reversible Video Watermarking Using Automatic Threshold Adjuster and Non-feedback-Based DE Method

Subhajit Das and Arun Kumar Sunaniya

Abstract For authentication and content protection, the act of hiding information in frames is done by reversible video watermarking. An automatic threshold adjuster-based reversible video watermarking (ATAVW) method using difference expansion (DE) for gray-scale video processing system is provided in this paper. To present the best watermarked video frames of an input video stream in terms of peak-signal-to-noise ratio (PSNR), the maximum payload size and the embedded threshold with respect to the embedding capacity is automatically calculated by the proposed algorithm. The main feature of ATAVW method is that the payload size and the embedding threshold are not needed to be specified by the user, like other feedback-based reversible video watermarking algorithms to restore the accurate original video. Software implementation results for an input video having 30 frames are explained based on their quality of services parameters. The ATAVW method is then compared with the non-feedback-based reversible watermarking algorithms. These results clearly demonstrated that the ATAVW method provides higher PSNR and embedding capability with respect to the DE method for reversible watermarking.

Keywords Difference expansion · Threshold adjuster · Video watermarking · PSNR · SSIM

1 Introduction

Nowadays, the distribution of video data is much easier due to the combination of the advance electronic and information technology [1–3]. But there is a need to improve the authentication of the digital video signal because editing tools-based applications made it easy to manipulate and destroy the copyright ownership. This concern turns to be further considerable when the video data are to be used second-hand as verification. Therefore, the authentication techniques are needed in order to keep up the authenticity, integrity, and security of digital video content. The drawback

S. Das (✉) · A. K. Sunaniya
National Institute of Technology, Silchar 788010, Assam, India
e-mail: Subhajitdas151@gmail.com

A. Elçi et al. (eds.), *Smart Computing Paradigms: New Progresses and Challenges*,
Advances in Intelligent Systems and Computing 766,
https://doi.org/10.1007/978-981-13-9683-0_9

of digital watermarking (DW) is that it can either recover the original multimedia data or watermark from the watermarked multimedia data [4, 5]. DW is the course of action to embed an additional, identifying information into a cover multimedia data, such as text, audio, image, or video. By adding a crystal clear watermark to the multimedia content, it is possible to detect intimidating alterations, as well as to authenticate the integrity and the possession of the digital media. Recently, DW methods are applied in various video signals-based applications [2–6]. By using DW, the original content is not distorted for video authentication. A foremost intuition of reversible image watermarking (RIW) is discussed by Honsinger [7–9]. RIW [10–13] is an attractive category for watermarking structure. Expanding the difference of two neighborhood pixels-based RIW method is proposed by Tian [14]. Some integer transform-based modification works over DE are found in [15–17]. Latter in 2017, Subhajit [17] provided a VLSI-based pipeline architecture for RIW by DE with high-level synthesis approach. By taking a constant user-specified threshold, the DE method is processed for data-embedding operation. For this reason, the number of embedding bits and the signal-to-noise ratio between original and watermarked images are not high at all possible simulation times. In 2015, Sudip [18] proposed an adaptively control-based algorithm over DE to detect the value of threshold in which the maximum embedding bit-based watermarked image has generated. The purpose of this algorithm is to adjust the threshold value automatically to generate user-specified embedding bits or payload capacity-based watermarked data. The distortion during data embedding can also be controlled by preserving the embedded bit and encryption keys. We applied the method provided by Sudip [18] for reversible video watermarking by taking a video stream as the input instead of an original image. In this paper we follow the method used in [18] to implement the data embedding and decoding processes of reversible video watermarking (RVW) by taking an input video stream. The algorithm is tested and verified in MATLAB environment. We have calculated the quality of service of each frame of the input video stream. Based on the result, we provided a comparative analysis with respect to other existing methods.

This paper is organized as follows: In Sect. 2, non-feedback-based difference expansion method is described in detail. The proposed ATAVW using DE for an input video stream is described in Sect. 3. In Sect. 4, details of software implementation of the proposed algorithm by taking 30 frames of input video stream are presented. Finally, the outcome results and conclusions are summarized in Sect. 5.

2 The Difference Expansion Method

The working principle of data embedding and decoding of Tian's [14] proposed DE method is mathematically illustrated in this section. The embedding and decoding processes are carried out in three steps. In the first step the average of neighbor pixels is calculated. The equation for average calculation is given in Eq. 1.

$$l = \left\lfloor \frac{m + n}{2} \right\rfloor \quad (1)$$

where (m, n) is denoted as original pair of pixel. The average of the two neighbor pixels is denoted as l. Next, the difference value h is calculated. Afterward, the randomly generated embedded bit is to be found in the obtained binary representation value of h at the LSB position by shifting 1-bit position of previous value from right to left. Precisely, it can be expressed as a result of Eq. 2.

$$h' = 2 * h + b \quad (2)$$

In the next step, some arithmetic calculations are made for integer transform. The watermarked pixels m' and n' are obtained by using the following Eq. 3.

$$m' = l + \left\lfloor \frac{h' + b}{2} \right\rfloor \text{and}\, n' = l - \left\lfloor \frac{h'}{2} \right\rfloor \quad (3)$$

In case of data extraction process, average (l') of m' and n' is first calculated, which is shown in Eq. 4.

$$l' = \left\lfloor \frac{m' + n'}{2} \right\rfloor \quad (4)$$

To recover the actual difference or the difference between input pixels m and n, the difference h' between watermarked pixels m' and n' is first calculated, as shown in Eq. 5. The key or the randomly generated bits are stored simultaneously, which is given in Eq. 6.

$$h' = m' - n' \quad (5)$$

$$b = h' - 2 * h \quad (6)$$

The arithmetic calculation for finding out h is given by the following Eq. 7.

$$h = \left\lfloor \frac{h'}{2} \right\rfloor \quad (7)$$

To recover input pixel pair values m and n, Eq. 8 has been followed.

$$m = l + \left\lfloor \frac{h' + b}{2} \right\rfloor \text{and}\, n = l - \left\lfloor \frac{h'}{2} \right\rfloor \quad (8)$$

h' is followed by boundary condition to overcome the problem of overflow and underflow [14], which is given by Eq. 9.

$$|h'| \leq \min(2(255 - l), 2l + 1) \tag{9}$$

Based on the measured difference values, four different sets have been created [14]. The payload size performs an important role to select some acceptable difference values for embedding. To control the embedding capacity, the payload size is controlled by a threshold value. But for the adaptive feedback-based RIW method [18], the threshold value T can be assorted to get the most wanted maximum payload size-based embedding. In the adaptive feedback-based RIW method, the l, h, l', and h' are denoted as Av, D, Av', and D', respectively. We applied this method for gray-scale video authentication. We take a gray-scale video as the original input video.

3 Automatic Threshold Adjuster-Based Reversible Video Watermarking

3.1 Embedding Part of Algorithm

The building structure for embedding part of ATAVW method is shown in Fig. 1a. The original input video is first processed into the frame maker block. The frame maker block reads each frame of the input video and then stores them with an appropriate address. One frame is processed in the embedding block at a time. The embedding block has three inputs. They are predefined threshold value, embedding key, and the host frame. Here the embedding key is a randomly generated key during embedding process. The process of data embedding is carried out by a constant threshold value which is initially specified by the user. The threshold value is used to overcome the overflow and underflow problems during encoding [14]. The initial payload capacity is calculated by using the user-specified threshold value. Then the threshold value is automatically adjusted by an adaptively controlled mechanism depending on the measured payload size of each frame of the input video. The threshold value is

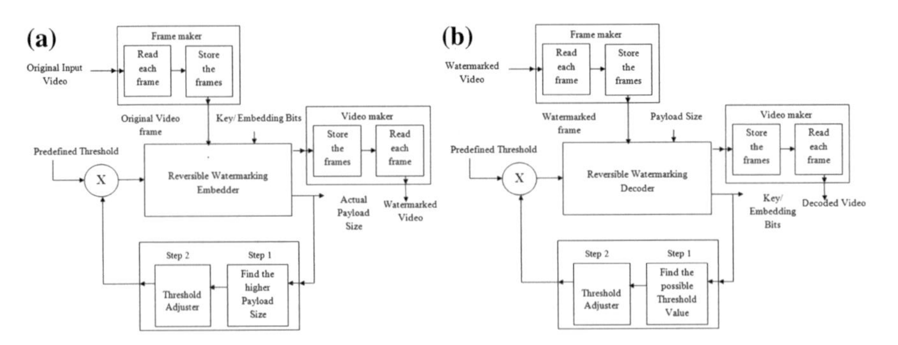

Fig. 1 Basic building structure of **a** the encoder and **b** the decoder of ATAVW method

varying for a limited range until an effectively higher PSNR-based watermarked frame with respect to the input video frame is detected.

The final selected threshold value is converted into 8-bit binary value, which is used as an encryption key for decoding process. The processes are applied for the frames of the input video signal. The watermarked frames are first stored by video maker. Then after the embedding process of all frames of input video, the video maker block reads each watermarked frames and makes the output watermarked video.

Algorithm 1: Data-Embedding Process

1: Read the frames of the input original gray-scale video.
2: Store the frames with an address.
3: Input of embedding process = original video's frame pixel pair (m, n).
4: To avoid overflow and underflow: $0 \leq m \leq 255$ and $0 \leq n \leq 255$.
5: Calculate the integer average of the two input image pixels m and n: $Av = \left\lfloor \frac{(m+n)}{2} \right\rfloor$
6: Calculate the difference: $D = m - n$.
7: Embedding of the bit b: $D' = 2 * D + b$; $b = 0$ or 1.
8: Embedded output pixel m′: $\mathrm{m}' = \mathrm{Av} + \left\lfloor \frac{(\mathrm{D}'+\mathrm{m})}{2} \right\rfloor$
9: Embedded output pixel n′: $\mathrm{n}' = \mathrm{Av} - \left\lfloor \frac{\mathrm{D}'}{2} \right\rfloor$
10: Output of embedding process = watermarked video frame pixel pair (m', n')
11: Adaptive feedback process is applied to find out possible highest threshold value and the payload. The adaptive feedback process is another function that runs along with the main code of watermarking to perform the additional steps over the Tian's proposed method.
12: Store the watermarked frames with an appropriate address.
13: Read the watermarked frames and make the output watermarked video.

In short, the encoding part of ATAVW consists of eight steps: (1) Read and store the frames of input original video, (2) check whether the gray-level intensity value of the input video is zero or not, (3) define the threshold value and calculate the difference values, (4) separate the difference values into four sets, (4) create a locality map, (5) collect the original LSB values, (6) data embedding by substitution, (7) an inverse integer transform is followed to perform an extra function to detect the maximum value of payload capacity, and finally (8) store and read the watermarked frames.

3.2 *Decoding Part of Algorithm*

The conceptual block diagram and the steps of the data extraction process of ATAVW are given in Fig. 1b and algorithm 2 in that order. The generated watermarked video from data-embedding process is taken as the input watermarked video for data extraction process. The process of data extraction starts by varying the encryption key. The possible threshold value is extracted from the encryption key by proper control mechanism. Using the threshold value, the algorithm is first verified, whether the payload capacity is higher than all possible threshold values with respect to the PSNR. When higher PSNR with respect to the higher payload capacity is detected this process starts the data extraction by following the inverse integer transform.

Algorithm 2: Data Extraction Process

1: Read the frames of the watermarked gray-scale video.
2: Store the watermarked frames with an address.
3: Input = watermarked image pixel pair (m', n')
4: Avoid overflow and underflow: $0 \leq m' \leq 255$ and $0 \leq n' \leq 255$
5: $Av' = \left\lfloor (m' + n')/2 \right\rfloor$
6: $D = \left\lfloor \frac{D'}{2} \right\rfloor$
7: Recovery of image pixel m: $\mathrm{m} = \mathrm{Av}' + \left\lfloor \frac{(\mathrm{D}+\mathrm{m}')}{2} \right\rfloor$
8: Recovery of image pixel n: $n = Av' - \left\lfloor \frac{\mathrm{D}}{2} \right\rfloor$
9: Output = Original image pixel pair (m, n)
10: Find out the possible threshold value.
11: Reverse adaptive feedback process. This process is required to find out the possible highest threshold value which is obtained during the encoding process for getting highest payload capacity-based watermarked image. The name reverse implies that this process take watermarked image data as input unlike adaptive feedback process during encoding where the input data is the original cover image.
12: Store the watermarked frames with an appropriate address.
13: Read the watermarked frames and make the output watermarked video.

4 Result of ATAVW Method

Both the processes of ATAVW are tested and verified. A video having 30 frames is taken as the input original video. The first four original video frames are shown in Fig. 2a. The watermarked frames of first frame are shown in Fig. 2b. Figure 2c shows the corresponding decoded frames. It is noted that the gray-level intensity value of the first frame is zero. As it does not satisfy the boundary condition, the embedding process is skipped for this frame.

We have calculated the quality factors to adjust the threshold value with high payload capacity. The main quality of services are mean squared error (MSE), normalized absolute error (NAE), average difference (AD), peak signal-to-noise ratio (PSNR), and structure similarity matrix index (SSIM) [19]. We have calculated the quality of services of each frame. The frame-wise respective values of the quality of services are shown in Fig. 3. As no embedding process has been made for the first frame, the PSNR and SSIM of first frame is 0 and 1, respectively.

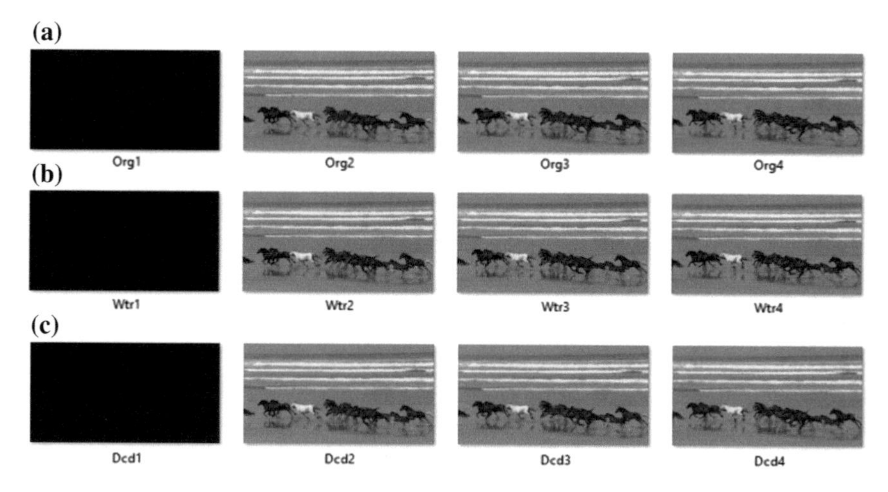

Fig. 2 **a** Input video frames, **b** watermarked video frames, and **c** decoded video frames

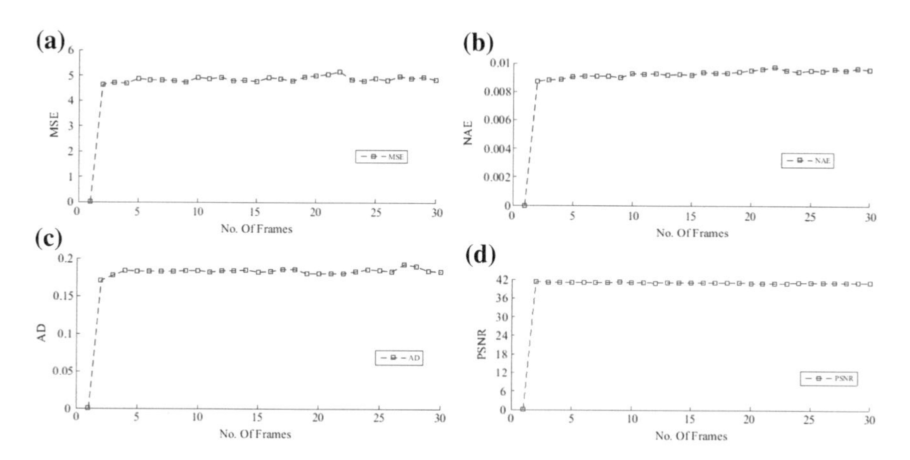

Fig. 3 **a** MSE versus number of frames, **b** NAE versus number of frames, **c** AD versus number of frames, and **d** PSNR versus number of frames

4.1 Comparative Analysis

The average PSNR with respect to the embedding capacity is first calculated by varying each frame size. Figure 4a shows comparative result of average PSNR using ATAVW method and non-feedback-based DE method. From Fig. 4a we conclude that the proposed ATAVW method provides highest PSNR of respective frames at all times over non-feedback-based DE method. The comparative SSIM between original and watermarked frames using those two methods is shown in Fig. 4b. We notice that we can embed more number of bits using ATAVW method. It is noticed that ATAVW method has higher embedding capability over non-feedback-based method.

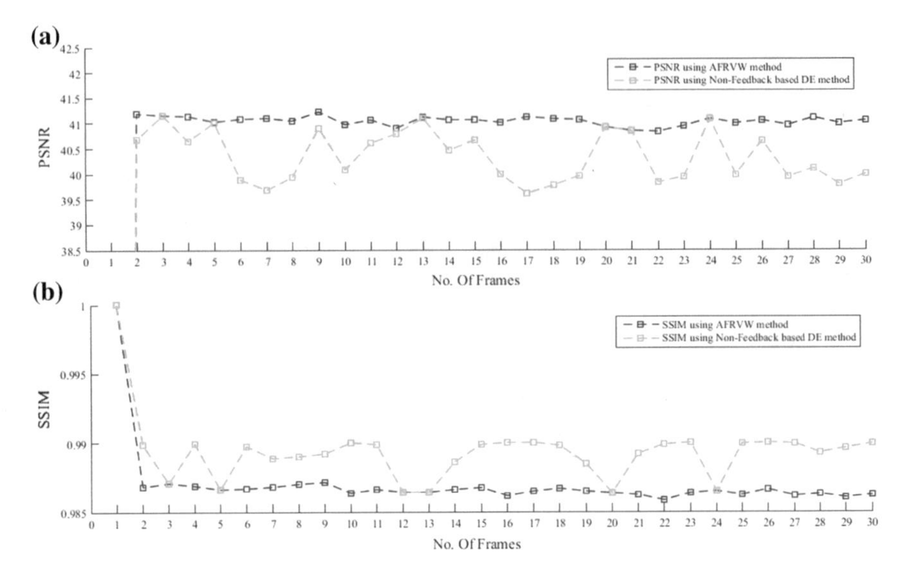

Fig. 4 Comparative analysis graphs based on **a** PSNR and **b** SSIM using ATAVW and non-feedback-based DE method

5 Conclusion

In this paper, an effective high-embedding capacity with higher PSNR based on the DE method is proposed. The name of the proposed method is automatic threshold adjuster-based reversible video watermarking (ATAVW). It is concluded from the comparative analysis between the proposed ATAVW and DE methods that, at the same time, the average embedding capacity with respect to the different size of frames by the ATAVW method is 32.64%, which shows improved value over DE method.

Acknowledgements The authors would like to show their appreciation to TEQUIP-III, NIT Silchar for providing financial assistance, and VLSI Research Lab, Department of Electronics, and Instrumentation Engineering, NIT Silchar to carry out the research work.

References

1. Potdar, V.M., Han, S., Chang, E.: A survey of digital image watermarking techniques. In: Proceedings of the IEEE International Conference on Industrial Informatics, pp. 709–716 Aug 2005
2. Gwenael, A.D., Dugelay, J.L.: A guide tour of video watermarking. Signal Process. Image Commun. **18**(4), 263–282 (2003)
3. Piva, A., Bartolini, F., Barni, M.: Managing copyright in open networks. IEEE Trans. Internet Comput. **6**(3), 18–26 (2002)

4. Shoshan, Y., Fish, A., Li, X., Jullien, G.A., Yadid-Pecht, O.: VLSI watermark implementations and applications. Int. J. Inf. Technol. Knowl. **2**(4), 379–386 (2008)
5. Li, X., Shoshan, Y., Fish, A., Jullien, G.A., Yadid-Pecht, O.: Hardware implementations of video watermarking. In: International Book Series on Information Science and Computing, no. 5, pp. 9–16. Institute of Information Theories and Applications, FOI ITHEA, Sofia, Bulgaria, June 2008 (supplement to the Int. J. Inform. Technol. Knowl. **2** (2008))
6. Cox, I.J., Kilian, J., Leighton, F.T., Shamoon, T.: Secure spread spectrum watermarking for multimedia. IEEE Trans. Image Process. **6**(12), 1673–1687 (1997)
7. Mohanty, S.P.:. Digital Watermarking: A Tutorial Review. http://www.linkpdf.com/download/dl/digital-watermarking-a-tutorial-review-.pdf (1999)
8. Barton, J.M.: Method and apparatus for embedding authentication information within digital data. US Patent 5646997 (1997)
9. Honsinger, C.W., Jones, P., Rabbani, M., et al.: Lossless recovery of an original image containing embedded data. US Patent 6278791 (2001)
10. Caldelli, R., Filippini, F., Becarelli, R.: Reversible watermarking techniques: an overview and a classification. EURASIP J. Inf. Secur. (2010). https://doi.org/10.1155/2010/134546
11. Feng, J.-B., Lin, I.-C., Tsai, C.-S., Chu, Y.-P.: Reversible watermarking: current states and key issues. Int. J. Netw. Secur. **2**, 161–171 (2006)
12. Coltuc, D., Tremeau, A., Delp, E.J., Wong, P.W.: Simple reversible watermarking scheme. In: SPIE: Security, Steganography, Watermarking Multimedia Contents, vol. 5681, pp. 561–568 (2005). https://doi.org/10.1117/12.585782
13. Coltuc, D., Chassery, J.M., Delp, E.J., Wong, P.W.: Simple reversible watermarking scheme: further results. In: SPIE: Security, Steganography, Watermarking Multimedia Contents VIII, vol. 6072, pp. 739–746 (2006). https://doi.org/10.1117/12.641376
14. Tian, J.: Reversible data embedding using a difference expansion. IEEE Trans. Circuits Syst. Video Technol. **13**(8), 890–896 (2003). https://doi.org/10.1109/tcsvt.2003.815962
15. Alattar, A.M.: Reversible watermark using the difference expansion of a generalized integer transform. IEEE Trans. Image Process. **13**(8), 1147–1156 (2004). https://doi.org/10.1109/tip.2004.828418
16. Liu, Y.-C., Hsien-Chu, W., Shyr-Shen, Yu.: Adaptive DE-based reversible steganographic technique using bilinear interpolation and simplified location map. Multimed. Tools Appl. **52**(2–3), 263–276 (2011). https://doi.org/10.1007/s11042-010-0496-0
17. Das, S., Maity, R., Maity, N.P.: VLSI-based pipeline architecture for reversible image watermarking by difference expansion with high-level synthesis approach. Circuits Syst. Signal Process. **37**(4), 1575–1593 (2018)
18. Ghosh, S., Das, N., Das, S., Maity, S.P., Rahaman, H.: An adaptive feedback based reversible watermarking algorithm using difference expansion. In: IEEE RETIS (2015)
19. Sasi Varnan, C., Jagan, A., Kaur, J., Jyoti, D., Rao, D.S.: Image quality assessment techniques in spatial domain. IJCST **2**(3) (2011)

Detection of White Ear-Head of Rice Crop Using Image Processing and Machine Learning Techniques

Prabira Kumar Sethy, Smitanjali Gouda, Nalinikanta Barpanda and Amiya Kumar Rath

Abstract Farmers in rural India have minimal access to agriculture aspect that can inspect paddy crop images and provide advice. Expert advice responses to queries often reach farmers too late. The disease in paddy crop mostly affects leaf and panicle. The disease that affects the panicle is more severe than the other parts of the paddy crop, as it directly hampers the production. Owing to the infestation of stem borer at the time of ear-head emergence, panicle gets dried and turns white in color, which is known as white ear-head. Automatic detection of white ear-head is done based on high-resolution images captured through mobile camera. In our proposed methodology, we analyze the image of defected panicle by using advanced image processing technique with machine learning to identify whether a panicle is white ear-head affected or a healthy one. This paper executes three machine learning techniques, that is PCA, Gabor filter and ANN, with an accuracy of 85, 90 and 95%, respectively.

Keywords PCA · Gabor filter · ANN · White ear-head

1 Introduction

Stem borer attacks paddy crop at different growth stages. At early stage when larvae come out of eggs it can damage the leaf or leaf sheath and middle-growing portion. As a result, the leaves get dried, which is known as "Dead Heart". Owing to its

P. K. Sethy (✉) · S. Gouda · N. Barpanda
Department of Electronics, Sambalpur University, Sambalpur 768019, Odisha, India
e-mail: prabirasethy@suniv.ac.in; prabirsethy.05@gmail.com

S. Gouda
e-mail: smitanjali.gouda@suiit.ac.in

N. Barpanda
e-mail: nkbarpanda@suiit.ac.in

A. K. Rath
Department of Computer Science and Engineering, VSSUT, Burla 768018, India
e-mail: amiyaamiya@rediffmail.com

A. Elçi et al. (eds.), *Smart Computing Paradigms: New Progresses and Challenges*, Advances in Intelligent Systems and Computing 766,
https://doi.org/10.1007/978-981-13-9683-0_10

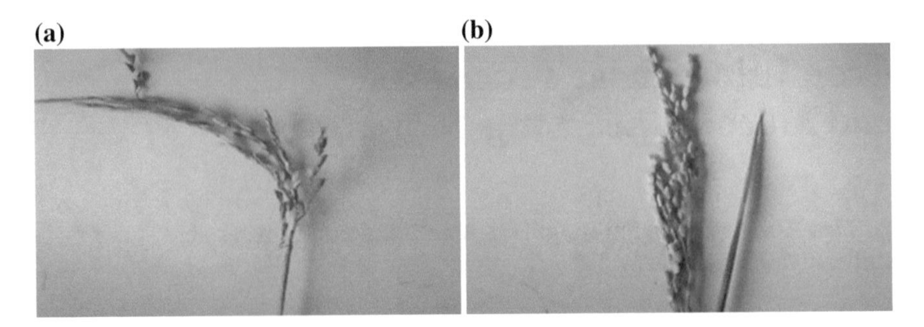

Fig. 1 **a** White ear-head panicle. **b** Healthy panicle

infestation at the time of ear-head emergence, ears get dried and turn white in color, which is known as "White Ear-Head" [1]. Ear-heads of affected plants can be easily pulled out and may cause heavy yield loss (Fig. 1).

Principal components analysis (PCA) is a digital image compression algorithm with a low level of loss that reduces dimensions of data, identifies standards in a set of data and expresses by emphasizing their similarities and differences. Patterns are recognized and compressed; that is their dimensions are reduced with negligible information loss. Such reduction is useful for image compression, data representation, and so on. PCA can discriminate multiple features; therefore, PCA feature extraction is a price mark choice to detect white ear-head. PCA transforms correlated variables into a smaller number of uncorrelated variables; that is into a set of additive orthogonal basis vectors or eigenvectors. For white ear-head detection, pixels of an image are converted into a number of Eigen-panicle feature vectors and are then compared with the selected trained image, and disease-affected panicles are detected [2]. Gabor filter is named after Dennis Gabor. It is a linear filter used for texture analysis which analyzes the presence of a particular frequency in an image in a specific direction in the region of analysis. It is mainly used for representing and discriminating texture. A two-dimensional (2D) Gabor filter is a Gaussian kernel function modulated by a sinusoidal plane wave in the spatial domain [3, 4]. Artificial neural networks (ANNs) or connectionist systems are based on biological neural networks. ANN consists of a collection of connected units or nodes called artificial neurons. These connected artificial neurons receive, process and transmit signals to others. In general, in ANN implementations, the signal is a real number, and a nonlinear function is used to calculate the output of artificial neuron. Weights are used to increase or decrease the strength of the signal. A signal is sent only if it crosses the threshold. Artificial neurons have different layers and they perform different kind of transformations on their inputs. ANN approaches were mainly used to solve problems as human brain would. Application of ANNs includes computer vision, machine translation, speech recognition, social network filtering, video games, medical diagnosis, and so on [5]. Liu Zhan-yu et al. introduced a methodology to measure the brown spot disease severity of rice crop using statistical methods as regression, principal component and partial least square regression. For their experimental purpose, a group of healthy and defected leaves were taken. The authors also compared the performance of the above three methods and concluded that the PLS regression analysis is better than stepwise regression and PCR to estimate brown spot disease [6]. Shuangping Huang et al. introduced hyperspectral image analysis based on BoSW model for rice panicle blast grading. Here BoSW model is used as the input to a χ^2 kernel support vector machine classifier to measure the severity of rice panicle blast. For their experimental purpose 312 rice panicles covering more than 50 varieties were taken which were collected under natural conditions. After validation they claimed that their proposed methodology successfully grade panicle blast with an accuracy of 81.41% for six-class grading and 96.40% for two-class grading [7].

2 Machine Learning Techniques

Here, the panicle images are captured from a smart phone in the month of July from paddy field. Panicle regions, that is the region of interest (ROI), are extracted using the method detailed below from the images. Using the proposed algorithms, features are extracted and matched with the corresponding template features to obtain a matching score that generates a class label either as genuine or as imposter.

2.1 PCA Algorithm

Principal component analysis follows the following steps:

Step-1: Take sample training images and compute the average panicle images.
Step-2: Calculate the deviation of each image from the mean image.
Step-3: Compute the difference between each image in the training set and merge all centered images.
Step-4: Using snapshot method of eigenface, sort and eliminate eigenvalues.
Step-5: Calculate the covariance matrix and its eigenvectors.
Step-6: Compare the image with a test image by using Euclidean distance and detect the disease-affected and healthy panicles.

PCA reduces dimensionality for computer vision and performs well while detecting objects. Here, a set of panicle images (training set) is taken and basis vector called Eigen-panicle is generated. Images are projected into the subspace that is spanned by the vector Eigen-panicle and are classified by comparison of its position in the panicle space with the position of known objects.

For generation Eigen-panicle, let a set of M panicle images $i_1, i_2, \ldots, i_M$ each of size N $\times$ N dimensions be considered, then the average panicle is given by

$$\bar{i} = \frac{1}{M}\sum_{j=1}^{M} i_j \tag{1}$$

All panicle images and the average panicle differ by vector $\varnothing_n = i_n - \bar{i}$. So covariance is used to calculate the eigenvector

$$C = AA^T \tag{2}$$

where $A = [\varnothing_1, \varnothing_2, \ldots, \varnothing_M]$ and is of $N^2 \times M$ dimensions. Hence, dimensions of C will be of $N^2 \times N^2$. So there will be N^2 eigenvectors with $N^2 \times 1$ dimension each. For large value of N, its computation time and memory occupied will be large and that is why, dimensionality reduction concept is used. So, the covariance matrix C is:

$$C = A^T A \text{ where, } A = [\varnothing_1, \varnothing_2, \ldots, \varnothing_M] \tag{3}$$

Now, covariance matrix dimension is $M \times M$. So, there are M eigenvectors having $M \times 1$ dimension, and $M \ll N^2$. For calculating lower dimensional subspace,

$$u_{i=A}v_i \tag{4}$$

where u_i is the ith eigenvector of original high-dimensional space and v_i is the ith eigenvector of lower dimensional subspace. Panicle images set, that is $\{i\}$ is transformed into Eigen-panicle components by,

$$w_{nk} = b_k\left(i_k - \bar{i}\right) \tag{5}$$

where $n = 1, 2, \ldots, M$ and $k = 1, 2, \ldots, K$

The weight vector $\tau_n = [w_{n1}, w_{n2}, \ldots, w_{nK}]$ represents each Eigen-panicle's contribution in representing input panicle image where Eigen-panicle is taken as a basis set for panicle images.

After projection of panicle images into the eigen space, similarity between panicle images is found by the Euclidean distance, that is, $\|y_1 - y_2\|$, where y_1 and y_2 are their corresponding feature vectors. The panicles are more similar if the distance is smaller. Similarity score $s(y_1, y_2)$, that is, inverse Euclidean distance is given by:

$$s(y_1, y_2) = \frac{1}{1 + |y_1 - y_2|} \in [0, 1] \tag{6}$$

For recognizing panicle, similarity score between input panicle image and each of the training images is calculated and the matched panicle image has the highest similarity [8–10].

2.2 Gabor Filter

Step-1: Take sample training images and check whether it is grayscale image.
Step-2: If it is a grayscale image then remove the noise using Gaussian low-pass filter and centralize it.
Step-3: Crop the image and locate its center point and divide the image into sectors such that the index value of the pixel represents sector number.
Step-4: Convert the image into polar form, perform convolution and obtain panicle code of the image.
Step-5: Compare the panicle code of the image with the test image database and detect the disease-affected and healthy panicles.

In 2D Gabor filter the time variable t is converted to spatial coordinates (x, y) and the frequency variable f is converted to (u, v) form in the frequency domain. Typically, a 2D Gabor function is defined as:

$$g(x,\ y) = g(x,\ y;\ f_o, \theta) = e^{-(\alpha^2 x_p^2 + \beta^2 y_p^2)} e^{i2\pi f_0 x_p} \tag{7}$$

where $x_p = x\ cos\ \theta\ +y\ sin\ \theta,\ y_p = -x\ sin\ \theta\ +y\ cos\ \theta$ and θ is angle of rotation of the Gaussian major axis and the plane wave. The normalized form of the 2D Gabor filter function is given by

$$g(x,\ y) = \frac{f_0^2}{\pi \gamma n} \mathrm{e}^{-(\frac{f_0^2}{r^2} x_p^2 + \frac{f_0^2}{n^2} y_p^2)} \mathrm{e}^{i2\pi f_0 x_P} \tag{8}$$

Let an arbitrary complex function f(x, y), that is centered at (x_0, y_0), then effective widths are given by:

$$\Delta x = \sqrt{\frac{\iint (x - x_0)^2 f(x, y) f * (x, y) dxdy}{\iint f(x, y) f * (x, y) dxdy}},\ \Delta y = \sqrt{\frac{\iint (y - y_0)^2 f(x, y) f * (x, y) dxdy}{\iint f(x, y) f * (x, y) dxdy}},$$

$$\Delta u = \sqrt{\frac{\iint (u - u_0)^2 F(u, v) F * (u, v) dudv}{\iint F(u, v) F * (u, v) dudv}},\ \Delta v = \sqrt{\frac{\iint (u - u_0)^2 F(u, v) F * (u, v) dudv}{\iint F(u, v) F * (u, v) dudv}}$$

$$\Delta x \Delta y \Delta u \Delta v \geq \frac{1}{16\pi^2} \tag{9}$$

In spatial–frequency domains feature extraction can be done with 2D Gabor filters by using the convolution

$$r_s(x,\ y;\ f, \theta) = g(x,\ y;\ f, \theta) \otimes s(x, y) = \iint g(x - x_r, y - y_r;\ f, \theta) s(x_r, y_r) dx_r dy_r \tag{10}$$

After panicle code is obtained it is compared with the test image database and the disease-affected panicles are detected [9].

2.3 Artificial Neural Network

Step-1: Take grayscale training input image.
Step-2: Detect its edges using Sobel filter.
Step-3: Divide the image into blocks and blocks into sub-blocks and determine the maximum pixel value.
Step-4: Calculate the mean square error, and if it is greater than 1e-5, then the panicle is disease affected.

Sobel filter detects edges and is a derivate mask that can detect two kinds of edges, that is, vertical direction and horizontal direction.

To compare image compression techniques, two error metrics are used, that is, mean square error (MSE) and the peak signal-to-noise ratio (PSNR). The cumulative squared error between the compressed and the original image is MSE, and PSNR is a measure of the peak error. The mathematical formula for MSE is

$$\frac{1}{MN}\sum_{y=1}^{M}\sum_{x=1}^{N}\left[I(x, y) - I'(x, y)\right]^{2} \tag{11}$$

After calculating MSE, it is compared with 1e-5 and if it is greater then the panicle is a disease-affected panicle [10].

3 Results and Discussion

Here, 20 sample test images and 10 train images are taken for all the three methods. In PCA the images taken are in the RGB format, whereas in Gabor filter and ANN grayscale images are taken.

$$\text{Accuracy} = \frac{\text{No. of correctly detected panicle by the algorithm}}{\text{Total no. of inputs}} \times 100\% \tag{12}$$

(a) Accuracy of PCA algorithm = (17/20) × 100% = 85%
(b) Accuracy of the proposed Gabor filter = (18/20) × 100% = 90%
(c) Accuracy of ANN algorithm = (19/20) × 100% = 95%

In demonstration, it is observed that the PCA algorithm correctly detects 17 number of panicles out of 20, whereas Gabor filter and ANN correctly detect 18 and 19 number of panicles, respectively, as shown in Table 1 and Graph 1. So the performance of ANN is better compared to PCA and Gabor filter.

The three techniques are also evaluated in terms of computational time. From our observation it is shown that the computation time of ANN is always less compared to PCA and Gabor filter. Again the Gabor filter has less computation time compared

Table 1 Performance analysis of PCA, Gabor filter and ANN

S. no.	Image dimensions (pixels × pixels)	Panicle type	Detected by PCA	Detected by Gabor filter	Detected by ANN
1	1080 × 1440	H.P.	H.P.	H.P.	H.P.
2	1080 × 1440	D.P.	H.P.	D.P.	D.P.
3	1080 × 1440	D.P.	H.P.	D.P.	D.P.
4	1080 × 1440	H.P.	H.P.	H.P.	H.P.
5	1080 × 1440	D.P.	D.P.	H.P.	D.P.
6	1080 × 1440	D.P.	D.P.	D.P.	D.P.
7	1080 × 1440	D.P.	D.P.	D.P.	D.P.
8	1080 × 1440	D.P.	D.P.	D.P.	D.P.
9	1080 × 1440	D.P.	D.P.	D.P.	D.P.
10	1080 × 1440	D.P.	D.P.	D.P.	D.P.
11	1080 × 1440	D.P.	D.P.	D.P.	D.P.
12	1080 × 1440	D.P.	D.P.	D.P.	D.P.
13	1080 × 1440	H.P.	H.P.	H.P.	H.P.
14	1080 × 1440	D.P.	D.P.	D.P.	D.P.
15	1080 × 1440	D.P.	D.P.	D.P.	D.P.
16	1080 × 1440	D.P.	D.P.	H.P.	D.P.
17	1080 × 1440	D.P.	D.P.	D.P.	D.P.
18	1080 × 1440	D.P.	H.P.	D.P.	D.P.
19	1080 × 1440	D.P.	D.P.	D.P.	H.P.
20	1080 × 1440	H.P.	H.P.	H.P.	H.P.
Accuracy			85%	90%	95%

H.P. Healthy Panicle, *D.P.* Defected Panicle

Graph 1 Performance analysis of PCA, Gabor filter and ANN

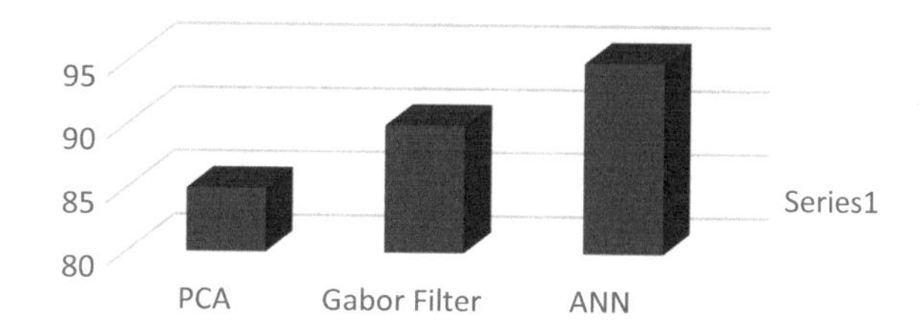

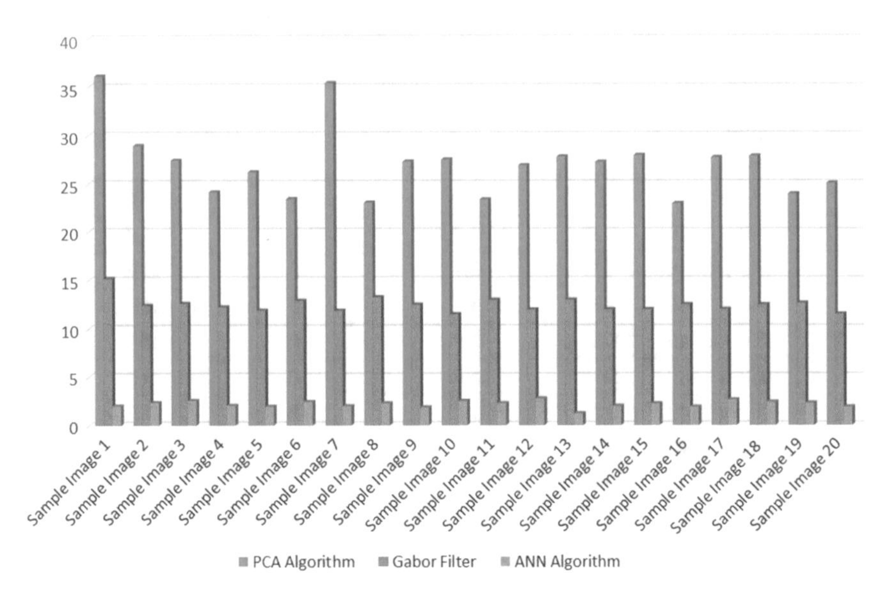

Graph 2 Computational time of PCA, Gabor filter and ANN

to PCA. So, in terms of accuracy and computational time, ANN is the best one, PCA is the worst and Gabor filter has average performance (Graph 2).

4 Conclusion

In this study we explored the performance of three machine learning techniques, that is, PCA, Gabor filter and ANN to estimate white ear-head detection using real-time captured sample images from rice field. Here the panicle of rice crop is considered as a case study, and according to the experimental observations, it has been suggested that the proposed three approaches, PCA, Gabor filter and ANN, are capable of detecting the white ear-head panicle with an accuracy of 85, 90 and 95%, respectively. This research can be extended to detect the defected grain within the panicle, so that accurate yield estimation prediction will be possible.

References

1. Hand Book on Rice Cultivation and Processing by NPCS Board of Consultants and Engineers. ISBN: 978-81-905685-2-4
2. do Espírito Santo, R.: Principal Component Analysis applied to digital image compression, Instituto do Cérebro - InCe, Hospital Israelita Albert Einstein – HIAE, São Paulo (SP), Brazil
3. Al-Kadi, O.S.: A Gabor Filter Texture Analysis Approach For Histopathological Brain Tumour Subtype Discrimination, King Abdullah II, School for Information Technology, University of Jordan Amman, 11942, Jordan

4. Grigorescu, S.E., Petkov, N., Kruizinga, P.: Comparison of texture features based on Gabor filters. IEEE Trans. Image Process. **11**(10), 1160–1167 (2002)
5. Dongare, A.D., Kharde, R.R., Kachare, A.D.: Introduction to artificial neural network. Int. J. Eng. Innov. Technol. (IJEIT) **2**(1) (2012)
6. Liu, Z.Y., Huang, J.F., Shi, J.J., Tao, R.X., Zhou, W., Zhang, L.L.: Characterizing and estimating rice brown spot disease severity using stepwise regression, principal component regression and partial least-square regression. J. Zhejiang. Univ. Sci. B. **8**(10), 738–744 (2007). https://doi.org/10.1631/jzus.2007.B0738
7. Huang, S., Qi, L., Ma, X., Xue, K., Wang, W., Zhu, X.: Hyperspectral image analysis based on BoSW model for rice panicle blast grading. Comput. Electron. Agricu. **118**, 167–178 (2015). https://doi.org/10.1016/j.compag.2015.08.031
8. Kumar, A., Zhang, D.: Personal authentication using multiple palm print representation. Pattern Recognit. **38**(10), 1695–1704 (2005). https://doi.org/10.1016/j.patcog.2005.03.012
9. Liu, Z., Shi, J., Zhang, L., Huang, J.: Discrimination of rice panicles by hyperspectral reflectance data based on principal component analysis and support vector classification. J. Zhejiang. Univ. Sci. **11**(1), 71–78 (2010). https://doi.org/10.1631/jzus.b0900193
10. Siddhichai, S., Watcharapinchai, N., Aramvith, S., Marukatat, S.: Dimensionality reduction of SIFT using PCA for object categorization. In: International Symposium on Intelligent Signal Processing and Communication Systems, May 2008

Hash Code Based Image Authentication Using Rotation Invariant Local Phase Quantization

Umamaheswar Reddy, Utkarsh Arya, Ram Kumar Karsh and Rabul Hussian Laskar

Abstract Image authentication remains a key issue in todays age of multimedia technology. With the easy availability of a multitude of hacking techniques as well as picture editing tools such as Photoshop, image security remains a key issue with a lot of research scope. As such, our paper presents a model to authenticate images transmitted across a non-secure channel. Cryptographic based hashing techniques have been shown to be more robust as compared to watermarking techniques. Therefore, we propose a model using Rotation invariant Local Phase Quantization (RI-LPQ) to generate a feature vector for the input image. LPQ is blur insensitive, and coupled with rotation invariance property, it provides an accurate representation of texture features of the image.

Keywords Multimedia security · RI-LPQ · Hash code · Image authentication

1 Introduction

Texture is defined as a continuously occurring pattern on a surface that defines the visual appearance of an object. Texture plays a crucial role in image processing, remote sensing [1] and machine vision applications where properties of texture can

U. Reddy (✉) · R. K. Karsh · R. H. Laskar
Department of Electronics and Communication Engineering,
NIT Silchar, Silchar 788010, Assam, India
e-mail: umamaheswardevireddy619@gmail.com

R. K. Karsh
e-mail: tnramkarsh@gmail.com

R. H. Laskar
e-mail: rabul2u4u@gmail.com

U. Arya
Department of Electronics and Communication Engineering,
NIT Uttarakhand, Srinagar 246174, Uttarakhand, India
e-mail: utkarsh.arya8@gmail.com

A. Elçi et al. (eds.), *Smart Computing Paradigms: New Progresses and Challenges*,
Advances in Intelligent Systems and Computing 766,
https://doi.org/10.1007/978-981-13-9683-0_11

be used to represent an object in the form of a feature vector. A feature vector may thus be defined as a set of values that represent an image in a unique way, with features like texture and contour being among the most common ones. Image blurring, which is a common transformation that images undergo, tends to degrade the texture quality. Therefore, a robust feature descriptor that is blur insensitive is of vital importance. There are only a handful of methods which are blur invariant, blur invariant moments [2] and modified Fourier phase [3] being some of them. But their advantages are limited as they consider only global features and not local features. Ojansivu et al. [4] proposed a new technique, local phase quantization, which is based on quantized phase of the discrete Fourier transform (DFT) computed in local image windows. Rotation, being one of the most common and difficult content preserving operations to model. While LPQ has been shown to be blur invariant, another update on this technique in the form of rotation-invariant LPQ has been proposed in [5].

A hexadecimal hash code is then generated from this feature vector and is sent along with the image across the non-secure channel. Pre-processing operations performed on the image render it invariant to other content preserving operation such as scaling, while being fragile to significant distortions in the form of object deletion and insertion in images. The receiver generates his own hash code on receiving the image, while might be either forged or operated on by content preserving operations. The two hash codes are then compared and on the basis of a sensitivity threshold parameter that is calculated empirically, the user can decide whether the image has been forged or not.

2 Related Work

There is a lot of work done conducted by experts in image hashing. Several earlier methods have used features either global [6–8], local [9–12] or combination of global and local features [13–16]. In [17], proposed that global features are extracted based on Zernike moments and local feature using shape-texture information of an image. For local feature there should be good saliency detection technique. Major drawback is rotation invariant up to 5° only. In [18–21], authors proposed hashing method using ring partition to obtain rotation robustness and better discriminative capability. They incorporated a secondary image by partitioning the original image in several rings. Major drawback of these methods is leaving around 20% visual information of an image.

3 Proposed Method

We begin by first computing the hash code for the input image. Hash code is a structured feature vector accurately able to represent the considered features, which in this case is texture. The input image is first converted into its YCbCr components.

Fig. 1 **a** RGB image **b** YCbCr image **c** Y component **d** Cb component **e** Cr component

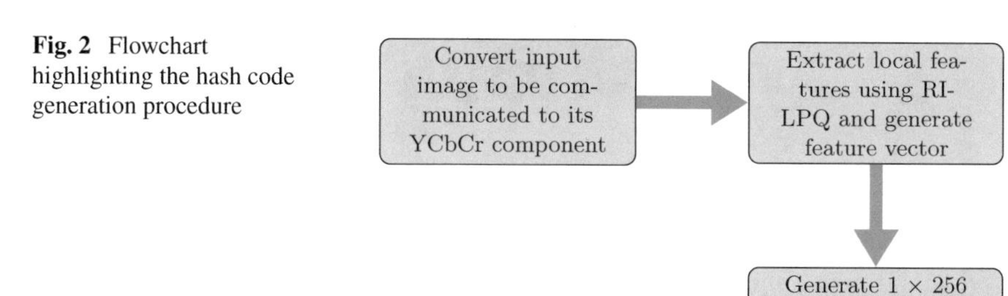

Fig. 2 Flowchart highlighting the hash code generation procedure

YCbCr color space is more preferred than RGB color space for forgery detection on account of ease of location of forged area. This is illustrated in Fig. 1.

As can be seen from Fig. 1, the input image has one of the monkeys head replaced with that of a cat. While there is no noticeable forge area that can be localized in either (a), (b), or (c), the chrominance red (Cr) and chrominance blue (Cb) components clearly show the head of the cat. YCbCr has therefore been used for analysis.

After this, RI-LPQ is applied and the corresponding hash code is generated. The process is summarized in Fig. 2.

After generating the hash code H1 for input image Im1. The image im1 and hash code are transmitted to receiver side. Hash code H2 is generated for Im2 (i.e., transformation form or forged/different corresponding to im1) and compared with H1, on the basis of sensitivity, which denotes the percentage of different bits in the two hash codes. A pre-computed threshold value is set and the obtained sensitivity is compared with this value. If the sensitivity is less than the selected threshold yields perceptually similar image pairs. Otherwise, the received image is considered to be a forged version of the transmitted image and the receiver can duly notify the sender of the same. The entire process is highlighted in Fig. 3. Details of our proposed method are described in the following sub-sections.

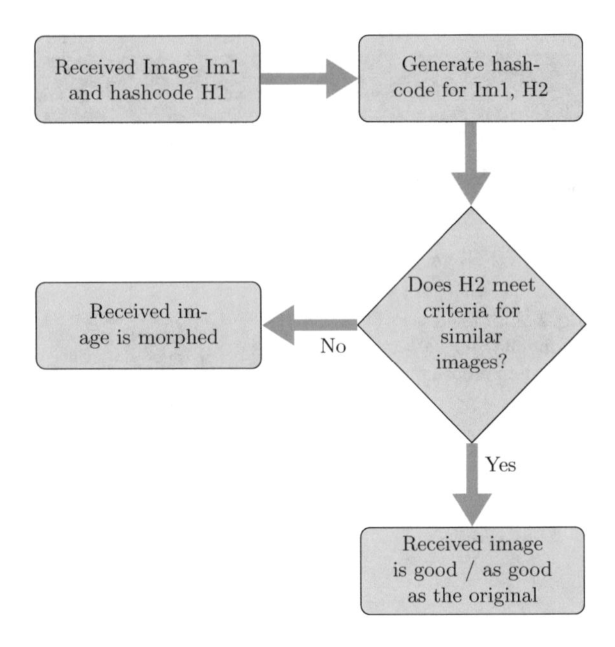

Fig. 3 Depicts Image authentication procedure

3.1 Rotation Invariant LPQ RI-LPQ

LPQ is developed as rotation invariant using following two stages: characteristic orientation estimation and directed descriptor extraction. These stages are described in details as follows.

3.1.1 Evaluation of Characteristic Orientation

Let us consider a two dimensional rotation matrix for an angle θ be defined as R_θ. Also, let $f'(x) = f(R_\theta^{-1}x)$ represent the rotated image. It can be shown that Fourier transform of f' is same as the Fourier transform of f rotated by R_θ. The similar approach may be applicable for the circular local neighbors N_x, in which the locations can be transformed to $x' = R_\theta x$. Using the same, the coefficients on the circular region of radius r with frequencies $v_i = r[cos(\phi_i)sin(\phi_i)]^T$ are evaluated, where $v_i = 2i/M$ and i = 0,..., M-1. The rectangular window is also changed with a circular Gaussian window as shown in (1).

$$W_G(x) = 1/(2\pi\sigma^2)exp\{-(x_1^2 + x_2^2)/(2\sigma^2)\}, \quad if\ |x_1|, |x_2| < N_G/2 \tag{1}$$
$$= 0, \quad otherwise$$

Based on the above statements, the vector $V(x) = [F(v_0, x), \ldots, F(v_{M-1}, x)]$ in case of rotation is obtained and relocated it to x'. Further, it process through a circular shift based on the rotation angle θ between the accuracy of discretization $2\pi/M$. Because of separability, $V(x)$ is estimated for every image locations x based on the convolutions of the rows and columns in single dimension. Only the phase of $V(x)$ has been considered in case to achieve blur insensitivity. This is done by getting down the signs of the imaginary part of $V(x)$ (i.e., $C(x) = Img\{V(x)\}$). The characteristic orientation is obtained from the quantized coefficients by putting the complex moment as in (2).

$$b(x) = \sum_{i=m}^{M-1} c_i e^{j\phi_i} \tag{2}$$

where c_i is the ith quantized element of $C(x)$. The characteristic orientation is then explained for each pixel position x as $\xi(x) = b(x)$. For a neighborhood N_x from f', the characteristic orientation will be $\xi(x)'$ which is approx. equal to $\xi(R_{-1}\,1\,x) + \theta$, where $\xi(y)$ is the characteristic orientation of N_y from f. By considering the discretization the equation is approximated. The direct calculation of $C(x)$ requires determination of frequency coefficients at M selected points.

But the direct procedure is not a best way of calculating the characteristic orientation. Therefore, an alternative scheme (i.e. an approximation method) is taken, where the shape of $C(x)$ is represented based on cosine $C^\wedge = A(x)\cos(\phi + \tau(x))$, where $\phi \in [0, 2]$, and $\tau(x)$ and $A(x)$ have been considered from two sample points i.e. $[F(v_0, x), F(v_{M/4-1}, x)]$.

Finally, $\xi(y)$ (i.e. characteristic orientation) is evaluated based on the maximum position value of $C^\wedge(x)$ i.e. $\xi(x) = -\tau(x)$, that refers to the representation of the complex moment (2). In the case of approximation scheme, error is inherent and produced using $C(x)$ which have do not have the sinusoidal shape. But, some of the $C(x)$ still follow sinusoidal shape for more than one period. The approximation produces adequately stable orientation for most of the test cases.

3.1.2 Evaluation of Oriented LPQ

In second stage, the binary descriptor vector is evaluated. The process is different from the original LPQ evaluation in that every location may be rotated similar to the characteristic orientation. This operation may be written mathematically by fixing the coefficients of oriented frequency as

$$F_\xi(u, x) = \sum_y f(y) w_R(R_{\xi(x)}^{-1}(y - x)) e^{-j2\pi u^T R_{\xi(x)}^{-1} y} \tag{3}$$

In the case of rotated image f' Eq. (3) can be written as

$$F_\xi(u, x)' = \sum_y f(y)' w_R(R_{\xi(x)}^{-1}(y - x)) e^{-j2\pi u^T R_{\xi(x)}^{-1} y} \tag{4}$$

$$F_{\xi}(u,x)' = \sum_{t} f(y) w_R (R^{-1}_{\xi(R_\theta^{-1}x)}(t - R_\theta^{-1}x)) e^{-j2\pi u^T R^{-1}_{\xi(R_\theta^{-1}x)} t} \tag{5}$$

We have considered the fact that $R_{\phi+\gamma} = R_\phi R_\gamma$ and given the values as $t = R_\theta^{-1} y$. For RI-LPQ, 256-dimensional feature descriptor vector is calculated based on $F_{\xi(x)}$. The histogram to be constructed later in the process is not affected by the rotation of f as only the coefficients $F_{\xi(u,x)}$ changes. In this way, the resulting feature vector is immune to blur and rotation. Besides, $F_{\xi(u,x)}$ (i.e. orientations) is quantized into different k values, and the resultant complex exponentials and set of window functions are computed. By doing so, the computational time does not differ much from that of the previous LPQ algorithm.

4 Experimental Analysis

For our experimental analysis, we have conducted tests on 1000 images, taken from public datasets [21, 22] and USC-SIPI [23]. Photoshop has been applied to the images and the manipulated images have been split into two equal parts: (a) forged images and (b) images either rotated, rescaled or blurred. On the basis of this, sensitivity parameter was calculated for each of the 1000 image pairs.

4.1 Threshold Extraction

Based on the experimental results on 1000 image pairs, the sensitivity threshold is calculated as follows (Fig. 4).

$$Threshold = x + 22.5$$

where, x = number of blue class image pairs/(number of blue class image pairs + number of red class image pairs. Therefore in our calculations, $x = 34/(34 + 150) = 0.1878$. Hence, $Threshold = 0.1878 + 22.5 = 22.6878$.

4.2 Qualitative Analysis

Case (1) Pairs in which one of the images has an object inserted or deleted (Class B) as shown in Fig. 5

It can be seen from Fig. 5 **that the sensitivity is greater than 22.6878**

Case (2) Pairs in which one of the images has undergone content preserving operations (Class A) as shown in Figs. 6 **and** 7

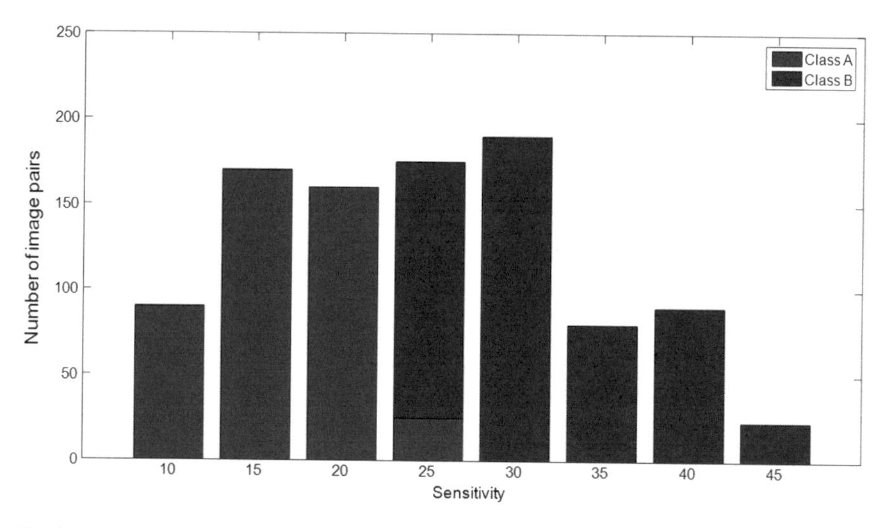

Fig. 4 Graphical evaluation of sensitivity threshold to separate the two classes of images

Operation	Original Input Image	Image with object deletion	Sensitivity parameter
Insertion			33.5611
Deletion			31.2500

Fig. 5 Object Insertion and Deletion

Case (3) Motion Blurring(Class A) as shown in Fig. 8

It can be observed that for case 2 and case 3, sensitivity values are less than threshold i.e., 22.6878

Input image	Rescaling%	Sensitivity
	90%	19.7841
	200%	20.4841

Fig. 6 Rescaling as content preserving operation

image1(With and without rotation)	Image2(With Rotation)	Sensitivity
		10.5496
	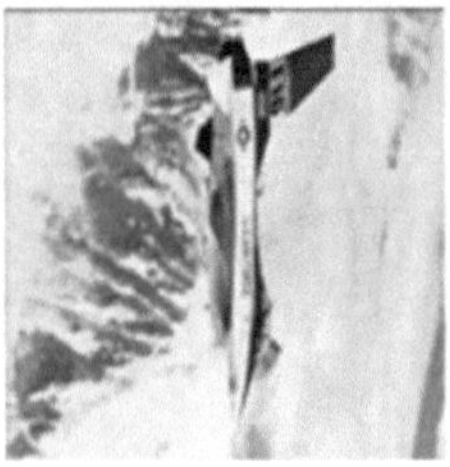	19.9219

Fig. 7 Rotation as content preserving operation

4.3 *Quantitative Analysis*

The algorithm was tested on MATLAB R2013a, on a computer running on 8 GB ram and Intel i7 processor. Accuracy is computed on the basis of total incorrect detections. Of the total images selected, 444 pairs were content preserved operated pairs while 556 of them were visually altered or forged pairs. Mathematically,

$$Accuracy = \frac{\text{total number of true detections}}{\text{total number of images}}$$

image without blurness	Blurred image	Sensitivity
		18.1612
		10.0050

Fig. 8 Motion blurring as content preserving operation

Table 1 Accuracy comparison of proposed method with other state-of-the-art methods

Method	LBP	Gabor	Proposed
Accuracy	90.2	91.6	94.8

where total number of false detections = total number of Class A pairs wrongly identified as Class B + Total number of class B pairs wrongly identified as Class A. Total number of class A pairs wrongly identified as Class B = 31 and Total number of class B pairs wrongly identified as Class A = 21 Total number of image-pairs = 1000.
Total number of incorrect detections = 31 + 21 = 52
Therefore, accuracy = (1000–52)/1000 = 94.8

In other words, for 94.8% of images taken, our scheme is able to detect forged images and content preserved images accurately. We have compared our method with two existing, state-of-the-art techniques, LBP [24] and Gabor wavelets [25]. Table 1 shows the efficacy of the proposed method with some state-of-the-art methods.

5 Conclusion

In this paper, we have proposed a new application of RI-LPQ texture analysis approach which works on Phase of the Fourier transform calculated locally based on window in every image position. Four low frequency coefficients of the phases are equally quantized into one of 256 hypercubes which consists eight-dimensional space and yields 8-bit code. All the LPQ codes collected from neighborhoods of image pixel form a histogram, which describes the texture and can be used for classification. Rotation invariant features have been additionally incorporated to make it rotation invariant along with blur insensitive. The efficacy of the proposed method has been shown by comparing with two related methods i.e. a Gabor filter bank based method and the LBP method. The experimental results show that the proposed LPQ method is more robust to blurring as compared to the state-of-the-art methods, shown by overall accuracy. Edge information however, can be incorporated to provide higher accuracy and our future work will include an edge information feature extractor to be fused with the existing texture feature extractor, RI-LPQ.

References

1. Hatt, M., Tixier, F., Pierce, L., Kinahan, P.E., Le Rest, C.C., Visvikis, D.: Characterization of PET/CT images using texture analysis: the past, the present any future? Eur. J. Nucl. Med. Mol. Imaging **44**(1), 151–165 (2017)
2. Flusser, J., Suk, T.: Degraded image analysis: an invariant approach. IEEE Trans. Pattern Anal. Machine Intell. **20**(6), 590–603 (1998)
3. Ojansivu, V., Heikkila, J.: A method for blur and similarity transform invariant object recognition. In: Proceedings International Conference on Image Analysis and Processing (ICIAP 2007), pp. 583–588. Modena, Italy (2007)
4. Ojansivu, V., Heikkila., J.: Blur insensitive texture classification using local phase quantization. In: Proceedings of International Conference on Image and Signal Processing (ICISP08), pp. 236–243 (2008)
5. Ojansivu, V., Rahtu, E., Heikkila. J.: Rotation invariant local phase quantisation for blur insensitive texture analysis. In: Proceedings of International Conference on Pattern Recognition, pp. 8–11 (2008)
6. Swaminathan, A., Mao, Y., Wu, M.: Robust and secure image hashing. IEEE Trans. Inf. Forensics Secur. **1**(2), 215–230 (2006)
7. Lei, Y., Wang, Y., Huang, J.: Robust image hash in radon transform domain for authentication. Signal Process. Image Comm. **26**(6), 280–288 (2011)
8. Saikia, A., Karsh, R.K., Laskar, R.H.: Image authentication under geometric attacks via concentric square partition based image hashing. In: Proceedings of TENCON 2017. IEEE Region 10 Conference, pp. 2214–2219 (2017)
9. Ahmed, F., Siyal, M.Y., Abbas, V.U.: A secure and robust hash-based scheme for image authentication. Signal Process. **90**(5), 1456–1470 (2010)
10. Tang, Z., Wang, S., Zhang, X., Wei, W., Zhao, Y.: Lexicographical framework for image hashing with implementation based on DCT and NMF. Multimed. Tools Appl. **52**(2), 325–345 (2011)
11. Monga, V., Mihcak, M.K.: Robust and secure image hashing via non-negative matrix factorizations. IEEE Trans. Inf. Forensics Secur. **2**(3), 376–390 (2007)
12. Lv, X., Wang, Z.J.: Perceptual image hashing based on shape contexts and local feature points. IEEE Trans. Inf. Forensics Secur. **7**(3), 1081–1093 (2012)

13. Karsh, R.K., Laskar, R.H., Richhariya, B.B.: Robust and secure hashing using Gabor filter and Markov absorption probability. In: Proceedings of IEEE Conference on Communication and Signal Processing (ICCSP), pp. 1197–1202 (2016)
14. Karsh, R.K., Laskar, R.H., Aditi.: Robust image hashing through DWT-SVD and spectral residual method. EURASIP J. Image Video Process. **2017**(1), 1–17 (2017)
15. Qin, C., Sun, M., Chang, C.: Perceptual hashing for color images based on hybrid extraction of structural features. Signal Process. **142**, 194–205 (2018)
16. Karsh, R.K., Saikia, A., Laskar, R.H.: Image authentication based on robust image hashing with geometric correction. Multimed. Tools Appl. 1–21 (2018). https://doi.org/10.1007/s11042-018-5799-6
17. Zhao, Y., Wang, S., Zhang, X., Yao, H.: Robust hashing for image authentication using zernike moments and local features. IEEE Trans. Inform. Forensics Secur. **8**(1), 55–63 (2013)
18. Tang, Z., Zhang, X., Zhang, S.: Robust perceptual hashing based on ring partition and NMF. IEEE Trans. Knowl. Data Eng. **26**(3), 711–724 (2014)
19. Karsh, R.K., Laskar, R.H.: Perceptual robust and secure image hashing using ring partition-PGNMF. In: Proceedings of TENCON 2015. IEEE Region 10 Conference, pp. 1–6 (2015)
20. Karsh, R.K., Laskar, R.H., Richhariya, B.B.: Robust image hashing using ring partition-PGNMF and local features. SpringerPlus **5**(1), 1–20 (2016)
21. Tang, Z., Zhang, X., Li, X., Zhang, S.: Robust image hashing with ring partition and invariant vector distance. IEEE Trans. Inform. Forensics Secur **11**(1), 200–214 (2016)
22. http://wang.ist.psu.edu/docs/home.shtml
23. http://sipi.usc.edu/database/
24. Ojala, T., Pietikainen, M., Maenpaa, T.: Multiresolution gray-scale and rotation invariant texture classification with local binary patterns. IEEE Trans. Pattern Anal. Machine Intell. **24**(7), 971–987 (2002)
25. Manjunathi, B.S., Ma, W.Y.: Texture features for browsing and retrieval of image data. IEEE Trans. Pattern Anal. Machine Intell. **18**(8), 837–842 (1996)

Novel Competitive Swarm Optimizer for Sampling-Based Image Matting Problem

Prabhujit Mohapatra, Kedar Nath Das and Santanu Roy

Abstract In this paper, a novel competitive swarm optimizer (NCSO) is presented for large-scale global optimization (LSGO) problems. The algorithm is basically motivated by the particle swarm optimizer (PSO) and competitive swarm optimizer (CSO) algorithms. Unlike PSO, CSO neither recalls the personal best position nor global best position to update the elements. In CSO, a pairwise competition tool was presented, where the element that fails the competition are updated by learning from the winner and the winner particles are just delivered to the succeeding generation. The suggested algorithm informs the winner element by an added novel scheme to increase the solution superiority. The algorithm has been accomplished on high-dimensional CEC2008 benchmark problems and sampling-based image matting problem. The experimental outcomes have revealed improved performance for the projected NCSO than the CSO and several metaheuristic algorithms.

Keywords Competitive swarm optimizer · Evolutionary algorithms · Large-scale global optimization · Particle swarm optimization · Swarm intelligence

1 Introduction

Maximum real-life optimization problems contract thru a huge number of choice variables, which are commonly known by means of Large-Scale Global Optimization (LSGO) problems. These problems are branded to face about harsh conditions such as high multimodality and robust communication between the parameters. Since the past few times, there has been a growing work in learning and examining the core performance of these problems. Numerous populations built metaheuristic algorithms such as Particle Swarm Optimization [1, 2], Evolutionary Algorithms (EAs) [3], Genetic Algorithms (GA) [4], and Differential Evolution (DE) algorithms [5, 6] have been proposed to solve them. Still, these algorithms undergo a lot of traumas while cracking the problems. The act of these algorithms depreciates quickly

P. Mohapatra (✉) · K. N. Das · S. Roy
National Institute of Technology, Silchar 788001, Assam, India
e-mail: prabhujit.mohapatra@gmail.com

A. Elçi et al. (eds.), *Smart Computing Paradigms: New Progresses and Challenges*, Advances in Intelligent Systems and Computing 766,
https://doi.org/10.1007/978-981-13-9683-0_12

by the rise of the dimension of the search space. There are two major causes of the performance decline of these algorithms. The initial, with the rise in the extent of the dimension the difficulty of the problem surges due to the adaptation of altered features. In addition, the search space rises exponentially by the problem magnitude, henceforth the algorithms are mandatory to do the tough job of discovering the whole search space in an effective method.

Therefore, large-scale optimization is the eventual task in diverse arenas of discipline and numerous procedures have been suggested to grip the problems. Particle Swarm Optimization (PSO) is one of such capable algorithms for cracking these problems. The algorithm is essentially motivated by the conducts of some creatures such as birds, fishes, and folks. In PSO, individual element discovers its individual finest location in the group, which is identified as an individual finest location while the entire group discovers the global best position. In order to catch the overall best of the optimization problem, the elements study from both the locations.

Though the algorithm has comprehended quick developments over the previous few times, still the show of the algorithm is faltering when the problem is high-dimensional and multimodal [7]. This weakness is attributed as premature convergence that commonly occurs in PSO [8]. In direction to increase the show of the PSO, numerous diverse reforms have been suggested. One key worry is about the tough influence of the global best position on the convergence speed, which is mainly responsible for the quick convergence. To check this problem Liang [9] proposed a different PSO alternative deprived of the global best term and the update approach trusts only on the particle best. An alternative way to handle the problem of premature convergence is to get free of together particle best and global best positions.

Another way to handle the complications over the swarm intelligence is the competition tool. This impression of the competition tool, which performances the cooperating character between the two inhabitants, has been offered in evolutionary algorithms [10–12]. Such competition tools might provide two likelihoods. Primary, weak solutions may able to absorb from strong ones in another population, and in next, the strong characters will be self-enthused by the earlier knowledge to harvest better solutions. Motivated by these mechanisms a multigroup evolutionary agenda aligned on a feedback mechanism [13] was presented in 2013. Following to the feedback mechanism brand, another practice called competitive swarm optimizer (CSO) [14] exposed the use of the competition mechanism between elements within one sole swarm. After respective competition, the element that loses will absorb from the winner element instead of from particle best or global best position. As the crucial asset behind the notion is the couple wise competition mechanism between different elements and neither particle best nor global best is involved in the search procedure, it brands the algorithm excellent, hypothetically unlike and intelligent to display a better overall show than other metaheuristic algorithms on large-scale optimization problems. It is the chief persuader behind the proposed algorithm.

This paper proposes a new PSO revised by altering the tool in CSO. Dissimilar to the original algorithm, here the winner group is also modernized along with the loser ones. The alteration not only extemporizes the show but also brands the algorithm to converge faster than the earlier one. In CSO, only part of the swarm population was proficient to update, but here the complete swarm catches the bet to renovate. The mathematical explanation is given as follows.

2 Algorithm

Let us consider the following problem:

$$min\ f = f(x)\ \text{such}\ x \in X.$$

Where $X \in R^n$ the feasible region and n is the exploration dimension. Primary the problem is resolved by randomly initializing a swarm $P(t)$ containing m particles, where m is known as the swarm size and t is known as the current generation. Each n dimensional position and velocity of ith particle is given by $x_i(t) = (x_{i,1}(t), x_{i,2}(t), \ldots, x_{i,n}(t))$ and $v_i(t) = \left(v_{i,1}(t), v_{i,2}(t), \ldots, v_{i,n}(t)\right)$ correspondingly. In each step, the elements in $P(t)$ are sorted into $m/2$ couples (presumptuous m is even). Then, a rivalry is made between the two elements in each couple and after the struggle, the element that won the competition recognized as winner is rationalized from the overall mean position of winners, whereas the loser will update its position and velocity by learning straight from the winner. It is clear that each element has to face one and only one competition. For a swarm size of m, total $m/2$ competitions happen to apprise all elements' position and velocity. The graphical drawing of the idea is presented in Fig. 1.

Let the position and velocity of the winner and the loser in the kth round of competition in iteration t be given by $x_{w,k}(t)$, $x_{l,k}(t)$, and $v_{w,k}(t)$, $v_{l,k}(t)$, respectively, where $k = 1, 2, \ldots., m/2$. After each competition, both the particle's velocity are updated using the mechanism as follows:

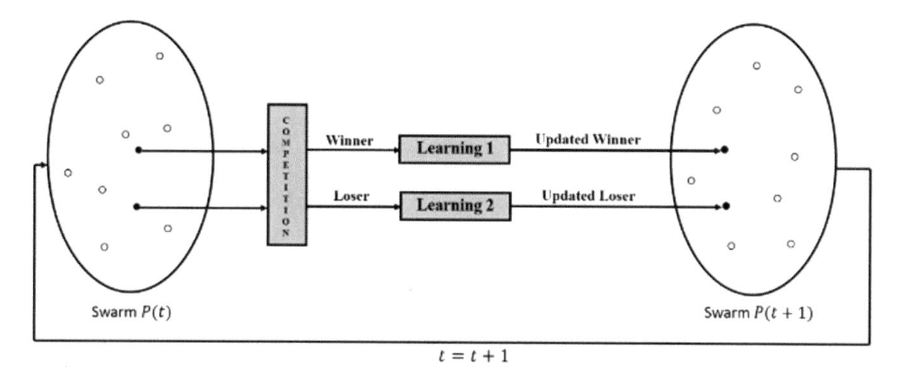

Fig. 1 The general idea of NCSO. For each iteration, particles are randomly selected in a pair from the population and after each competition, the loser and the winner are updated

$$v_{l,k}(t+1) = C_1(k,t)v_{l,k}(t) + C_2(k,t)\big(x_{w,k}(t) - x_{l,k}(t)\big) + \alpha_1 C_3(k,t)\big(\overline{x_k(t)} - x_{l,k}(t)\big) \tag{1}$$

$$v_{w,k}(t+1) = C_4(k,t)v_{w,k}(t) + \alpha_2 C_5(k,t)\big(\overline{x_w(t)} - x_{w,k}(t)\big) \tag{2}$$

By means of the above outcome, the positions of both the losers are rationalized with the new velocity as follows:

$$x_{l,k}(t+1) = x_{l,k}(t) + v_{l,k}(t+1) \tag{3}$$

$$x_{w,k}(t+1) = x_{w,k}(t) + v_{w,k}(t+1) \tag{4}$$

Where $C_1(k,t)$, $C_2(k,t)$, $C_3(k,t)$ and $C_4(k,t)$, $C_5(k,t)\epsilon[0, 1]^n$ are five arbitrarily generated vectors after the rivalry and learning approaches in generation t. Here, $\overline{x_w(t)}$ and $\overline{x_w(t)}$ are the mean position value of the applicable elements and winners correspondingly. The α_1 and α_2 are the parameters that controls the influence of the mean positions. The first part of both the equations (1) and (2), $C_1(k,t)v_{l,k}(t)$, $C_4(k,t)v_{w,k}(t)$ are comparable to the inertia term in the canonical PSO. But here in the planned method, the inertia term ω is substituted by randomly created vectors. The second part $C_2(k,t)\big(x_{w,k}(t) - x_{l,k}(t)\big)$ is named the cognitive component after Kennedy and Eberhart. Here the element that fails the competition absorbs from its competitor dissimilar to its personal best in PSO. It does not remember the personal best or global best location. The pseudo code of the NCSO is outlined in Algorithm 1.

Algorithm 1 The pseudo code of the Improved Competitive Swarm Optimizer (NCSO). R Represents a set of particles. The terminal condition is the total number of fitness evaluations.

1. $t = 0$;
2. arbitrarily reset $P(0)$;
3. **while** *terminal condition* is not satisfied **do**
4. compute the fitness of all elements in $P(t)$;
5. $R = P(t), P(t+1) = \emptyset$
6. **while** $R \neq \emptyset$ **do**
7. arbitrarily pick two elements $X_1(t)$ and $X_2(t)$ from R ;
8. Such that $X_1(t) \leq X_2(t)$
9. allocate $X_w(t) = X_1(t)$ and $X_l(t) = X_2(t)$;
10. apprise $X_{w,k}(t)$ and $X_{l,k}(t)$ using eqn. (1-4);
11. add the rationalized $X_{w,k}(t+1), X_{l,k}(t+1)$ to *P (t + 1)*;
12. eliminate $X_1(t)$ and $X_2(t)$ from R ;
13. **end while**
14. t = t + 1;
15. **end while**

3 Experimental Setup

Here, we will achieve a set of trials on the set of benchmark functions offered in CEC 2008 special session on LSGO problems. There are seven benchmark functions accessible in the CEC 2008 from which the functions f_1, f_4 and f_6 are separable functions and the rest are nonseparable ones. But as the dimension surges f_5 converts to separable ones. In all the trials, the population size m is wide ranging from 200 to 1000 while the social components α_1 and α_2 are diverse in the variety 0 and 0.3. The ranges of these parameters are engaged similar to the CSO [14].

The trials are replicated on a PC with an Intel i5 2.00 GHz CPU, Microsoft Windows 7 64-bit operating system and Matlab R2013a. The outcomes are averaged over 25 different rounds and for each independent run, the maximum number of function evaluations (FEs) is set to be $5000*n$ for the CEC 2008 benchmark problems. It is inessential to mention that n is the dimension of the search space.

The trial is exhibited to match the projected method NCSO with CSO and other newly proposed algorithms for large-scale optimization on 1000D CEC2008 benchmark functions. The compared algorithms for the first benchmark include CCPSO2 [15], multilevel cooperative co-evolution (MLCC) [16], separable covariance matrix adaption strategy (sep-CMA-ES) [17], efficient population utilization strategy for particle swarm optimizer (EPUS-PSO) [18] and DMS-PSO [19]. The measures suggested in the individual benchmarks have been acknowledged in this trial too.

4 Analysis of the Results

For respective benchmarks, the mean and standard deviation of the optimization error of each function is described. The cells that symbolize the best statistical outcomes for one particular function is highlighted. The t-test has been used to crisscross the significant variance between the algorithms. A negative t value signifies comparative small optimization error for the projected algorithm, whereas positive t value characterizes the relative high-optimization error.

Table 1 display the outcomes of each the algorithms on all the functions of the CEC 2008 benchmark functions. An examination is employed with the t-test, for that the t values are recorded in the last row of each algorithm. The negative t value is decorated only when the algorithm's performance is significantly healthier than that of the other algorithm and it is considered as a win. Similarly, the ties and loses are also measured. In case of a tie, the corresponding t value is also painted in blue. The last column counts the number of wins, ties and losses and is expressed as $w/t/l$. The higher number of wins than the number of ties and losses represent a very significant difference between the proposed algorithm and the competitor algorithm. From these statistical results, it is clear that the proposed algorithm NCSO has a significantly better overall show in comparison with all other algorithms on 1000D CEC 2008 benchmark functions. The dominant performance of NCSO and DMS-

Table 1 Comparison of all the algorithms on the CEC 2008 benchmark (f_1–f_7) of 1000 dimensions

		f_1	f_2	f_3	f_4	f_5	f_6	f_7	$w/t/l$
NCSO	Mean	4.01E–23	**3.45E+01**	9.53E+02	5.50E+02	2.22E–16	8.32E–13	**–1.41E+04**	
	Std.	2.40E–24	**3.52E–01**	9.38E–02	1.54E+01	0.00e+00	1.78E–14	6.54E+01	–
	t -Values	–	–	–	–	–	–	–	
CSO	Mean	1.66E–22	3.76E+01	9.80E+02	5.21E+02	2.22E–16	**5.306E–13**	– 1.38E+04	
	Std.	1.18E–23	1.18E+00	6.48E-01	2.95E+01	0.00E+00	**1.673E–14**	3.37E+02	**4/1/2**
	t -Values	**–5.23E+01**	**–1.26E+01**	**–2.06E+02**	4.36E+00	**0.00E+00**	6.17E+01	**–4.37E+00**	
CCPSO2	Mean	5.18E–13	7.82E+01	1.33E+03	1.99E–01	1.18E–03	1.02E–12	– 1.43E+04	
	Std.	9.61E–14	4.25E+01	2.63E+02	4.06E–01	3.27E–03	1.68E–13	8.27E+01	**5/0/2**
	t -Values	**–2.70E+01**	**–5.14E+00**	**–7.17E+00**	1.78E+02	**–1.80E+00**	**–5.56E+00**	9.48E+00	
MLCC	Mean	8.45E–13	1.087E+02	1.79E+03	**1.37E–10**	4.18E–13	1.06E–12	**–1.47E+04**	
	Std.	5.00E–14	4.754E+00	1.58E+02	**3.37E–10**	2.78E–14	7.68E–14	**1.51E+01**	**5/0/2**
	t -Values	**–8.45E+01**	**–7.78E+01**	**–2.65E+01**	1.79E+02	**–7.51E+01**	**–1.45E+01**	4.47E+01	
Sep-CMA-ES	Mean	7.81E–15	3.65E+02	**9.10E+02**	5.31E+03	3.94E–04	2.15E+01	– 1.25E+04	
	Std.	1.52E–15	9.02E+00	**4.54E+01**	2.48E+02	1.97E–03	3.19E–01	9.36E+01	**5/1/1**
	t -Values	**–2.57E+01**	**–1.83E+02**	4.74E+00	**–9.58E+01**	**–1.00E+00**	**–3.37E+02**	**–7.01E+01**	
EPUS-PSO	Mean	5.53E+02	4.66E+01	8.37E+05	7.58E+03	5.89E+00	1.89E+01	– 6.62E+03	
	Std.	2.86E+01	4.00E–01	1.52E+05	1.51E+02	3.91E–01	2.49E+00	3.18E+01	**7/0/0**
	t -Values	**–9.67E+01**	**–1.14E+02**	**–2.75E+01**	**–2.32E+02**	**–7.53E+01**	**–3.80E+01**	**–5.14E+02**	
DMS-PSO	Mean	**0.00E+00**	9.15E+01	8.98E+09	3.83E+03	**0.00E+00**	7.75E+00	– 7.50E+03	
	Std.	**0.00E+00**	7.13E–01	4.38E+08	1.70E+02	**0.00E+00**	8.92E–02	1.63E+01	**6/0/1**
	t -Values	**8.35E+01**	**–5.63E+01**	**–1.03E+02**	**–9.61E+01**	7.65E+45	**–4.34E+02**	**–4.90E+02**	

PSO are very much similar on CEC 2008 1000D functions. In general, regardless of search dimension, DMS-PSO always provides the best results for the functions f_1 and f_5, whereas for the function f_2 NCSO algorithm is best suitable. Similarly, MLCC is perfectly matchable for the functions f_4 and f_7 in most cases and sep-CMA-ES is for the function f_3. Another observation is that, although the algorithms are specialized to produce optimum solution for some particular functions but under some conditions, they also produce optimum solution for others too. But as the dimension surges to 500 or more, these algorithms switch to their committed functions. It is an additional observation that algorithms such as CCPSO2 and EPUS-PSO are incompetent to produce optimal solutions for any of the benchmark functions. Hence, finally, we can conclude that the proposed algorithm NCSO is significantly better than all the other remaining algorithms.

5 Sampling-Based Image Matting Problem

Image matting is a noteworthy procedure in the picture and video altering and furthermore assumes an essential part in picture preparing. The essential target of the system is to smooth-and-correct concentrate of the frontal area district from a picture and to decide if the undetermined pixel is in the closer view locale of the picture. When all is said in done, the data about the endeavored picture can be accomplished by a resultant tri-map picture [20]. The tri-map picture parts the procured picture into three locales, for example, frontal area district, foundation locale and undetermined locale. In the writing, different picture tangling techniques [21–23] have been proposed to fulfill the concentrate frontal area. These techniques are arranged into inspecting-based picture tangling innovation and spread based picture tangling innovation. This present examination concentrates on inspecting-based picture tangling innovation, where the specimen set is worked from the gathered closer view and foundation pixels of the known areas. In inspecting-based picture image matting, it is assumed that each undetermined pixel's area can be correctly surveyed by a couple of closer view foundation pixels from known locales. Notwithstanding, the best specimen match of each undermined pixel is not generally in the example set because of the accessibility of a set number of tests. The precision of tangling results vitiates when the best example combine is not picked in the specimen set. With a specific end goal to dodge this trouble, a vast scale testing strategy is required for packaging more pixels of the known locale. The global sampling-based technique [23] is one such strategy, which assembles every known pixel on the limit of the undermined district to frame the specimen set. This specimen set is large to the point that, the best example match for each undermined pixel can be picked from it. In a worldwide testing-based technique, random search [24] was connected to locate the best arrangement. However, the technique neglects to locate the best example matches because of the absence of investigation and power property. Keeping in mind the end goal to look for best example sets for all undermined pixels, the expansive scale numerical streamlining issue was detailed and unraveled with another DE variation (called CC-DE-S) in [25].

When all is said in done, for each specific undermined pixel, the problem can be considered as a searching in two-dimensional pixel area. Keeping in mind, the end goal to build the effectiveness, the two-dimensional advancement issue is demonstrated into a huge scale improvement issue in multidimensional inquiry space. The target of this new advancement issue now is to look for best forefront foundation for all undermined pixel on the double. Here, each position describes a gathering of closer view foundation test sets for all undermined pixels and each measurement implies a frontal area test or foundation test. The best position is what gets the best forefront foundation test sets for all undermined pixels on the double. This LSGO issue is presently being illuminated by MCSO calculation; whose issue definition is as per the following.

Problem definition

The image matting presented by Porter and Duff can be numerically shown as a direct blend condition, where the shade of an undetermined pixel is computed as the mix of a frontal area shading and foundation shading. For kth undermined pixel, this equation can be written as

$$I_k = \alpha_k F_k + (1 - \alpha_k) B_k \tag{5}$$

where I_k is the color of the kth undetermined pixel, F_k is the color of foreground pixel and B_k is the color of background pixel. The varying value of the α_k in the range [0, 1] determines whether the undetermined pixel is a foreground pixel or background pixel of the given image. Hence, the variables F_k and B_k need to be optimized for the kth undetermined pixel.

In past, various assessment criteria have been proposed to decide the best frontal area foundation test combine for each undetermined pixel. In this investigation, the assessment display [28] in light of spatial and shading wellness of chose test match is utilized. The evaluation value of the selected foreground–background sample pair $\left(F^i, B^j\right)$ is given by

$$f\left(F^i, B^j\right) = f_c\left(F_k^i, B_k^j\right) + f_s\left(F^i\right) + f_s\left(B^j\right), \tag{6}$$

where F_k^i is the color of ith foreground sample and B_k^j is the jth background sample in built sample set. The color fitness function f_c of the selected sample pair describes the quality of the sample pair as follows.

$$f_c\left(F_k^i, B_k^j\right) = \left\| I_k - \left(\hat{\alpha} F_k^i + \left(1 - \hat{\alpha} B_k^j\right)\right)\right\|, \tag{7}$$

$$\hat{\alpha} = \frac{\left(I_k - B_k^j\right)\left(F_k^i - B_k^j\right)}{\left\| F_k^i - B_k^{j2}\right\|}, \tag{8}$$

The smaller f_c value better explains the selected sample pair with the corresponding color of the undermined pixel. However, there always exists a possibility of well explanation of a bad sample pair to the corresponding undermined pixel's color. Therefore, the spatial information $f_s(F^i)$ and $f_s(B^j)$ are also reflected in f. The $f_s(F^i)$ and $f_s(B^j)$ values calculate the spatial distance between the foreground sample, background sample and the corresponding undermined pixel, respectively, as follows:

$$f_s(F^i) = \frac{\|X_{F^i} - X_k\|}{D_F}, \; f_s(B^j) = \frac{\|X_{B^j} - X_k\|}{D_B} \tag{9}$$

Here, X_{F^i},X_{B^j} and X_k are the spatial coordinates of the ith foreground sample, jth background sample and the corresponding undermined pixel. The factors $D_F = min_i \|X_{F^i} - X_k\|$ and $D_B = min_j \|X_{B^j} - X_k\|$ are the nearest distance of the corresponding undermined pixel to the foreground and background boundary. A smaller $f_s(F^i)$ and $f_s(B^j)$ indicate that the selected foreground and background are very closer to the corresponding undermined pixel. Therefore, the smaller f value represents a high-quality sample pair for one undetermined pixel.

Remembering the ultimate objective to perceive the idea of the general population, an appraisal work described in Eq. 10 in perspective of the above appraisal show has been proposed in [25]. This has considered to settle by NCSO figuring with a particular ultimate objective to find the best course of action. The assessment work for the ith individual at gth generation is figured as takes after.

$$\varphi(X_{i,g}) = \sum_{k=1}^{N_u} f\left(x_{i,g}^{2k-1}, x_{i,g}^{2k}\right) \tag{10}$$

Here, $x_{i,g}^{2k-1}$ and $x_{i,g}^{2k}$ represent a selected foreground and background sample for an undermined pixel. Hence clearly, the minimization of φ value results in improving the overall quality of all the undermined pixels in an individual.

Experiments and results

The inspecting-based picture tangling issue is a distinct LSGO issue [25]. NCSO is utilized to take care of this issue and the got consequence of mean square error (MSE) for each of the 27 preparing pictures [26] are thought about in Table 2, with random search [24], PSO [2], DE [27] and CC-DE-S [25]. Keeping in mind the end goal to abstain from missing the best example combines, the global sampling method [24] is connected to assemble the specimen sets. Amid reproduction, the most extreme capacity assessment is set to 5000.

From Table 2, it is discovered that the proposed calculation NCSO acquires less MSE esteem than both of the arbitrary scan and DE calculations for every one of the pictures. It is important that, out of 27 preparing pictures, NCSO calculation outflank CC-DE-S (in [25]) and PSO for 16 and 20 pictures individually. Subsequently, the prevalence of the NCSO calculation is guaranteed.

Table 2 Comparison of mean square error of different algorithms in solving sampling-based image matting problem

Image No.	Random Search	DE	PSO	CC-DE-S	NCSO
Image01	13.9×10^{-3}	1.27×10^{-3}	1.25×10^{-3}	1.18×10^{-3}	$\mathbf{1.16 \times 10^{-3}}$
Image02	7.02×10^{-3}	3.12×10^{-3}	8.23×10^{-3}	$\mathbf{2.72 \times 10^{-3}}$	3.02×10^{-3}
Image03	13.0×10^{-3}	12.34×10^{-3}	$\mathbf{4.21 \times 10^{-3}}$	10.02×10^{-3}	11.15×10^{-3}
Image04	100.17×10^{-3}	17.45×10^{-3}	20.10×10^{-3}	$\mathbf{13.16 \times 10^{-3}}$	13.9×10^{-3}
Image05	15.80×10^{-3}	2.54×10^{-3}	2.20×10^{-3}	$\mathbf{2.01 \times 10^{-3}}$	2.20×10^{-3}
Image06	10.22×10^{-3}	1.96×10^{-3}	2.67×10^{-3}	1.95×10^{-3}	$\mathbf{1.93 \times 10^{-3}}$
Image07	15.65×10^{-3}	2.29×10^{-3}	2.57×10^{-3}	2.06×10^{-3}	$\mathbf{2.05 \times 10^{-3}}$
Image08	57.26×10^{-3}	26.47×10^{-3}	$\mathbf{12.32 \times 10^{-3}}$	23.23×10^{-3}	13.21×10^{-3}
Image09	12.54×10^{-3}	4.58×10^{-3}	$\mathbf{2.50 \times 10^{-3}}$	4.58×10^{-3}	3.51×10^{-3}
Image10	8.12×10^{-3}	3.40×10^{-3}	3.11×10^{-3}	3.08×10^{-3}	$\mathbf{2.98 \times 10^{-3}}$
Image11	12.97×10^{-3}	5.51×10^{-3}	4.04×10^{-3}	4.26×10^{-3}	$\mathbf{3.95 \times 10^{-3}}$
Image12	4.86×10^{-3}	2.72×10^{-3}	2.82×10^{-3}	2.33×10^{-3}	$\mathbf{2.11 \times 10^{-3}}$
Image13	57.17×10^{-3}	24.03×10^{-3}	21.58×10^{-3}	19.74×10^{-3}	$\mathbf{18.91 \times 10^{-3}}$
Image14	4.60×10^{-3}	2.36×10^{-3}	2.00×10^{-3}	$\mathbf{1.94 \times 10^{-3}}$	2.00×10^{-3}
Image15	9.25×10^{-3}	6.13×10^{-3}	3.88×10^{-3}	4.33×10^{-3}	$\mathbf{3.33 \times 10^{-3}}$
Image16	80.49×10^{-3}	63.10×10^{-3}	62.30×10^{-3}	$\mathbf{61.85 \times 10^{-3}}$	63.25×10^{-3}
Image17	10.92×10^{-3}	2.48×10^{-3}	2.58×10^{-3}	$\mathbf{2.32 \times 10^{-3}}$	2.35×10^{-3}
Image18	11.64×10^{-3}	3.82×10^{-3}	3.81×10^{-3}	3.61×10^{-3}	$\mathbf{3.60 \times 10^{-3}}$
Image19	6.18×10^{-3}	1.25×10^{-3}	$\mathbf{0.94 \times 10^{-3}}$	0.97×10^{-3}	0.95×10^{-3}
Image20	4.45×10^{-3}	2.24×10^{-3}	1.94×10^{-3}	2.15×10^{-3}	$\mathbf{1.78 \times 10^{-3}}$
Image21	22.50×10^{-3}	4.69×10^{-3}	4.84×10^{-3}	$\mathbf{4.56 \times 10^{-3}}$	4.60×10^{-3}
Image22	11.15×10^{-3}	1.86×10^{-3}	1.94×10^{-3}	1.84×10^{-3}	$\mathbf{1.80 \times 10^{-3}}$
Image23	4.49×10^{-3}	3.60×10^{-3}	3.76×10^{-3}	3.15×10^{-3}	$\mathbf{3.04 \times 10^{-3}}$
Image24	6.29×10^{-3}	5.06×10^{-3}	4.45×10^{-3}	4.76×10^{-3}	$\mathbf{4.43 \times 10^{-3}}$
Image25	24.96×10^{-3}	15.98×10^{-3}	15.50×10^{-3}	$\mathbf{15.20 \times 10^{-3}}$	15.23×10^{-3}
Image26	47.48×10^{-3}	29.41×10^{-3}	25.44×10^{-3}	24.87×10^{-3}	$\mathbf{24.18 \times 10^{-3}}$
Image27	20.88×10^{-3}	16.15×10^{-3}	16.33×10^{-3}	$\mathbf{15.90 \times 10^{-3}}$	16.11×10^{-3}

6 Conclusion

In this paper, an innovative algorithm called as novel competitive swarm optimizer (NCSO) is offered. The algorithm neither follows particle best or global best in the method to renovate the positions of the elements. However, it enhances with an added policy to the CSO algorithm by apprising the winner elements. The experiential tests and inquiries are conveyed on high dimension of CEC2008 benchmark problems and sampling-based image matting problem. The algorithm has exposed sovereignty show on CSO and numerous metaheuristic algorithms predominantly meant to resolve the large-scale optimization problems.

References

1. Kennedy, J.F., Kennedy, J., Eberhart, R.C., Shi, Y.: Swarm Intelligence, 1st edn, Morgan Kaufmann (2011)
2. Kennedy, J., Eberhart, R.: Particle swarm optimization. In: Proceedings of the IEEE International Conference on Neural Networks, pp. 1942–1948. IEEE (1995)
3. Back, T.: Evolutionary Algorithms in Theory and Practice: Evolution Strategies, Evolutionary Programming, Genetic Algorithms. Oxford University Press, Oxford (1996)
4. Goldberg, D.E., Holland, J.H.: Genetic algorithms and machine learning. Mach. Learn. **3**(2), 95–99 (1988)
5. Price, K.V.: An introduction to differential evolution. New Ideas in Optimization, pp. 79–108. McGraw-Hill Ltd., England (1999)
6. Storn, R., Price, K.: Differential evolution–a simple and efficient heuristic for global optimization over continuous spaces. J. Global Optimiz. **11**(4), 341–359 (1997)
7. Yang, Y., Pedersen, J.O.: A comparative study on feature selection in text categorization. In: Proceedings of International Conference on Machine Learning, Morgan Kaufmann Publishers, pp. 412–420 (1997)
8. Chen, W.N., Zhang, J.: Particle swarm optimization with an aging leader and challengers. IEEE Trans. Evol. Comput. **17**(2), 241–258 (2013)
9. Liang, J.J., Qin, A.: Comprehensive learning particle swarm optimizer for global optimization of multimodal functions. IEEE Trans. Evol. Comput. **10**(3), 281–295 (2006)
10. Goh, C., Tan, K.: A competitive and cooperative co-evolutionary approach to multi-objective particle swarm optimization algorithm design. Eur. J. Oper. Res. **202**(1), 42–54 (2010)
11. Hartmann, S.: A competitive genetic algorithm for resource-constrained project scheduling. Naval Res. Logist. (NRL) **45**(7), 733–750 (1998)
12. Whitehead, B., Choate, T.: Cooperative-competitive genetic evolution of radial basis function centers and widths for time series prediction. IEEE Trans. Neural Netw. **7**(4), 869–880 (1996)
13. Cheng, R., Jin, Y.: A multi-swarm evolutionary framework based on a feedback mechanism. In: Proceedings of IEEE Congress on Evolutionary Computation, pp. 718–724. IEEE ()
14. Cheng, R., Jin, Y.: A competitive swarm optimizer for large scale optimization. IEEE Trans. Cybernet. **45**(2), 191–204 (2014)
15. Li, X., Yao, X.: Cooperatively coevolving particle swarms for large scale optimization. IEEE Trans. Evol. Comput. **16**(2), 210–224 (2012)
16. Yang, Z., Tang, K.: Multilevel cooperative coevolution for large scale optimization. IEEE Congress on Evolutionary Computation (IEEE World Congress on Computational Intelligence), pp. 1663–1670. IEEE, Hong Kong (2008)
17. Ros, R., Hansen, N.: A simple modification in cma-es achieving linear time and space complexity. Parallel Problem Solving from Nature–PPSN X, pp. 296–305. Springer, Germany (2008)
18. Hsieh, S.-T., Sun, T.-Y.: Solving large scale global optimization using improved particle swarm optimizer. In: IEEE Congress on Evolutionary Computation, 2008. CEC 2008 (IEEE World Congress on Computational Intelligence), pp. 1777–1784. IEEE (2008)
19. Zhao, S.-Z., Liang, J.J.: Dynamic multi-swarm particle swarm optimizer with local search for large scale global optimization. In: IEEE Congress on Evolutionary Computation, 2008. CEC 2008 (IEEE World Congress on Computational Intelligence), pp. 3845–3852. IEEE (2008)
20. Wang, J., Cohen, M.F.: An iterative optimization approach for unified image segmentation and matting. In: Proceedings of Tenth IEEE international conference on computer vision, pp. 936–943 (2005)

21. Wang, J., Cohen, M.F.: Optimized color sampling for robust matting. In: Proceedings of IEEE Computer Society Conference on Computer Vision and Pattern Recognition, pp. 1–8 (2007)
22. Gastal, E.S.L., Oliveira, M.M.: Shared sampling for real-time alpha matting. Comput. Gr Forum **29**(2), 575–584 (2010)
23. He, K., Rhemann, C., Rother, C., Tang, X., Sun, J.: A global sampling method for alpha matting. In: Proceedings of IEEE Conference on Computer Vision and Pattern Recognition, pp. 2049–2056 (2011)
24. Barnes, C., Shechtman, E., Finkelstein, A., Goldman, D.B.: Patchmatch: a randomized correspondence algorithm for structural image editing. ACM Trans. Gr. **28**(3), 24 (2009)
25. Cai, Z.-Q., Lv, L., Huang, H., Hu, H., Liang, Y.-H.: Improving sampling-based image matting with cooperative coevolution differential evolution algorithm. Soft Comput. 1–14 (2016)
26. Rhemann, C., Rother, C., Wang, J., Gelautz, M., Kohli, P., Rott, P.: A perceptually motivated online benchmark for image matting. In: Proceedings of IEEE Conference on Computer Vision and Pattern Recognition, pp. 1826–1833 (2009)
27. Storn, R., Price, K.: Differential evolution—a simple and efficient heuristic for global optimization over continuous spaces. Glob. Optim. **11**(4), 341–359 (1997)

Sound, Voice, Speech, and Language Processing

Note Transcription from Carnatic Music

S. M. Suma, Shashidhar G. Koolagudi, Pravin B. Ramteke and K. S. Rao

Abstract In this work, an effort has been made to identify note sequence of different *ragas* of Carnatic Music. The proposed heuristic method makes use of standard just-intonation frequency ratios between notes for basic transcription of music piece into written sequence of notes. The notes present in a given piece of music are obtained using pitch histograms. The normalized pitch contour of the music piece is segmented based on detection of the note boundaries. These segments are labeled using note information already available. Without prior knowledge of *raga*, 30 out of 64 sequences are identified accurately and additional 18 sequences are identified with one note error. With the prior *raga* knowledge 76.56% accuracy is observed in note sequence identification.

Keywords Carnatic Music · Note transcription · Pitch · *Raga*

1 Introduction

Technically, a note may be understood as an identifiable fundamental frequency component (pitch) in an audio signal with a clear beginning and an ending time [1]. The ratio of fundamental frequencies of two notes is referred to as an interval [1]. Carnatic Music, one of the popular categories of Indian Classical Music (ICM), may

S. M. Suma · S. G. Koolagudi · P. B. Ramteke (✉)
Department of Computer Science and Engineering, National Institute of Technology Karnataka, Surathkal 575025, Karnataka, India
e-mail: ramteke0001@gmail.com

S. M. Suma
e-mail: isumabhat@gmail.com

S. G. Koolagudi
e-mail: koolagudi@nitk.edu.in

K. S. Rao
Indian Institute of Technology Kharagpur, Kharagpur, West Bengal, India
e-mail: ksrao@sit.iitkgp.ernet.in

A. Elçi et al. (eds.), *Smart Computing Paradigms: New Progresses and Challenges*, Advances in Intelligent Systems and Computing 766,
https://doi.org/10.1007/978-981-13-9683-0_13

be represented using both 12 interval and 22 interval system [2, 3]. Since 12 interval system is widely used for teaching and understanding of ICM, it has been considered for note transcription in this work. In ICM the basic notes or *Swaras* are identified as *Shadja, Rishaba, Gandhara, Madhyama, Panchama, Dhaivatha* and *Nishadha* popularly known as *Sa, Ri, Ga, Ma, Pa, Dha* and *Ni*. Out of 7 basic notes, *Sa* is called as base note or *Adhara Shadhja* denoted as *S0*. The frequencies of all other notes depend on the frequency of the base note *S0*. Singer of Carnatic Music varies one's pitch with respect to the base note for rendering other notes [2]. Though these frequency ratios are ideal, while singing usually pitch may deviate from these ratios within some limits. For example, the standard ratio of *R1* to *S0* is 1.067, but in our dataset, it is observed that the variation is from 1.05 to 1.1. These characteristics of the note differ in different *ragas*. Hence melody of musical piece in ICM depends on the *raga* it is rendered in. Note transcription refers to the process of converting a music segment into series of notes present within. Note transcription from a music piece, that leads to automatic *raga* identification has wide range of applications in ICM such as automatic annotating of large music corpora, teaching and/or learning basics of ICM, music indexing/retrieval, *raga* identification and so on [4–6]. In this work, a method has been proposed to identify the notes present in given segment of monophonic music using estimated probability density function (pdf) of pitch values. The transcription is performed in two steps: segmenting and labeling. The segmentation is performed using heuristic methods and segments are labeled using just-intonation frequency ratios as given in [2].

The remaining paper is organized as follows. Section 2 reviews related work. Section 3 explains the proposed method. Experimentation and results are given in Sect. 4. Section 5 concludes the work with some further research directions.

2 Literature Review

Since notes correspond to the fundamental frequency of a singer, many research contributions use pitch and its derivative as the major features for note transcription. In [1], various methods used for Western music transcription, in particular singing transcription are explained. Various heuristics and statistical methods such as Hidden Markov Models are used for Western Music Transcription [1]. Since there are many differences between Western Music (WM) and ICM, the transcription methods that are proposed for ICM are quite different from that of WM. Pitch tracker is proposed for analyzing pitch variations for various notes and their intonations [7]. Various inflexions in the notes, just-intonation frequency ratios for different notes and the application of pitch trackers while processing Carnatic Music are also explained. The spectral features of Carnatic notes such as study of energy, frequency and time spectra was carried out and signature formants are identified for the signal [8]. The characteristics of notes described in [1, 7, 8] are helpful in identifying notes. Pitch features are used for note transcription of Hindustani Music as well [4]. Two heuristics that make use of variations in pitch and time segments are used as features. In the results, more number of note insertions and note displacement were observed. Approach based on frequency analysis use method for identifying note boundaries based on pitch values and slope of pitch contour [9]. It also tries to associate Indian musical notes with equal tempered scale of western musical notes which is not appropriate as ICM notes are just-intoned [9]. Rajeswari and Geetha have proposed a method to segment given music piece based on *Tala* (rhythmic pattern) and then tried to fit one or more notes within each segment [10].

3 Proposed Method

Singers change the pitch of their voice while singing different notes. Hence pitch is the important feature that helps in identification of notes. The process involves 3 stages, namely, identifying notes present, note boundaries and then sequence of notes. At first, the pitch contour is extracted from the music signal using probabilistic YIN (PYIN), a modified autocorrelation method for pitch estimation [11]. The average pitch values are obtained for each frame (25 ms with 50% overlapping to avoid transition cliques). From the pitch values, pdf is obtained using kernel density estimation method (Gaussian kernel is assumed). Bandwidth range is taken into account from minimum and maximum pitch values. The pdf curve obtained is smoothened using 5 point running averaging technique to remove small glitches. The prominent peaks representing the high probable frequencies of music piece sung in *Kalyani Raga* are identified as shown in Fig. 1. The peaks in the Fig. 1 represent the prominent frequency components rendered through the song. It may be observed that M2 and N3 are very near to P and S (higher octave) respectively, which is the characteristic of this *raga*. From the prominent peaks of pdf, base note is identified. Variance of base note is very less and its frequency lies in between 100 Hz and 250 Hz [12]. Using these characteristics, we first find all potential candidates to be a S0. From these potential notes, the frequency with the least variance and highest mean value is selected as S0. The peaks of pdf are normalized by taking the ratio of peak frequency to the base note. Further, the standard note ratios are used to associate each normalized peak value to the nearest note value.

In ICM, notes do not have predefined duration. Therefore, detecting the boundary of a note becomes difficult task. Here, first derivative of pitch contour is considered for finding the note onset. From the pitch contour one can observe that whenever the transition happens from one note to another note, the difference between adjacent pitch values will be considerably high (refer Fig. 2b). Note boundary (NB) may be identified by $NB = \{t|DP(t) > threshold\}$, where threshold is set to the average of derivative values. The pitch contour of a music piece is shown in the Fig. 2b and the corresponding first derivative is shown in Fig. 2c. One can observe the note transition only in the regions with sharp non-zero changes in the derivative signal. This property is used to find note boundaries. Finally, labelling of each segment is done to obtain the sequence of notes from pitch contour. The sequence of labeled

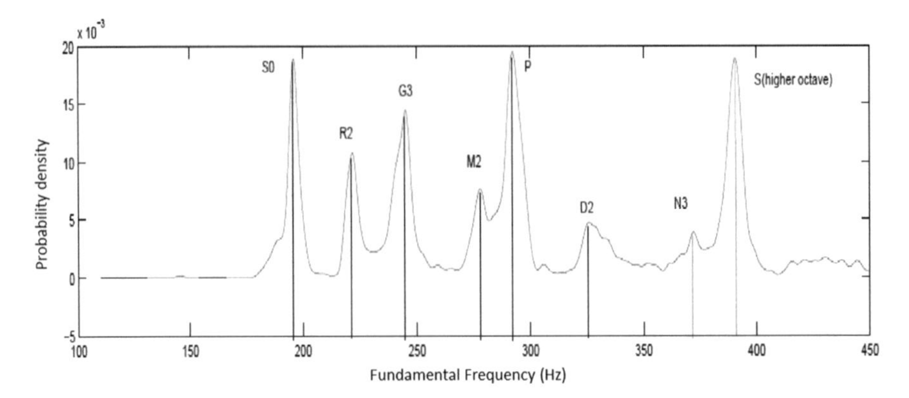

Fig. 1 Probability density function of pitch (*Raga* Kalyani)

segments gives the note sequence. Figure 2 illustrates all the steps involved in the labeling process.

4 Experiments and Results

The standard and common database for ICM processing is not available, therefore suitable database is collected to provide different variations in ICM. The database used in this study is monophonic in nature consisting of 8 *ragas* (total 64 music clips). Each *raga* contains 6–12 clips recorded from artists of both genders which includes both *Sampurna ragas* and *ragas* with missing notes. The intention of selecting different *ragas* is to include all variations of notes present in Carnatic Music. The performance evaluation of note transcription system is conducted for the testing clips

Table 1 Note Identification accuracy without raga knowledge

Raga name	Sequences accurately identified	Sequences with 1 note error	Sequences with 2 note error
Chandrajyoti	50.00	16.67	8.30
Harikhambojhi	66.67	33.33	0.00
Kalyani	30.00	30.00	20.00
Malayamarutha	50.00	37.50	0.00
Mayamalavagowla	30.00	30.00	20.00
Nattai	60.00	20.00	0.00
RasikaPriya	54.54	27.27	0.00
Varali	44.44	22.22	22.22
Average recognition	46.87	28.12	9.38

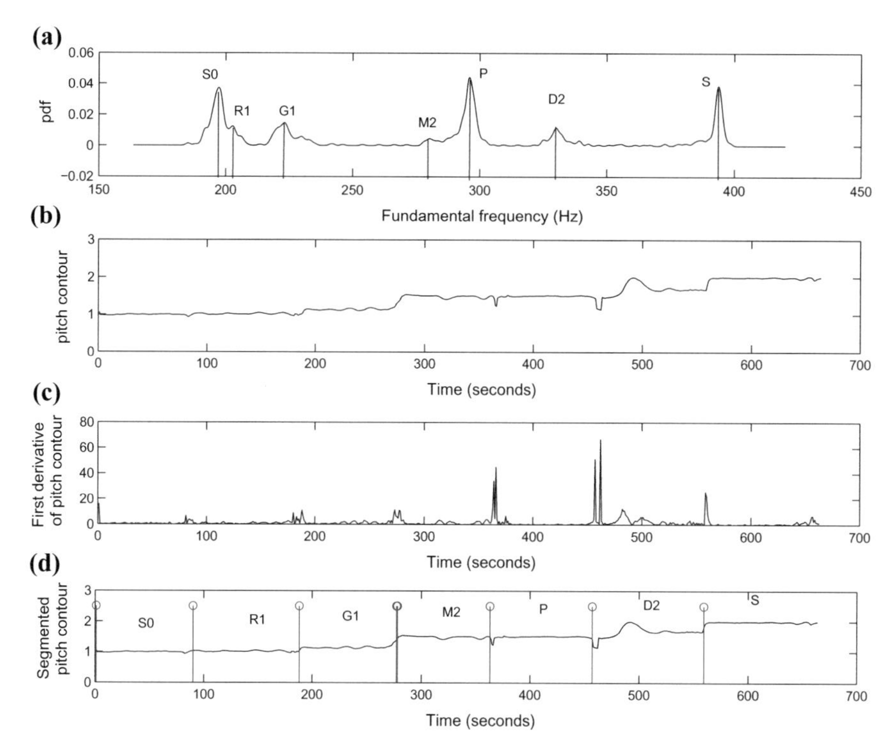

Fig. 2 Note Boundary detection for music piece of *Raga Chandrajyoti*. **a** pdf of pitch contour. **b** Pitch contour of the input music piece. **c** First derivative of pitch contour. **d** Segmented pitch contour

with no prior knowledge and with prior knowledge of their *raga*. The system generated sequence is tested against the manual transcription done by the music experts. The note identification percentage without the prior knowledge of *raga* are given in Table 1. *Kalyani raga* has more note identification errors due to the oscillatory movement of M2 around P and frequency of note N3 is very nearer to *Shadja* of second octave. The 1 note and 2 note errors are mainly influenced by the presence of *gamakas* in the music clips. The other factors that influence the error are the errors in *Shadhja* identification and boundary detection. A marginal error in these two steps contributes heavily to the erroneous note and note sequence identification. In note sequence identification without the prior *raga* knowledge, 30 out of 64 sequences are identified accurately and 18 additional sequences are identified with one note error.

The same experiment is conducted with the prior knowledge of *raga*. Since we already know the notes that should be present in a given *raga*, we associate the most probable note to the peaks obtained from pdf. All remaining procedure and the parameters remain the same. The note sequence identification results are tabulated in the Table 2. Using prior *raga* knowledge, 49 out of 64 music clips are identified with correct note sequences. The accuracy achieved is 76.56%. Similar issues in ICM

Table 2 Note Sequence Identification results (in percentage)

Raga name	Accuracy (%) (with no *raga* knowledge)	Accuracy (%) (with *raga* knowledge)
Chandrajyoti	50.00	75.00
Harikhambojhi	66.67	100.00
Kalyani	30.00	70.00
Malayamarutha	50.00	87.50
Mayamalavagowla	30.00	83.33
Nattai	60.00	100.00
RasikaPriya	54.54	81.81
Varali	44.44	55.56
Average recognition	46.87	76.56

are addressed in [4–6]. However, the results are not directly comparable because of differences in datasets used, such as number of *ragas*, singers. Pandey et al., considered only two *ragas* for experimentation [4].

5 Conclusion and Future Work

In this paper, we have proposed a straight forward method for identifying the notes and the note sequence in the given piece of Carnatic Music. The experiment is conducted on different *ragas* to test if the note transcription is *raga* independent. The experiments were conducted without and with the prior knowledge of *raga*. The results show that the knowledge of *raga* improves the note transcription accuracy. The study may be extended for polyphonic database as well. The system is tested for Carnatic Music, the same can be extended to Hindustani Music. The same task can also be modeled using HMMs or other statistical methods and results can be compared.

References

1. Klapuri, A., Davy, M.: Signal Processing Methods for Music Transcription. Springer, New York (2006)
2. Krishnaswamy, A.: On the twelve basic intervals in South Indian classical music. In: Audio Engineering Society Convention 115. Audio Engineering Society (2003)
3. Avtar Vir, R.: Theory of Indian Music. Pankaj Publications (1999)
4. Pandey, G., Mishra, C., Ipe, P.: Tansen: A system for automatic raga identification. In: IICAI, pp. 1350–1363 (2003)
5. Bhattacharjee, A., Srinivasan, N.: Hindustani raga representation and identification: A transition probability based approach. IJMBC **2**(1–2), 66–91 (2011)

6. Shetty, S., Achary, K.K.: Raga mining of Indian music by extracting arohana-avarohana pattern. Int. J. Recent. Trends Eng. **1**(1), 362–366 (2009)
7. Krishnaswamy, A.: Application of pitch tracking to South Indian classical music. In: International Conference on Multimedia and Expo., Proceedings (ICME'03), vol. 3, pp. III–389. IEEE (2003)
8. Chandrasekaran, J., Jina Devi, H., Swamy, N.V.C., Nagendra, H.R.: Spectral analysis of Indian musical notes. Indian J. Tradit. Knowl. **4**(2), 127–131 (2005)
9. Prashanth, T.R., Venugopalan, R.: Note identification in Carnatic music from frequency spectrum. In: International Conference on Communications and Signal Processing (ICCSP), pp. 87–91. IEEE (2011)
10. Sridhar, R., Geetha, T.V.: Swara indentification for South Indian classical music. In: 9th International Conference on Information Technology (ICIT'06), pp. 143–144. IEEE (2006)
11. Mauch, M., Dixon, S.: pyin: a fundamental frequency estimator using probabilistic threshold distributions. In: International Conference on Acoustics, Speech and Signal Processing (ICASSP), pp. 659–663. IEEE (2014)
12. Ranjani, H.G., Arthi, S., Sreenivas, T.V.: Carnatic music analysis: Shadja, swara identification and raga verification in alapana using stochastic models. In: IEEE Workshop on Applications of Signal Processing to Audio and Acoustics (WASPAA), pp. 29–32. IEEE (2011)

Sentence-Based Dialect Identification System Using Extreme Gradient Boosting Algorithm

Nagaratna B. Chittaragi and Shashidhar G. Koolagudi

Abstract In this paper, a dialect identification system (DIS) is proposed by exploring the dialect specific prosodic features and cepstral coefficients from sentence-level utterances. Commonly, people belonging to a specific region follow a unique speaking style among them known as dialects. Sentence speech units are chosen for dialect identification since it is observed that a unique intonation and energy patterns are followed in sentences. Sentences are derived from a standard Intonational Variations in English (IViE) speech dataset. In this paper, pitch and energy contour are used to derive intonation and energy features respectively by using Legendre polynomial fit function along with five statistical features. Further, Mel frequency cepstral coefficients (MFCCs) are added to capture dialect specific spectral information. Extreme Gradient Boosting (XGB) ensemble method is employed for evaluation of the system under individual and combinations of features. Obtained results have indicated the influences of both prosodic and spectral features in recognition of dialects, also combined feature vectors have shown a better DIS performance of about 89.6%.

Keywords Dialect identification system · IViE speech corpus · Prosodic features · Spectral features · Sentence segmentation · XGB model

---

N. B. Chittaragi (✉)
Department of Information Science and Engineering, Siddaganga Institute of Technology, Tumkur, India
e-mail: nbchittaragi@gmail.com

N. B. Chittaragi · S. G. Koolagudi
Department of Computer Science and Engineering, National Institute of Technology Karnataka, Surathkal 575025, Karnataka, India
e-mail: koolagudi@nitk.edu.in

A. Elçi et al. (eds.), *Smart Computing Paradigms: New Progresses and Challenges*, Advances in Intelligent Systems and Computing 766,
https://doi.org/10.1007/978-981-13-9683-0_14

1 Introduction

Over the last few years, speech community is targeting for automatic speech recognition (ASR) systems inherited with automatic recognition system to handle speech variabilities exists with natural speech. Due to this, dialect identification from speech has become increasingly essential. Every language has several dialectal variations. Regional dialectal variations mainly exist due to geographical location, socioeconomic status, cultural background, education details and so on [1].

Identification of dialects from speech plays a significant role in speaker profiling and identification of speaker specific details in forensic applications. Speech variabilities of an utterance are the main cause for degradation of the performance of ASR system. Embedding DIS within ASR makes the system more robust to address speech variability issues. DIS can be used in routing of calls at call centers so that the human operators who understand the specific dialect can assist the caller efficiently. The performance of spoken document retrieval system and multi-language translation system can be enhanced.

Every dialect of a language shows unique distinguishing characteristics such as speech rate, levels of stress, rising and fall of pitch pattern, rhythmic patterns, beginning and ending patterns in sentence utterance, and so on. Indeed, all these factors assist in capturing the unique speaking patterns for each dialect [2]. Sometimes it is a very challenging task to capture dialect specific properties from shorter utterances (may in terms of Milliseconds or seconds). It is also observed that few languages comprise of prosodic variations lie over longer utterances such as sentences instead of phonemes or words [1]. However, work done in this regard is rarely observed, as identifying the exact meaningful and complete sentence boundaries from the continuous speech is a challenging task. Hence, in this paper, an attempt is made to propose a DIS for nine dialects of British English. Sentences are extracted through an automatic segmentation algorithm. DIS is built by exploring prosodic, statistical, and spectral features from spoken sentences. Decision tree-based XGB ensemble algorithm is used for evaluation and significant improvement is observed.

The remaining part of the paper is organized as follows: Existing work for DIS is given in Sect. 2. Dataset details are provided in Sect. 3. Proposed DIS is discussed in Sect. 4, along with sentence segmentation algorithm, feature extraction, and XGB method. Section 5 comprises of discussions of the obtained results. Section 6 presents conclusions of the present work and future works.

2 Existing Works

Literature shows varieties of techniques proposed for dialect identification through which humans and machines make distinction across dialects of a language. This is also being addressed for the prominent languages of the world. Practically, dialects of the majority of languages have shown phonological, lexicon, morphological, and structural differences between them [3]. Several approaches are proposed by considering the dialectal cues existing at various levels of utterances [4, 5]. Majority of the existing systems processes complete speech utterances available for a longer duration for dialect classification [6]. Apart from these, few works address DIS from different speech units such as phoneme, syllable level [2] and at word level [4]. Usually, sentence utterances are considered better choices, since they convey specific style or pattern followed among specific dialect speakers. Also, to model DIS, features explored from sentences are more significant evidence at the suprasegmental level processing of speech.

Acoustic and phonotactics are two important models followed commonly for dialect recognition. Acoustic model-based dialect systems are proposed by extracting spectral characteristics such as MFCCs and its derivatives [6]. Prosodic features namely, pitch, intonation, intensity, rhythmic features, stress patterns are derived for dialect identification [2]. Several authors have attempted by combining both features and observed improvement. Apart from these language modeling techniques are also being applied for DIS, as these can discriminate dialects similar to languages. Applying Phone Recognition-Language Modeling (PRLM) and its variants namely, parallel PRLM (PPRLM) and Parallel Phone Recognition (PPR) models can also contribute to DIS [7].

In addition to these, extraction of sentence-level features for dialect identification is also seen in literature for Arabian and Chinese languages. However, these are applied to the text-based dataset and processed using n-gram features [8]. Unavailability of standard systems for sentence-based dialect identification from speech has motivated to take this problem. When literature is reviewed through classification methods point of view, commonly Gaussian mixture models, support vector models and hidden Markov models, and artificial neural networks are used most extensively for evaluating DIS [6]. Due to rare use of ensemble algorithms for dialect recognition tasks, in this paper, an attempt is made by employing XGB algorithm with decision tree classifiers as base learners [9].

3 Dataset Details

IViE speech dataset is available with nine dialectal varieties of British English. It has been recorded from native speakers belonging to nine urban areas across the British Isles. Before the recording begins, speakers are asked to read the Cinderella story and later to narrate the story in their own sentences. In this way, recording has captured sentences with pauses, repetitions, etc. Spontaneous dataset is recorded from 12 speakers for each dialect (6F + 6M adolescents) [10].

4 Proposed Dialect Identification System

Many times a sentence spoken is sufficient to make decisions about the languages and dialects [8]. Sentences carry a majority of distinct dialectal cues and these remain same across every sentences [1]. Also, sentence-level processing leads to a reduction

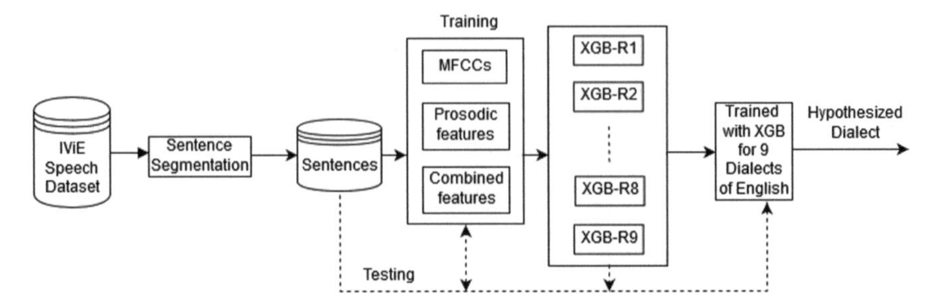

Fig. 1 XGB-based dialect identification system

of computational resources and even beneficial under limited data availability constraints. This section includes brief details of the sentence segmentation algorithm used and extraction of spectral and prosodic features. Later, details of ensemble algorithm are provided. Block diagram of the proposed sentence-level DIS is presented in Fig. 1.

4.1 Sentence Segmentation Algorithm

The process of identification of voiced regions, which are assumed to be complete sentences is separated from spontaneous utterances by using a dynamic threshold-based algorithm [11]. Algorithm followed for sentence segmentation is as follows. Two acoustic features namely short-term energy and spectral centroid are derived from spontaneous utterances of the IViE dataset. These two features combination has recognized clear sentence boundaries through silent regions which are characterized by lower energy and centroid features. Here, centroid feature represents the center of gravity of its spectrum and it will be lower when silent regions exist. Further, every segmented unit is listened manually to retain only meaningful and complete sentence units. 200 sentences are taken for each dialects leading to $200 * 9 = 1800$ sentences in the dataset.

4.2 Feature Extraction

Spectral features play a prominent role in capturing the essential differences exist between languages and dialects. MFCC features highly correlated with human speech perception process, hence are commonly used for modeling spectral cues [12]. In this work, RASTA (Relative Spectra)-based 13 MFCCs features with derivatives such as delta and delta-delta giving 39 features are extracted from 40 filter banks [13]. Distinct prosodic variations have been observed across nine British English dialect speakers

[10]. Prosodic features duration, pitch (F0) and energy features are extracted. Indeed, the F0 change, produced due to rise and fall of the pitch while speaking, recognizes a unique phonological pattern followed in a sentence of a specific dialect.

Prosodic features reflecting intonation patterns from pitch and energy contours, and statistical features from sentence utterances have been explored. Pitch features are computed using a subharmonic-to-harmonic ratio-based pitch estimation algorithm with 30 ms frame size and 15 ms step. From each sentence, the pitch value is computed to draw pitch contour. Further, a Legendre polynomial fit function of order 15 is used. 15 coefficients of Legendre polynomials have produced a good fit and these forms a feature vector. Later, five statistical features namely; min, max, mean, standard deviation, and variance are extracted from pitch contour forming a total 20-dimensional feature vector. Similarly, 15 coefficients and five statistical features are explored from energy contour. Finally, 20 features are added to form the total 40-dimensional prosodic feature vector.

4.3 XGB Ensemble Algorithm

Ensemble algorithms are found to be very powerful prediction and classification techniques in enhancing the performance with a combination of multiple classifiers over single classification methods. Recently, the concept of combining multiple weak classifiers and aggregating their results are found to improve the robustness and performance. However, use of these methods for speech processing in particular for dialect identification is rarely found [9]. Hence, XGB method is used for classifying nine dialects of the English language.

Gradient boosting involves three important steps: First, selection of a suitable loss function, multi-class *logloss* is used, since it is a classification problem. Second, choosing decision trees as base learners where trees are constructed greedily. Few parameters such as best split, number of leaf nodes, maximum levels are fine-tuned to produce a better performance. In the third step, trees are added one at a time; a gradient descent procedure is used for minimization of loss during the addition of trees. The XGBoost library is used for implementation for handling nine dialect classes [14].

5 Experiments and Results

The performance of the proposed system is evaluated on sentence speech corpus with XGB method. Simple validation (SV) and cross fold validation (CV) configurations are used for analysis. Series of experiments are carried out with individual and in combination.

In SV configuration, the dataset is divided into training (80%) and testing (20%) data. Due to the biased division of data, the results are considered to be less significant.

Table 1 Dialect recognition performance in % using XGB algorithm

Sl. no.	Features	80:20		50:50		30:70	
		SV	CV	SV	CV	SV	CV
1	MFCCs	76.32	75.46	73.51	71.48	70.63	68.51
2	Prosodic + statistical features	66.43	63.08	62.59	61.85	59.92	60.03
3	Combination of features	89.58	87.38	85.00	84.16	83.92	82.86

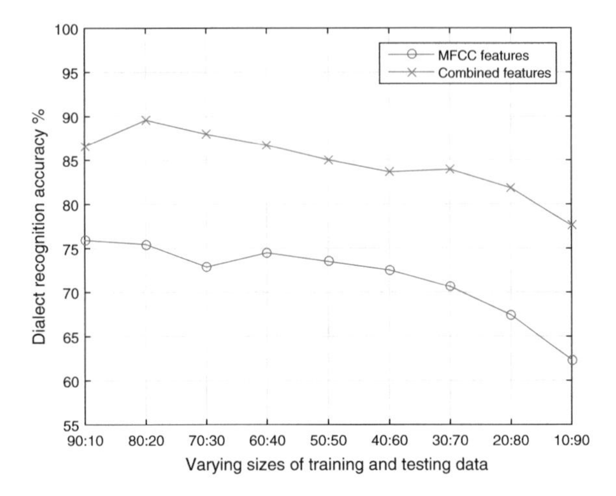

Fig. 2 Nine different ratios of train and test data

To overcome this, CV is used with 5-folds of the dataset. Five different combinations of input data are used for training and testing to measure the average behavior of the system and hence CV results are robust. Further, experiments are conducted by considering varying sizes for training and testing data. These experiments assist in verifying the robustness and generalization of system performance. In the area of forensic science, it may be required to make a compelling prediction even with lesser data during investigation [15]. Table 1 shows recognition results for the proposed DIS for three important combinations. Figure 2 gives the comparison of results obtained with SV configuration for nine different ratios of the train and test data with MFCCs and combined feature vectors.

MFCCs features extracted have resulted in an accuracy of about 76.32% with 80% training data. Similarly, series of experiments are conducted for varying sizes of train data (no. of sentences). These have indicated the slight reduction in recognition rate, as the size of training data for each dialect class is reduced. This is because tree-based models specifically XGB algorithm learns even from less sized data. Along with these, prosodic features explored along with statistical features have achieved 66.43% of recognition performance. Results have indicated the existence of distinct prosodic features among nine dialects. Further, with combination of features highest recognition rate of 89.58% is obtained with 80% training data. Hence, prosodic cues have shown complimentary cues to spectral information. Similarly, for varying sizes of training data, reduction in performance is observed, whereas reduction is comparatively smaller. This is due to generalization performance of XGB model when compared to single classifier-based machine learning algorithms. XGB algorithm has played a significant role in distinguishing nine dialects based on weighted vote of weak classifiers. This is found to be successful in construction of a strong learner for classification of dialects.

6 Conclusions and Future Work

In this paper, an attempt is made to recognize the dialect from the sentence utterances using an XGB ensemble algorithm. Standard IViE dialect dataset of British English is used for evaluation. The dynamic threshold-based segmentation algorithm is used for automatic sentence extraction. Both spectral and prosodic features are used for assessment of system behaviors. Highest recognition of about 89.58% is achieved with combined feature set. Further, the proposed system has shown effective results when the machine is trained with 80% of data and also given slight reduction with smaller training data. In future, accurate language specific dialect discriminating prosodic and spectral features can be explored.

References

1. Chambers. J.K., Trudgill, P.: Dialectology, 2 edn. Cambridge University Press (1998)
2. Rouas, J.L.: Automatic prosodic variations modeling for language and dialect discrimination. IEEE Trans. Audio Speech Lang. Process. **15**(6), 1904–1911 (2007)
3. Mehrabani, M., Hansen, J.H.L.: Automatic analysis of dialect/language sets. Int. J. Speech Technol. **18**(3), 277–286 (2015)
4. Huang, R., Hansen, J.H.L., Angkititrakul, P.: Dialect/accent classification using unrestricted audio. IEEE Trans. Audio Speech Lang. Process. **15**(2), 453–464 (2007)
5. Chen, N.F., Shen, W., Campbell, J.P.: A linguistically-informative approach to dialect recognition using dialect-discriminating context-dependent phonetic models. In: ICASSP, IEEE International Conference on Acoustics, Speech and Signal Processing, pp. 5014–5017 (2010)
6. Biadsy, F.: Automatic dialect and accent recognition and its application to speech recognition. PhD thesis (2011). Columbia University
7. Zissman, M.A., Gleason, T.P., Rekart D.M., Losiewicz, B.L.: Automatic dialect identification of extemporaneous conversational, Latin American Spanish speech. In: ICASSP, pp. 777–780 (1996)
8. Xu, F., Wang, M., Li, M.: Sentence-level dialects identification in the Greater China region. Int. J. Nat. Lang. Comput. (IJNLC) **5**(6), 9–20 (2016)
9. Chittaragi, N.B., Prakash, A., Koolagudi, S.G.: Dialect identification using spectral and prosodic features on single and ensemble classifiers. In: Arabian Journal for Science and Engineering (2017, November)
10. Grabe, E., Post, B.: Intonational variation in the British Isles. In: Speech Prosody, International Conference (2002)

11. Giannakopoulos, T.: Study and Application of Acoustic Information for the Detection of Harmful Content, and Fusion with Visual Information. University of Athens, Greece, Department of Informatics and Telecommunications (2009)
12. Chittaragi, N.B., Koolagudi, S.G.: Acoustic features based word level dialect classification using SVM and ensemble methods. In: 2017 Tenth International Conference on Contemporary Computing (IC3), pp. 1–6 (2017)
13. Hermansky, H., Morgan, N.: Rasta processing of speech. IEEE Trans. Speech Audio Process. **2**(4), 578–589 (1994)
14. Chen, T., Guestrin, C.: Xgboost: a scalable tree boosting system. In: Proceedings of the 22nd ACM SIGKDD International Conference on Knowledge Discovery and Data Mining, pp. 785–794 (2016)
15. Harris, M.J., Gries, S.T., Miglio, V.G.: Prosody and its application to forensic linguistics. LESLI: Linguist. Evid. Secur. Law Intell. **2**(2), 11–29 (2014)

Characterization of Consonant Sounds Using Features Related to Place of Articulation

Pravin Bhaskar Ramteke, Srishti Hegde and Shashidhar G. Koolagudi

Abstract Speech sounds are classified into 5 classes, grouped based on place and manner of articulation: velar, palatal, retroflex, dental and labial. In this paper, an attempt has been made to explore the role of place of articulation and vocal tract length in characterizing the different class of speech sounds. Formants and vocal tract length available for the production of each class of sound are extracted from the region of transition from consonant burst to the rising profile of the immediate following vowel. These features along with their statistical variations are considered for the analysis. Based on the non-linear nature of the features Random Forest (RF) is used for the classification. From the results, it is observed that the proposed features are efficient in discriminating the class of consonants: velar and palatal, palatal and retroflex and palatal and labial sounds with an accuracy of 92.9%, 93.83 and 94.07 respectively.

Keywords Formants · Manner of articulation · Place of articulation · Random forest · Vocal tract length

1 Introduction

Phonemes are the smallest and indivisible unit of speech [1]. They can be classified based on a number of parameters such as place and manner of articulation [2]. For a given phoneme, place of articulation refers to the narrowest part of the vocal tract

P. B. Ramteke (✉) · S. G. Koolagudi
National Institute of Technology Karnataka, Surathkal, Karnataka, India
e-mail: ramteke0001@gmail.com

S. G. Koolagudi
e-mail: koolagudi@nitk.edu.in

S. Hegde (✉)
Nitte Mahalinga Adyanthaya Memorial Institute of Technology,
Karkala, Karnataka, India
e-mail: srishti.hegde.sh@gmail.com

A. Elçi et al. (eds.), *Smart Computing Paradigms: New Progresses and Challenges*,
Advances in Intelligent Systems and Computing 766,
https://doi.org/10.1007/978-981-13-9683-0_15

involved in the production of the speech sound [3]. In most of the Indian languages, consonants are classified into 5 groups based on place of articulation, namely velar, palatal, retroflex, dental and labial [4]. Velar: the articulation is at the back of the tongue against soft palate, e.g. $/k/$, $/g/$ and $/\eta/$. Palatal consonants are pronounced by raising the tongue against hard palate, e.g. $/ch/$ and $/j/$. The tip of the tongue is curled at the back against the palate results in retroflex consonants like $/t/$, $/d/$. Labial sounds such as $/p/$, $/b/$, are pronounced by sudden release of constriction at the lips. Dental consonants are pronounced by tongue touch to the upper teeth as in $/th/$ and $/dh/$ [5].

Phonological processes are the frequently observed patterns of pronunciation errors in children from are $2\frac{1}{2}$ years to $6\frac{1}{2}$ years [6]. Persistence of the phonological processes beyond the specified age may indicate developing phonological disorder [7]. Identification of class of sound been mispronounced is an effective way to study, detect and hence correct in the early stages. In language learning, identification of class of sound usually being mispronounced may help in improving the pronunciation [8]. Consonants burst regions are very difficult to recognize, if the properties efficient in characterizing each class of sound are used, it may help in improving the performance of the system [4]. In this work, an attempt has been made to explore the features efficient in discriminating different class of sounds. The spectral features such as formants and vocal tract length available for the production of each class of sound are analyzed.

Rest of the paper is organized as follows. Section 2 discuss the related work done in the past. Section 3 provides implementation details of the proposed approach. Section 4 explains the results along with the performance metrics. Section 5 gives the conclusion along with some future directions.

2 Literature Survey

Significant amount of research has been done in the field of phoneme recognition to improve the speech recognition accuracy. In phoneme recognition and speech recognition systems, the vocal tract parameter estimation using spectral analysis has been employed. This involves critical band filter which extracts the information related to auditory system by decomposing the speech signal into discrete set of spectral samples [4]. Cepstral analysis decompose the speech signal into excitation source and vocal tract response, hence provide the details of vocal tract configuration involved in the production of different speech sounds [9]. This task is computationally intensive, hence a compact representation of the same is achieved with the help of LPC analysis [10]. Various combinations of spectral representations of human speech production and perception mechanisms have been explored for the accurate recognition of phonemes for speech recognition and speech to text conversion [11]. The effectiveness of different prosodic features namely fundamental frequency ($F0$), energy and duration, in automatic speech recognition is put forward [12]. The orthographic accentuation of prosodic features is achieved and incorporated with the hidden Markov model (HMM) based speech recognizer. The word recognition error is observed to reduce by 28.91% for Spanish language [12]. With an approach towards speech recognition based on excitation source, discussions were made on the contribution of articulatory and excitation source information in discriminating sound units. Since the performance of the speech recognition highly rely on the efficiency of the phoneme recognition. The objective is to uplift the phoneme recognition performance using combination of spectral features, articulatory features-manner, place, roundness, frontness, height- and excitation source features [13]. HMM trained with articulatory and spectral features is observed to achieve little high accuracy compared to the model trained with excitation and spectral features.

From the literature, it is observed that, various combinations of the spectral, prosodic and excitation source features have been explored to improve the performance of phoneme recognition and hence the speech recognition. The features specific to the place and manner of articulation of the phoneme may help in improving the performance of the system. Hence, the proposed approach explores the role of formants and vocal tract length in discriminating class of consonants.

3 Proposed Methodology

Formants and vocal tract length features are extracted from each pair of speech classes and random forest (RFs) are used to test the efficiency of discrimination of the proposed features.

3.1 Dataset

TIMIT acoustic-phonetic speech corpus is designed to fulfill the needs of speech data for acquisition of acoustic-phonetic knowledge along with the evaluation of speech related tasks [14]. It consists of recordings of 630 speakers from both genders in 8 major dialects of American English with 10 phonetically rich sentences. The consonant sounds for each class: velar, palatal, retroflex, dental and labial with the vowel followed are manually segmented from the database, as shown in Figure 1.

3.2 Feature Extraction

Formants: Formants characterize the vocal tract resonance during pronunciation of the different speech units. Formants can be extracted from linear prediction by solving the poles of the filter by setting the denominator to 0 [15]. Features are extracted from the transition region of the consonant burst and the immediate following vowel. 4 formants along with their statistical variations are considered. $F1$ represents the

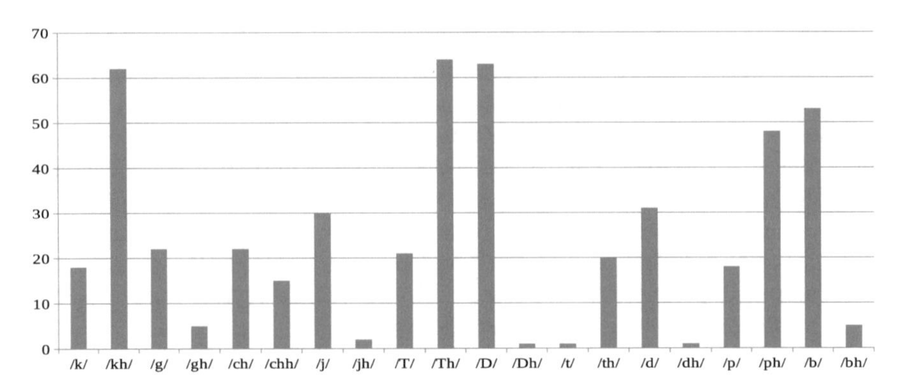

Fig. 1 Distribution of the data considered for the experimentation

Table 1 Details of the features and their statistical variations

Features	Features and their statistical
Formants	Formant1 (F1), Formant2 (F2), Formant3 (F3), Formant4 (F4), minimum F1, minimum F2, minimum F3, minimum F4, maximum F1, maximum F2, maximum F3, maximum F4, mode F1, mode F2, mode F3, mode F4, median F1, median F2, median F3, median F4
Vocal tract length	Vocal tract length 1 (V1), Vocal tract length 2 (V2), Vocal tract length 3 (V3), Vocal tract length 4 (V4), minimum V1, minimum V2, minimum V3, minimum V4, maximum V1, maximum V2, maximum V3, maximum V4, mode V1, mode V2, mode V3, mode V4, median V1, median V2, median V3, median V4

volume of the pharyngeal cavity during pronunciation where increase in volume results in lower value of $F1$ [16]. It also gives a measure of change in the tongue height and degree of lip opening. The degree of lip opening is directly proportional to $F1$ value. $F2$ approximates the length of oral cavity, where lower $F2$ value represents longer oral cavity [16]. It correlates with the tongue retraction function. The amount of constriction is given by $F3$, where if the constriction happens at the back gives high values of $F3$ [16].

Vocal tract length: Vocal tract is assumed to be a uniform lossless acoustic tube which is closed at one end and open at the other end. The tube generates resonant frequencies uniformly spaced in nature. The vocal tract length is given by equation, $L = (2n - 1)\frac{C}{4F_n} \quad n = 1, 2, 3, \ldots$ [17]. Where C = 35,300 cm/s is the speed of sound at 35 °C and L is the length of uniform vocal tract in centimeter. F_n is the nth formant frequency. From the estimated vocal tract lengths, statistical variations are derived as given in Table 1.

3.3 *Classifiers Used*

Random Forests (RFs): Random forests are the class of ensemble learning methods generally used for the classification, regression, etc. It is constructed by combining the decision tree based classifiers, where each classifier use an independent random subset of input vector for training [18]. Class label is assigned to a test sample by taking the most popular class voted by all the tree predictors of the forests. Random forest tries to overcome the internal unbiased generalization error, hence efficient in estimation of missing data [18]. It inherits an ability to achieve good accuracy even when the large proportion of the dataset is unrecoverable and the dataset with unbalanced class population.

4 Results

The work mainly focuses on the role of the place of articulation in characterizing the classes of speech unit. Formants, vocal tract lengths and their statistical variations are extracted from 5 classes of sounds grouped based on place of articulation: velar, palatal, retroflex, dental and labial. This results in features vector of size 40 (refer Table 1). Based on the non-linear nature of the feature vector, random forests are used for the classification. Beginning with 2 classes at a time, the classifier is trained and tested to obtain the accuracy of the classification with 10-fold cross-validation. The accuracy of random forest is given in Table 2. From the results, it can be clearly observed that the proposed features are efficient in discriminating the class of sound palatal against the velar, retroflex and labial. The accuracy of the distinction between the palatal and velar sounds is 92.90%. The results achieved for the palatal and retroflex, and palatal and labial are 93.83% and 94.07% respectively. This shows that there is a significant difference in the place of articulation of palatal-velar, palatal-retroflex and palatal-labial speech sounds. The proposed features are not efficient in discriminating other classes of sounds. An average accuracy of 70% is achieved with the proposed features for the remaining pairs of class of consonant sounds. This raises the question on the generalization of the applicability of the proposed features

Table 2 Accuracies of classification of speech classes using random forest

Sl. No	Consonant classes	Accuracy (%)		Sl. No	Consonant classes	Accuracy (%)
1	Velar - Palatal	**92.90**		6	Palatal - Dental	79.41
2	Velar - Retroflex	76.40		7	Palatal - Labial	**94.07**
3	Velar - Dental	74.10		8	Retroflex - Dental	77.43
4	Velar - Labial	70.36		9	Retroflex - Labial	72.18
5	Palatal - Retroflex	**93.83**		10	Dental - Labial	77.91

in discriminating all classes of speech sound. The reason behind the low accuracy for the other classes may come from the position of the tongue hump, wherein some of these classes though the obstruction of the air happens at the different places, the position of the tongue may be similar in nature. This may affect the performance of the classification.

5 Conclusion

Here, an attempt has been made to analyze the role of place of articulation in characterizing the speech sounds. The classification is performed between each pair of 5 classes: velar, palatal, retroflex, dental and labial sounds based on place of articulation. Formants and vocal tract length are extracted with the assumption that these are similar within a particular class of speech unit. These features along with their statistical variations formed the feature vector of 40. Based on the non-linear nature of the features Random Forest (RF) is used for the classification. From the results, it is observed that, the proposed features are fairly efficient in discriminating palatal against the velar, retroflex and labial sounds. However it is possible that with a detailed study of each class of sounds along the same line, the spectral features that are characteristic to each of these sound classes may provide a better representation of the sound classes and hence improve the classification performance.

References

1. Jones, D.: The phoneme: its nature and Use. Cambridge, England, Heffer (1950)
2. Denes, P.B.: On the statistics of spoken English. J. Acoust. Soc. Am. **35**(6), 892–904 (1963)
3. Clements, G.N.: Place of articulation in consonants and vowels: a unified theory. Work. Pap. Cornell Phon. Lab. **5**, 77–123 (1991)
4. Rabiner, L. R., Juang, B. H.: Fundamentals of speech recognition. Tsinghua University Press (1999)
5. Hogg, R.M.: Phonology and morphology. Camb. Hist. Engl. Lang. **1**, 67–167 (1992)
6. Grunwell, P.: Phonological Assessment of Child Speech (PACS). College Hill Press (1985)
7. Shriberg, L.D., Kwiatkowski, J.: Phonological disorders I: A diagnostic classification system. J. Speech Hear. Disord. **47**(3), 226–241 (1982)
8. Eskenazi, M.: Using automatic speech processing for foreign language pronunciation tutoring: Some issues and a prototype. Lang. Learn. Tech. **2**(2), 62–76 (1999)
9. Fukada, T., Tokuda, K., Kobayashi, T., Imai, S.: An adaptive algorithm for mel-cepstral analysis of speech. In: 1992 IEEE International Conference on Acoustics, Speech, and Signal Processing, 1992. ICASSP-92, vol. 1, pp. 137–140. IEEE (1992)
10. Shrawankar, U., Thakare, V.M.: Techniques for feature extraction in speech recognition system: A comparative study arXiv:1305.1145 (2013)
11. Zue, V.W.: The use of speech knowledge in automatic speech recognition. Proc. IEEE **73**(11), 1602–1615 (1985)
12. Milone, D.H., Rubio, A.J.: Prosodic and accentual information for automatic speech recognition. IEEE Trans. Speech Audio Process. **11**(4), 321–333 (2003)

13. Manjunath, K.E., Sreenivasa Rao, K.: Articulatory and excitation source features for speech recognition in read, extempore and conversation modes. Int. J. Speech Technol. **19**(1), 121–134 (2016)
14. Garofolo, J.S., Lamel, L.F., Fisher, W.M., Fiscus, J.G., Pallett, D.S.: Darpa timit acoustic-phonetic continous speech corpus. NASA STI/Recon technical report n **93**, (1993)
15. McCandless, S.: An algorithm for automatic formant extraction using linear prediction spectra. IEEE Trans. Acoust., Speech, Signal Process. **22**(2), 135–141 (1974)
16. Stevens, K.N.: Acoustic Phonetics, vol. 30. MIT Press, Cambridge (2000)
17. Paige, A., Zue, V.: Calculation of vocal tract length. IEEE Trans. Audio Electroacoust. **18**(3), 268–270 (1970)
18. Breiman, L.: Random forests. Mach. Learn. **45**(1), 5–32 (2001)

Named Entity Recognition Using Part-of-Speech Rules for Telugu

SaiKiranmai Gorla, Sriharshitha Velivelli, Dipak Kumar Satpathi, N L Bhanu Murthy and Aruna Malapati

Abstract Part-of-Speech (POS) plays an important role in identifying and classifying Named Entities (NEs) for any language, especially, for inflected and agglutinating languages like Telugu. The main objective of our work is to generate POS-based rules to identify named entities in Telugu language using Decision Tree classifier. The corpus is generated by crawling through Telugu newspaper websites, which consists of 54457 words and each of these words are manually annotated with NEs tags, namely person, location, organization, and others. We have achieved competent performance by the generated POS rules and they help to get deeper insights into NER for Telugu.

Keywords Named entity recognition · Decision tree · Part-of-speech tagging

1 Introduction

Named Entity Recognition (NER) is identification and classification of textual elements (words or sequences of words) into a predefined set of categories called Named Entities such as the names of a person, organization, location, expressions of the time, quantities, monetary values, percentages. The term Named Entity (NE) was first introduced in the Sixth Message Understanding Conference (MUC-6)[12]. The

S. Gorla (✉) · S. Velivelli · D. K. Satpathi · N. L. B. Murthy · A. Malapati
Birla Institute of Technology and Science Pilani, Hyderabad Campus, Hyderabad, India
e-mail: p2013531@hyderabad.bits-pilani.ac.in

S. Velivelli
e-mail: f20130847@hyderabad.bits-pilani.ac.in

D. K. Satpathi
e-mail: dipak@hyderabad.bits-pilani.ac.in

N. L. B. Murthy
e-mail: bhanu@hyderabad.bits-pilani.ac.in

A. Malapati
e-mail: arunam@hyderabad.bits-pilani.ac.in

A. Elçi et al. (eds.), *Smart Computing Paradigms: New Progresses and Challenges*,
Advances in Intelligent Systems and Computing 766,
https://doi.org/10.1007/978-981-13-9683-0_16

NER task has important significance in many of the NLP applications such as machine translation [6], question answering [19], automatic summarization [3], social media analysis [15], and semantic search [13] in biomedical domain [2].

Telugu is a Dravidian language in the southern part of India and ranks third among the mostly spoken Indian languages. It is highly inflected and agglutinating language providing rich and challenging sets of linguistic and statistical features resulting in long and complex word forms [4]. There has been not much of research work on NER for Telugu language and we would attempt to propose learning models. There has been evolved models for NER in English with good prediction accuracies [10, 16]. Though we may get some insights from the learning models developed for NER in English, the language-dependent features make it difficult to use similar models for Telugu. For the purpose of analysis of such inflectionally rich languages, the root and the morphemes of each word have to be identified. Telugu is having a large number of morphological variants for a given root. The primary word order of Telugu is SOV (Subject–Object–Verb). The word order of subjects and objects is largely free. For example, రాముడు సీతాకు హారాన్ని పంపాడు (Ram sent a necklace to Sita) or రాముడు హారాన్ని సీతాకు పంపాడు (Ram sent a necklace to Sita). Internal changes in the sentences or position swap between various word group or phrases will not affect the meaning of the sentence.

The main reason for the richness in morphology of Telugu (and other Dravidian languages) is that the significant portion of grammar that is handled by syntax in English is being handled with morphology in Telugu language. Each inflected word in Telugu starts with root and has many suffixes. The word suffix used here is to refer to inflections, post-positions, and markers which indicate tense, number, person, gender, negatives, and imperatives. These suffixes are affixed with each root word to generate word forms. In English, generally, phrases include several words and for most of the cases, such phrases are mapped to a single word in Telugu. For example, *(he came and went)*, *"gelavalEDanukonnAvA" (do you think he will not win?)*, *"rAjamaMDrovaipu" (toward rajahmundary)* are single words in Telugu which makes NER task complex.

Named Entity Recognition in Indian languages, especially, in Telugu is difficult due to the following challenges:

1. Absence of capitalization: In English, the NE start with a capital letter which places an important role for identification, but there is no concept of capitalization in Telugu.
2. Telugu is highly inflectional and agglutinating language: The way lexical forms get generated in Telugu are different from English. Words are formed by productive derivation and inflectional suffixes to roots or stems. *For example*, *hyderabad*, *hyderabad ki*, *hyderabadki*, *hyderabad lo*, etc. all refer to Location Hyderabad where *lo*, *ki*, and *ni* are all post-position markers in Telugu. All the post-positions get added to the root word Hyderabad.
3. Recognizing acronyms: There are different ways to represent acronyms. *Example*: ***bi.je.pi*** *and* ***bajapa*** *both are acronyms of* ***BhAratIya janatA pArTI***. Sometimes letters in acronyms could be the English alphabet or in native language.

4. Resource-poor language: unavailability of corpus, name dictionaries (gazetteers), morphological analyzers, Part-of-Speech (POS) taggers, etc.
5. Relatively free order.

In this paper, we attempt to find rules around POS to recognize named entities and classify them to one of the categories: name of person or location or organization. We explore decision tree learning to get the rules and make use of annotated data as described in Sect. 4.1.

The rest of the article is organized as follows. We discuss related work in Sect. 2 and illustrate features, dataset, and algorithm in Sect. 3. The experiments and results are discussed in Sect. 4 and conclusion of our study is summarized in Sect. 5.

2 Related Work

NER task has been investigated extensively for English language. The earlier approaches for NER can be grouped into rule-based, lookup from large gazetteer lists and machine learning approaches. Rule-based NER system uses handcrafted rules based on pattern matching or regular expressions[1]. The second approach recognizes named entities that are available in the gazetteer list [11]. The disadvantages of these approaches are maintaining the lists are expensive, intensive, and inflexible and conversion from domain to another is difficult. The NER systems which are based on machine learning approaches are Support Vector Machine (SVM) [22], Maximum Entropy Model (ME) [7], Hidden Markov Model (HMM)[5], Conditional Random Field (CRF) [17], and Decision Trees (DT) [14].

Telugu NER developed by Srikanth and Murthy [21] has used part of LERC-UoH Telugu corpus where CRF-based Noun Tagger is built using 13,425 words. This has been considered as one of the features for rule-based NER system for Telugu mainly focusing on identifying person, location, and organization without considering POS tag or syntactic information. This work is limited to only single word NEs. Praneeth et al. [20] built CRF-based NER system with language- independent and dependent features. They have conducted experiments on data released as a part of NER for South and South East Asian Languages (NERSSEAL)[1] competition with 12 classes and obtained F1-score of 44.89%.

In other Indian languages, a considerable amount of work has been done in Bengali and Hindi. Ekbal et al. [8] developed an NER system for Bengali and Hindi using SVM. The system uses different contextual informations of words in predicting four NE classes, such as person, location, organization, and miscellaneous. They have used annotated corpora of 122,467 tokens for Bengali and 502,974 tokens for Hindi. The system has been tested with 35K and 60K tokens for Bengali and Hindi with an F1-score 84.15% and 77.17% respectively. In this [9] paper, they developed the NER system using CRF for Bengali and Hindi using contextual features with an F1-score of 83.89% for Bengali and 80.93% for Hindi.

[1] http://ltrc.iiit.ac.in/ner-ssea-08/.

3 Methodology

The decision tree learning is widely used for its ability to visually and explicitly express rules. They are supervised learning models which are predominantly used for classification tasks. The decision tree learnt from the data can be re-represented as a set of if-then rules.

One of the most primary decision tree learning algorithms is ID3 by Quinlan. The decision trees are constructed in a top-down manner. The attribute/feature that has got the most information about the target label assumes root of the decision tree and the same principle is followed for constructing the decision tree.

Entropy measures the purity/impurity of the data. Data is pure if it is nonhomogeneous in the target label. It is impure if it is a mix of target variable. Entropy is zero when data is pure and high when it is impure. ID3 uses Information Gain (IG) as the split function to decide decision node. Information gain is the expected reduction in entropy caused by partitioning the data according to an attribute. IG is applied on all the available attributes and the attribute which gives the maximum value is the next node in the tree.

$$Gain(S, A) = Entropy(S) - \sum_{v \in Values(A)} \frac{|S_v|}{|S|} Entropy(s_v) \tag{1}$$

where S is data, A is the attribute on which we are splitting, Values(A) is the set of all possible values for attribute A, S_v is the subset of S for which A has value v.

Each path from root node to the leaf node is a conjunction of attributes and the tree is a disjunction of these conjunctions. These disjunctions are the rules which we get from decision tree. Each path from root node to the leaf node is a rule in our rule set.

The algorithm is implemented using C50[2] package in R. This package implements C5.0 decision tree learning algorithm, which is an extension of the basic ID3.

4 Experiment and Results

4.1 Corpus

We have generated data for our experiments by crawling through Telugu Newspaper websites. We have manually annotated each word of the corpus as belonging to one of the NEs, i.e., PERSON, LOCATION, and ORGANIZATION. The data consists of 54557 tokens words out of which 16829 are unique word forms. The number of NEs in the corpus are 2658 persons, 2291 locations, and 1617 organizations.

[2]https://cran.r-project.org/web/packages/C50/index.html.

Table 1 POS tag set for Telugu language

Category	POS tag	Example
Noun	NN	vimAnam (aeroplane), vanamu (forest)
Proper noun	NNP	jAnaki (Janaki), amErikA(America)
Noun denoting spatial and temporal expressions	NST	vemTanE (quickly), akkaDa (there)
Pronoun	PRP	amdaru (all), Ame(she)
Demonstratives	DEM	I (these), A (that)
Verb	VM	tinaDam (eating)
Post-position	PSP	lO, tO, kOsam
Conjunctions	CC	ayitE, kAgA, amtEkAka
Question words	WQ	evaru, E
Quantifiers	QF	anni, komtamamdi
Cardinals	QC	remdu (two), vamDa (hundred)
Ordinals	QO	remDO (second), okkO
Classifiers	CL	mamdi (persons)
Intensifier	INTF	marimta, atyamta
Interjection	INJ	pApam, abbO, hAmmayya
Reduplication	RDP	peDDa peDDA, kumcham kumcham
Adjective	JJ	peDDa, cinna
Adverb	RB	amtagA, haTTigA
Particle	RP	tappa, okavELa
Quotative	UT	ani, anna
Special symbol	SYM	"!", "?", "."

4.2 *Features*

Reddy et al. has developed TnT Part-of-Speech (POS) tagger that can classify a Telugu word into one of the 21 POS tags as shown in Table 1. In order to obtain POS tag, we have used TnT POS tagger [18] on our data.

The context word feature (window size) is a previous (w_{-1}) and next (w_{+1}) word of a particular word (w_0) have been commonly used as a one of the features for NER. For example, in the sentence, రాముడు సీతా ని ఇష్టపడ్డారు the context word feature of window size 3 for a word సీతా (w_0) will be రాముడు (w_{-1}) and ని (w_{+1}). If the context word feature of a current word itself is used as a feature, then most likely, it emerges as the root with 16K (refer Sect. 4.1) unique words as its children. The Decision Tree learns from training data turns into a lookup table, instead of generating meaningful rules. If a new word appears in the test data, then the model fails to classify the new word. Hence, the context word feature is not considered as a feature for building a decision tree model.

In this paper, POS tag information of the current and surrounding words of window size 3 is considered for NER. For the first word of a sentence, the previous word POS is taken as "NIL" and for the last word of a sentence, the next word POS is taken as "EOS". "NIL"'and "EOS" are used to represent the beginning and end word in the sentence.

For example, consider the sentence: సీత తన దుస్తులు ఇష్టపడ్డారు (Sita likes her dress). POS tags for the sentence(in order) : {NNP PRP NN VM}. Features for each word in the sentence: సీత—{NIL, NNP, PRP}, తన—{NNP,PRP, NN}, దుస్తులు—{PRP, NN, VM}, ఇష్టపడ్డారు—{NN, VM, EOS}.

4.3 Evaluation Metrics

The standard evaluation measures like Precision, Recall, and F1-score are considered to evaluate our experiments.

$$Precision(P) = \frac{c}{r} \tag{2}$$

$$Recall(R) = \frac{c}{t} \tag{3}$$

$$F1 - score = \frac{2 * P * R}{P + R} \tag{4}$$

where r is the number of NEs predicted by the system, t is the total number of NEs present in the test set, and c is the number of NEs correctly predicted by the system.

4.4 Results

Srikanth et al. [21] has implemented CRF-based noun tagger for finding whether a given word is a noun or not. This has been considered as one of the features along with Suffix/Prefix, Gazetteers, and Context word. They proposed linguistic rules, as advised by Telugu linguists, by making use of these features to classify NEs in Telugu. This approach has achieved an F1-score of 81.55, 81.45, and 79.89% for person, location, and organization, respectively.

Shishtla et al. [20] has developed a learning model by implementing CRF classifier for NER in Telugu. They have used context words, POS, and chunk of a current word as features and achieved F1-score of 44.99%.

By making use of decision tree learning, we propose to generate rules around POS of contextual words from historical data without using the intricate deeper features

Table 2 Evaluation results using window size 3

NE class	Precision	Recall	F1-score
Location	60.53	78.88	68.49
Person	67.22	76.84	71.71

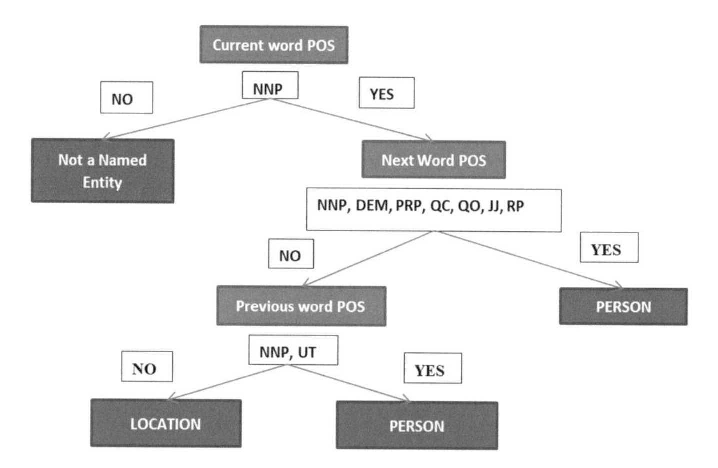

Fig. 1 Decision tree generated using POS tags

and algorithms. In our experiments, the corpus is randomly split 70:30 for training and testing. Table 2 summarizes Precision, Recall, and F1-score of window size 3 and averaged on 50 runs. The experiments were also repeated for window size 5. The Precision, Recall, and F1-score remains more or less the same. Hence, we can conclude that POS tag for Telugu NER is influenced by only surrounding words. The Decision Tree generated using POS tag is shown in Fig. 1.

Out of the three NE classes experimented our model failed to classify organization class due to following two reasons:

1. The organization is a multi-word entity,
2. Each word in the organization has different POS tags.

Our results outperform Shishtla et al. [20] and the learning model adopted is much simpler with only POS tag of context words as a feature.

4.5 Rules Generated From the Decision Tree

Rules generated from the decision tree are shown below with examples in English and their translation in Telugu. In the examples given below, the **underlined words** represent the **named entities** and **bold** represent the **POS** tag under consideration.

- If *current word POS* is not *proper noun* then *current word* is not a *named entity*.
  - As name of person and name of location are both proper nouns anything other than proper noun is not an entity.
- If *current word POS* is *proper noun* and *next word POS (or previous word POS)* is also *proper noun* then *current word* is name of a *person*.
  - Often person names consist of two or more words than the name of location. Further, locations in English are written using two words while they are expressed as a one word in Telugu. Hence if two consecutive words have POS as proper noun, then it is the name of the person.
  - *Examples of two words English locations are written as one word in Telugu (transliteration):*
    New Zealand—న్యూయార్క్ (*nyUjilAmD*)
    New Delhi—న్యూఢిల్లీ (*nyUDhillI*)
    Andhra Pradesh—ఆంధ్రప్రదేశ్ (*AmdhrapradESI*).
- If *current word POS* is *proper noun* and *next word POS* is *pronoun* then *current word* is name of a *person*.
  - Pronouns are commonly used as substitutions for person names than locations. In Telugu, person name is followed by a pronoun unlike in English.
  - *Examples*:
    English sentence " *Seetha likes* ***her*** *dress*" in Telugu written as " *sIta* ***tana*** *dustulu ishTapaDDAru*". Here the pronoun *tana* is followed after *sIta* the name of person unlike English where the pronoun is after the verb.
    Rahul is taller than ***me*** is translated in Telugu as *rAhul* ***nA*** *kmTE poDuvugA unADu*.
- If *current word POS* is *proper noun* and *next word POS* is *Demonstratives*, then it is name of a *person*.
  - Person and Location can form a sentence with demonstratives in a similar way. Often Person is followed by demonstratives then location. Hence, we conclude that the current word is a name of a person.
  - For example, *rOjA A vidamgA mATlaDatumDi*, *sIta I plETu lO BHOjanam cEsimdi*. As the set of sentences possible are more for a person, it is inferred that the word is likely to be the name of a person.
  - *Examples*:
    English sentence *Sanjay denied* ***these*** *allegations* in Telugu is *samjay* ***I*** *ArOpaNlanu khamDimcAru*. Name of the person *samjay* is followed by the demonstrative ***I***.
    British Prime Minister David Cameron announced ***these*** *changes—briTish pradhAnamamtri DEviD kAmerUn* ***I*** *mArpulanu prakaTimcAru*.
- If *current word POS* is *proper noun* and *next word POS* is *Cardinal/Ordinal*, then *current word* is name of a *person*.

- Cardinals followed by person names represent age/experience/position of a person. In contrast, the location followed by a Cardinal represent the position of the place. Hence, it is more likely to be a person name.
- *Examples:*

  Mary Kom won ***5*** *world championships—mErI kOm* ***5*** *prapamca cAmpiyass hiplanu gelcukunaru.* Here *mErI kOm* is followed by ***5*** representing how many times she got world championship.

  Spinner Ravichandran Ashwin is the ***third*** *runner up—spinnar ravicamdran aSvin* ***mUDO*** *rkAmk nulabeTTUkunADu.*

- If *current word POS* is *proper noun* and *previous word POS* is *Cardinal*, then *current word* is name of the *location.*
  - A number before a word indicate the number of items of the object. But it cannot occur before name of a person. Postal address for location has door number, street number before the location name. Number can also come before location in the form of year indicating events held in that location during the specified year.
  - *Examples:*

    2016 *American presidential elections—****2016*** *anerikA adyaksh ennikalu.*

    Millions *of Syrian people are going to Europe -****lakshalAdi*** *siriyA prakalu yUrOpA bAri paTTAru.*

- If *current word POS* is *proper noun* and *next word POS/previous word POS* is *NST (Noun denoting spatial and temporal expression)*, then the *current word* is name of a *location.*
  - Here, the POS tag refers to "spatial" terms which tend to co-occur before/after a location.
  - *Example:*

    India has enough foreign currency reserves—bhArat ***vadda*** *saripaDinamta vedESImAraka ravya niluvalu unnayi.*

- If *current word POS* is *proper noun* and *previous word POS* is *Quotative*, then the *current word* is name of a *person.*
  - Quotative words are used when we refer to what people say. In Telugu when we are quoting a statement by some person, we use a quotative term followed by the person who said that.
  - *Example:*

    Modi ***said*** *that he will win the elections in Himachal Pradesh—himAcal pradES ennikalalO mEmu gelistAmu* ***ani*** *mODi cepperu.*

5 Conclusion

The POS rules are generated for NER from an annotated corpus of Telugu text documents obtained by crawling through newspaper websites. The performance of POS rules, with average F1-score of 71.71% for name of person and 68.50% for name of location, is competitive and comparable to the more generic models built with advanced features and algorithms. It is interesting to observe that the POS rules are in tune with the Telugu linguist view of recognizing NEs. It is also exciting to note that NER for Telugu is influenced by the context of immediate neighboring words but not beyond them.

References

1. Appelt, D.E., Hobbs, J.R., Bear, J., Israel, D., Kameyama, M., Martin, D., Myers, K., Tyson, M.: SRI international FASTUS system: MUC-6 test results and analysis. In: Proceedings of the 6th Conference on Message Understanding, pp. 237–248. Association for Computational Linguistics (1995)
2. Atkinson, J., Bull, V.: A multi-strategy approach to biological named entity recognition. Expert. Syst. Appl. **39**(17), 12968–12974 (2012)
3. Baralis, E., Cagliero, L., Jabeen, S., Fiori, A., Shah, S.: Multi-document summarization based on the yago ontology. Expert. Syst. Appl. **40**(17), 6976–6984 (2013)
4. Bharadwaja Kumar, G., Murthy, K.N., Chaudhuri, B.: Statistical analyses of telugu text corpora. Int. J. Dravidian Linguist. (IJDL) **36**(2), 71–99 (2007)
5. Bikel, D.M., Miller, S., Schwartz, R., Weischedel, R.: Nymble: a high-performance learning name-finder. In: Proceedings of the 5th Conference on ApplIed Natural Language Processing, pp. 194–201. Association for Computational Linguistics (1997)
6. Chen, Y., Zong, C., Su, K.Y.: A joint model to identify and align bilingual named entities. Comput. Linguist. **39**(2), 229–266 (2013)
7. Chieu, H.L., Ng, H.T.: Named entity recognition: a maximum entropy approach using global information. In: Proceedings of the 19th International Conference on Computational Linguistics, vol. 1, pp. 1–7. Association for Computational Linguistics (2002)
8. Ekbal, A., Bandyopadhyay, S.: Bengali named entity recognition using support vector machine. In: IJCNLP, pp. 51–58 (2008)
9. Ekbal, A., Bandyopadhyay, S.: A conditional random field approach for named entity recognition in Bengali and Hindi. Linguist. Issues Lang. Technol. **2**(1), 1–44 (2009)
10. Finkel, J.R., Manning, C.D.: Joint parsing and named entity recognition. In: North American Association of Computational Linguistics (NAACL) (2009), pubs/joint-parse-ner.pdf
11. Grishman, R.: The NYU system for MUC-6 or where's the syntax? In: Proceedings of the 6th Conference on Message Understanding, pp. 167–175. Association for Computational Linguistics (1995)
12. Grishman, R., Sundheim, B.: Message understanding conference-6: A brief history. In: COLING 1996 Volume 1: The 16th International Conference on Computational Linguistics. vol. 1 (1996)
13. Habernal, I., KonopíK, M.: SWSNL: semantic web search using natural language. Expert. Syst. Appl. **40**(9), 3649–3664 (2013)
14. Isozaki, H.: Japanese named entity recognition based on a simple rule generator and decision tree learning. In: Proceedings of the 39th Annual Meeting on Association for Computational Linguistics, pp. 314–321. Association for Computational Linguistics (2001)

15. Jung, J.J.: Online named entity recognition method for microtexts in social networking services: a case study of twitter. Expert. Syst. Appl. **39**(9), 8066–8070 (2012)
16. Klein, D., Smarr, J., Nguyen, H., Manning, C.D.: Named entity recognition with character-level models. In: 7th Conference on Natural Language Learning (2003). https://nlp.stanford.edu/pubs/klein2003ner.pdf
17. McCallum, A., Li, W.: Early results for named entity recognition with conditional random fields, feature induction and web-enhanced lexicons. In: Proceedings of the 7th Conference on Natural Language Learning at HLT-NAACL 2003, vol. 4, pp. 188–191. Association for Computational Linguistics (2003)
18. Reddy, S., Sharoff, S.: Cross language pos taggers (and other tools) for Indian languages: an experiment with Kannada using Telugu resources. In: Proceedings of the 5th International Workshop on Cross Lingual Information Access, pp. 11–19 (2011)
19. Rodrigo, Á., Pérez-Iglesias, J., Peñas, A., Garrido, G., Araujo, L.: Answering questions about European legislation. Expert. Syst. Appl. **40**(15), 5811–5816 (2013)
20. Shishtla, P.M., Gali, K., Pingali, P., Varma, V.: Experiments in Telugu ner: a conditional random field approach. In: Proceedings of the IJCNLP-08 Workshop on Named Entity Recognition for South and South East Asian Languages (2008)
21. Srikanth, P., Murthy, K.N.: Named entity recognition for Telugu. In: IJCNLP, pp. 41–50 (2008)
22. Takeuchi, K., Collier, N.: Use of support vector machines in extended named entity recognition. In: Proceedings of the 6th Conference on Natural Language Learning, vol. 20. pp. 1–7. Association for Computational Linguistics (2002)

Computational Intelligence: Algorithms, Applications and Future Directions

Objective Assessment of Pitch Accuracy in Equal-Tempered Vocal Music Using Signal Processing Approaches

Roshni Biswas, Y. V. Srinivasa Murthy, Shashidhar G. Koolagudi and Swaroop G. Vishnu

Abstract This paper presents an approach for assessing the pitch in vocal monophonic music objectively using various signal processing techniques. A database has been collected with 250 recordings containing both *arohan* and *avarohan* patterns rendered by 25 different singers for 10 Hindustani classical ragas. The fundamental frequency ($F0$) values of the user renditions are estimated and analyzed with the original pitch values to quantify the level of variations in pitch initially the five-point moving window has been considered to smoothen the contour. Later, first order and second order differential techniques are applied to estimate the note onset. This process is computationally economical when compared with the available approaches. The technique of cents has been used to evaluate the variation among the target and singing pitch as *cent* is a unit of the most common tuning system for quantifying intonation in equal tempered music. From this analysis, it is observed that singers with professional training have deviations within 15–20 cents, and non-musicians have deviations above 50 cents. Five expert singers rated the global pitch accuracy from the recordings and these results were found to exhibit high correlation with the system's assessments. Such an evaluation system with quantitative analysis coupled with visual representation will greatly aid the training process of singers.

Keywords Pitch detection · Singing accuracy

R. Biswas
National Institute of Technology Rourkela, Rourkela 769 008, Odisha, India
e-mail: 114CS0011@nitrkl.ac.in

Y. V. S. Murthy · S. G. Koolagudi (✉) · S. G. Vishnu
National Institute of Technology Karnataka, Surathkal 575 025, Karnataka, India
e-mail: koolagudi@yahoo.com

Y. V. S. Murthy
e-mail: urvishnu@gmail.com

S. G. Vishnu
e-mail: vishnuswaroop21@gmail.com

A. Elçi et al. (eds.), *Smart Computing Paradigms: New Progresses and Challenges*,
Advances in Intelligent Systems and Computing 766,
https://doi.org/10.1007/978-981-13-9683-0_17

1 Introduction

Accuracy in pitch is a criterion of paramount importance to every singer. Notes can be "in tune" or "out of tune", in majority of the times. The definition of pitch accuracy is measured by variations in the pitch value from target to singing pitch. The process of assessing the decision manually could be subjective and needs high expertise. For those who are learning it is not always possible to get expert opinion all the time. An automated system to asses the deviation in pitch would be quite helpful in such cases. This motivated us to develop an automated system to assess the pitch accuracy in monophonic clips. This work attempts to objectively analyze pitch accuracy in a monophonic sample of sung music with efficiency, precision and certainty. The system not only quantifies the accuracy level note-by-note, it also graphically depicts the sung pitch against an assumed target pitch contour.

It is difficult to quantify error in pitch using the qualitative terms that are used traditionally in music. However, if the accuracy could be assessed with precision, it would be extremely helpful to practice and master pitch-precision without depending on the human ear or a teacher. A tool providing instant feedback to aid a singers learning will refine the sound production to a great extent and avoid the limitation of possible variabilities in human judgment [1–3].

There are several applications possible by developing a system to assess singer performance automatically. The singer can practice anytime without the support of an expert singer. The carnatic music of ICM is effectively taught at some places of karnataka (a state in india). It is not possible for every one to travel all the way to karnataka to learn it hence automation of assessment can help us overcome such barriers and spread the essence of such regional musics world wide. The system helps us to develop an online tutorials and provide an online assessment of the performance.

There are two main challenges in pitch determination: precision and speed. An efficient pitch detection algorithm will estimate between successive frames and produce a smooth high-resolution contour. The problem of octave errors seems to be present in all pitch tracking algorithms known so far, and they have to be manually dealt with to some extent. Along with accuracy, it is very important to extract the pitch values almost instantaneously to provide the feedback soon enough. Another challenge is to correctly determine the exact onset point of each note.

In this work, a system has been developed to assess the corrections of 25 different songs. A simple method has been proposed to locate the mote onsets to make it useful for real-time applications. The concept of differential equation in time domain is used instead of complex spectrogram analysis. Ten different ragas have been tested from 25 singers of different expertise levels.

The rest of this paper is organized as follows: Sect. 2 critically evaluates the existing related work, Sect. 3 describes the framework and methodology of the proposed system, Sect. 4 describes the experimentation and analysis of results and Sect. 5 concludes the paper by highlighting the scope of future work.

2 Literature Review

The importance of fundamental frequency in various speech and music processing applications. Has led to the invention of various pitch detection algorithms. As pitch and note onset detection are two important sub-tasks of pitch assessment, the literature is again subdivided into Pitch detection algorithms and locating onset of a note.

There are several pitch detection algorithms were introduced in the literature in both time and frequency domains. The first technique used in time domain is zero crossing rate (ZCR). After certain analysis it is found that ZCR is not accurate to identify the cycles and noise cancellations. Later, the concept of correlation technique has been introduced on the same signals as the correlation value would be higher for every cycle. The peak-to-peak difference of the correlation signal has been computed for to estimate the pitch period. This approach is suitable for voiced sound signals without noise. In the case of fricatives and noisy signals, it gives wrong information [4]. Further, maximum likelihood (ML) is considered by dividing a frame into 't' equal parts. The left signal has been equally added to all parts. This approach is could practically fail due to many issues. One such issue is existence of silence.

The fundamental or first harmonic of any tone is perceived as its pitch. The pitch contour is a curve that tracks the perceived pitch of sound over time. Pure tones have a clear pitch but complex sounds in music typically have intense peaks at many different frequencies. Pitch Detection Algorithms (PDAs) estimate the pitch or fundamental frequency of an oscillating signal, usually a digital recording of speech or music. There is no single ideal PDA—a variety of algorithms exist, most falling broadly into the time, frequency or both the two domains.

No particular algorithm is the best, the various approaches have different kinds of drawbacks and thus the algorithm most suited for the application is used. The difficulty with Autocorrelation techniques has been that peaks occur at sub-harmonics as well, and it is sometimes difficult to determine which peak is the fundamental frequency and which represent harmonics. Average Magnitude Difference Function (AMDF) suffers from incorrect pitch detection in noisy conditions, hence it might not work correctly for music with accompaniment or a drone in the background. The Cepstrum method assumes that the signal has regularly-spaced frequency partials, thus for inputs other than speech it might provide erroneous results. YIN is very immune to amplitude changes resulting in sub-harmonic errors [4–6].

The Sawtooth Waveform Inspired Pitch Estimator (SWIPE) addresses the problem causing features in the above and is seen to work the best, except for the few octave errors in case of male voices that can manually be corrected quite easily.

In this work, an effort has been made to estimate the accuracy of a singer performance over the original note values. Pitch information has been considered since it is the only effective feature in order to estimate the singer performance. Hindustani Classical Music has been observed to have Equal Tempered influences, unlike Carnatic Music which is more inclined to the Just Intonation tuning system [7]. The twelve-tone Equal Temperament divides the octave into 12 parts, all of are equal

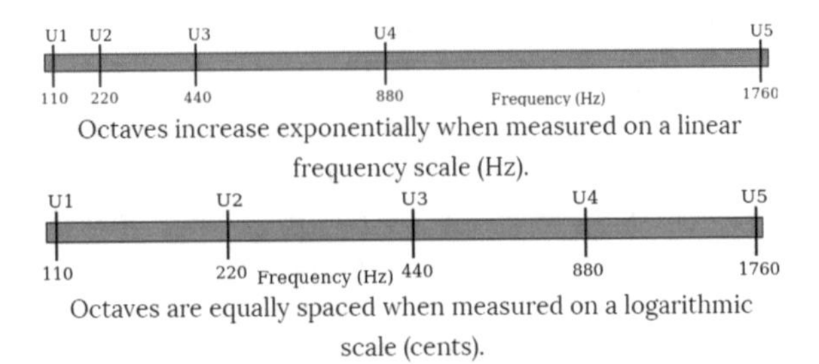

Fig. 1 Equal-tempered octave representation

(100 cents each) on a logarithmic scale, as shown in Fig. 1. It is the most convenient and thus widely used music tuning system in the world.

3 Proposed Methodology

Database Collection: The database contains 50 recordings of the arohan and avarohan of 10 basic Hindustani Classical ragas, by 8 singers having varied levels of expertise in singing. All recodings are done at a sampling frequency of 44100 Hz sampling rate, mono channels and 16 bits per sample.

The list of ragas' used for this work are Raga Bhupali, Raga Bhairav, Raga Bhairavi, Raga Bilawal, Raga Kafi, Raga Khamaj, Raga Asavari, Raga Marwa, Raga Vasant, and Raga Todi.

The proposed methodology has three stages, namely:

- Segmentation of the auditory signal, pitch (F0) estimation and plotting of sung pitch against correct pitch
- Visual inspection of the pitch contour and identification of the steady-state phase of each sung note
- Comparison of mean-estimated pitch with the target note in cents.

3.1 Pitch Estimation

The Pitch Detection Algorithm should not only estimate the fundamental frequency as accurately as possible, but also detect if the section of speech is voiced or unvoiced. Sawtooth Waveform Inspired Pitch Estimator (SWIPE) estimates the pitch of the vector signal X every dt seconds with minimum error. The spectrum is computed using a Hann window with an overlap WOVERLAP between 0 and 1. The spectrum is sampled uniformly in the ERB scale with a step size of DERBS ERBs. The pitch is searched within the range [PMIN PMAX] (in Hertz) with samples distributed every DLOG2P units on a base-2 logarithmic scale of Hertz. The pitch is fine-tuned using parabolic interpolation with a resolution of 1 cent. Pitch estimates with strength lower

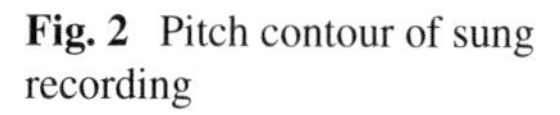

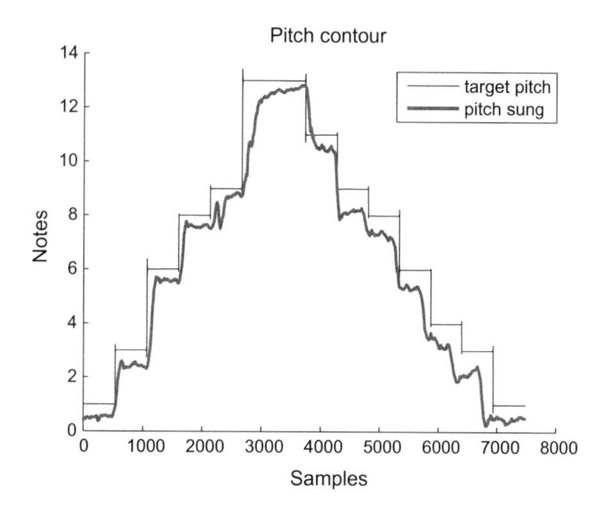

Fig. 2 Pitch contour of sung recording

than STHR are treated as undefined [8]. The pitch detected by SWIPE is shown in Fig. 2.

The F0 values are converted to note degrees in the scale sung. The ascending and descending steps of note changes can be identified by visual inspection here.

3.2 Note Onset Detection

The contour is slightly mean-smoothened by taking the nearest 5 values of each bin (two pitch samples before and after) to even out minute variations. This helps to find the boundary between consecutive notes easily. The 1st order derivative doesnt have peaks sharp enough. However, the note onset points can be clearly identified in the 2nd order derivative of the sung pitch contour. This has been depicted in Fig. 3. Steady state regions occur immediately after the note onsets. After locating the note onsets and allowing a nominal space for note change error, mean pitch of each steady note region is calculated. This gives the pitch of notes which have been sung [9].

3.3 Error Estimation

The unit of cents is the standard measure for representation and comparison of musical pitches and intervals. An equally-tempered semi-tone spans 100 cents. An octave spans twelve semitones, and thus makes 1200 cents. Each sung-note pitch value is compared with the respective target-note pitch value and represented in cents. Likewise, the error for each of the notes sung is computed. An error within 15–20 cents is slightly off but cannot be perceived by the human hear, beyond 50

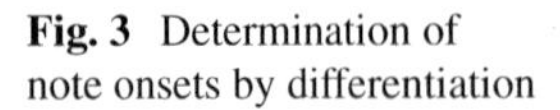
Fig. 3 Determination of note onsets by differentiation

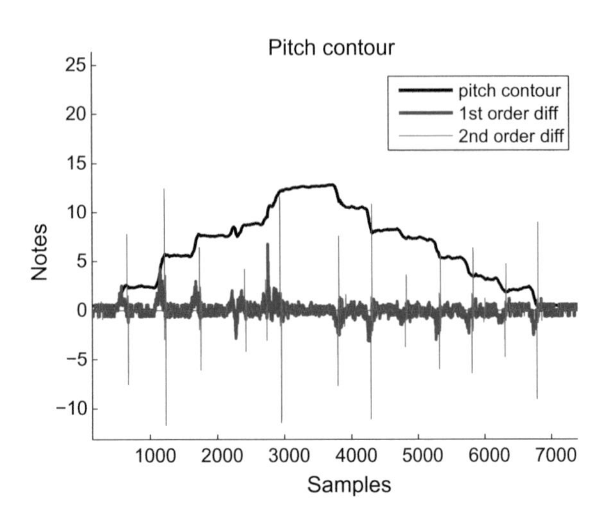

cents is bordering on a wrong note, and in between 20 and 50 sounds 'out-of-tune'. So the average level of accuracy by a singer can be measured by taking mean of errors in all notes sung by him/her [1, 2, 10].

4 Result and Observations

A singer has to concentrate on pitch, articulation and dynamics at the same time, making it possible for unsatisfactory aspects of the sound to slip by without noticing. Different exercise can be used to improve vocal accuracy—involving reproducing sounds, single pitch-matching, imitation of intervals, short pitch patterns or melodies. But, scales being the foundation of all music, working on them is one of the most important parts of learning how to control the voice. Being able to sing all the basic scales proficiently and confidently, gives any singer a head start on learning almost any composition [2, 11].

In Indian classical vocal music, singing the arohan and avarohan of ragas as a part of a daily practice routine is essential. Not only for warming up the voice but it is the baseline for improving singing techniques. Because there are no words to be memorized, melodies to remember, nor any complicated rhythms to maneuver, these exercises can better address things like breath control, extending the vocal range, developing a variety of vocal timbres and colors, perfecting pitch and intonation, and controlling mental and emotional focus.

The pitch glitches that are observed to happen can be categorized into : errors in melodic contour, pitch interval errors and tonal centre errors. Problems in gauging direction or distance of intervals and drift in the tonal centre or reference pitch are common.

Table 1 Rank of tested singers

Singer rank (System generated)	Singer rank (By 5 expert singers)	Mean error in pitch (in cents)
1	1	14.5
2	1	15.0
3	1	17.0
4	2	27.5
5	2	29.0
6	2	37.5
7	3	86.5
8	3	116.0

Five expert singers were additionally asked to judge the singers by ear. Their ranking was in congruence with the system but not as precise. Out of the 8 singers assessed, the ones with more experience had more accuracy. Non musicians were seen to be grossly off-tune. The results are shown in Table 1.

5 Conclusion and Futurework

In this work, objective assessment of pitch has been done. The challenges in pitch extraction techniques have been reported. The problem of detecting note onset points has been addressed and pitch precision has been calculated note-by-note. It has been concluded that pitch error above 20 cents are made by learners and this should be addressed with the help of pictorial tools and feedback, as has been described in this work. It has been concluded that such objective assessment of pitch is thus reliable for use by learners. This method is independent of a teachers' qualitative evaluation and the variabilities of it.

References

1. Larrouy-Maestri, P., Lévêque, Y., Schön, D., Giovanni, A., Morsomme, D.: The evaluation of singing voice accuracy: a comparison between subjective and objective methods. J. Voice **27**, 259-e1 (2013)
2. Lal, P.: A comparison of singing evaluation algorithms. In: INTERSPEECH (2006)
3. McLeod, P., Wyvill, G.: Visualization of musical pitch. In: Computer Graphics International, vol. 2003 (2003)
4. Kasi, K., Zahorian, S.A.: Yet another algorithm for pitch tracking. In: Acoustics, Speech, and Signal Processing (ICASSP), vol. 1, pp. I–361. IEEE (2002)
5. Hanžl, J.B.V.: Comparing pitch detection algorithms for voice applications

6. McLeod, P., Wyvill, G.: A smarter way to find pitch. In: Proceedings of International Computer Music Conference (2005)
7. Serra, J., Koduri, G.K., Miron, M., Serra, X.: Assessing the tuning of sung Indian classical music. In: ISMIR, pp. 157–162 (2011)
8. Camacho, A., Harris, J.G.: A sawtooth waveform inspired pitch estimator for speech and music. J. Acoust. Soc. Am. **124**, 1638–1652 (2008)
9. Datta, A., Sengupta, R., Dey, N.: Objective analysis of srutis from the vocal performances of Hindustani music using clustering algorithm. EUNOMIOS Open Online J. Theory Anal. Semiot. Music **25** (2011)
10. Larrouy-Maestri, P., Dominique, M.: Criteria and tools for objectively analysing the vocal accuracy of a popular song. Logop. Phoniatr. Vocology **39**, 11–18 (2014)
11. Larrouy-Maestri, P., Magis, D., Morsomme, D.: The evaluation of vocal pitch accuracy. Music Percept. Interdiscip. J. **32**, 1–10 (2014)

Gibbs Sampled Hierarchical Dirichlet Mixture Model Based Approach for Clustering Scientific Articles

G. S. Mahalakshmi, G. MuthuSelvi and S. Sendhilkumar

Abstract Biomedical research is progressing remarkably and there arises a necessity for identifying most interested sub-research discipline in biomedicine. This sounds similar to identifying the popular research sub-filed which is growing in fast pace under biomedical research. Application of topic models upon research articles derives to better clustering algorithms which reveal interesting insights to the underlying research problem. This paper proposes a new clustering algorithm which is a fusion of GSDMM and HDP. The resulting scientific article clusters are compared with K-means clustering, which reveals interesting results.

Keywords GIBBS sampling · Dirichlet multinomial mixture · K-means clustering

1 Introduction

Document clustering [1] has been widely explored in the past decades. Word based clustering approaches [2] are primitive and are lost in research arena after the introduction of topic based clustering algorithms [3]. Word based clustering algorithms are unable to cluster text based on context [1]. Text clustering approaches which includes context as a prominent feature for clustering are the need of the hour [2, 4, 5]. Alternatively, candidate documents are enriched with external knowledge using a lexicon or taxonomy for arriving at better clustering. Because the candidate elements are scientific articles, to understand the existing patterns is not that difficult [6]. However, the issue is quite opposite; since the scientific articles have rich semantic information, there are many such overlap patterns in untagged scientific article

G. S. Mahalakshmi · G. MuthuSelvi (✉)
Department of Computer Science and Engineering, Anna University, Chennai, Tamil Nadu, India
e-mail: gmuthuselvi16@gmail.com

G. S. Mahalakshmi
e-mail: gsmaha@annauniv.edu

S. Sendhilkumar
Department of Information Science & Technology, Anna University, Chennai, Tamilnadu, India
e-mail: ssk_pdy@yahoo.co.in

A. Elçi et al. (eds.), *Smart Computing Paradigms: New Progresses and Challenges*,
Advances in Intelligent Systems and Computing 766,
https://doi.org/10.1007/978-981-13-9683-0_18

dataset [7], therefore, one is confused as to use the most prominent and suitable feature for clustering. The effective representation of documents, use of similarity measure among the document pairs, and the nature of clustering algorithm, have a definite role in obtaining efficient clusters.

Clustering is an unsupervised approach by itself. Using unsupervised modelling of text to be represented as topics prior to clustering, improves the performance of clustering algorithm [8]. Topic Models work by defining a probabilistic representation of the latent factors of corpora called topics.

Latent Dirichlet Allocation (LDA) (refer Fig. 1) is a mixture membership model where a document is described by a particular topic proportion, and a topic is defined as a distribution over words [9]. The Hierarchical Dirichlet Process (HDP) (refer Fig. 2) is another alternative topic modeling approach [10] which is based on Naive Bayes.

HDP is based on a Dirichlet process for each group of data. In HDP, groups share statistical strength by sharing of clusters across groups. HDP provides more accurate topic distribution over LDA [11] because it derives an unlimited number of topics as well as the relation between topics based on the content of the document; but, LDA derives only specified number of topics. In traditional clustering, each document is only associated with one cluster. However, this is not convincing as a scientific article is multi-topic yet diverse in nature. Explicit Semantic Analysis (ESA) is robust in the above context. It is used to cluster documents by using their vector representations [12]. This paper proposes a topic based clustering based on Gibbs Sampled Dirichlet Multinomial Mixture (GSDMM) and Hierarchical Dirichlet Process (HDP). This is a fusion attempt where Gibbs sampling is combined with hierarchical Dirichlet processes which is the novelty of the research idea.

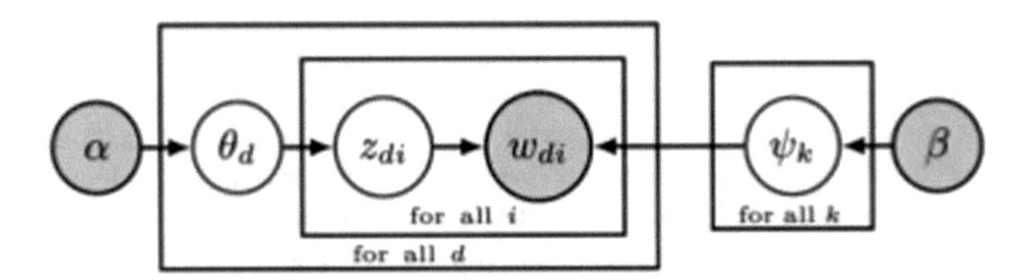

Fig. 1 The process behind LDA

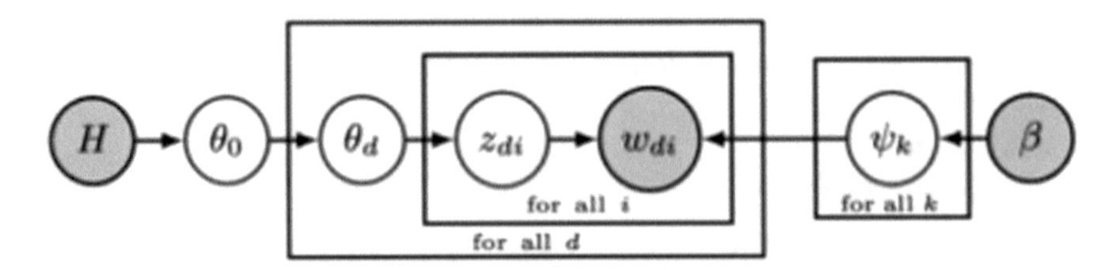

Fig. 2 The process behind HDP

2 Gibbs Sampled Hierarchical DMM (GSHDMM) Clustering

DMM is a probabilistic based generative model for documents [7], and employs the following assumptions: (1) the documents are generated by a mixture model [13], and (2) there is a one-to-one correspondence between mixture components and clusters. Following this, Gibbs Sampling algorithm for the Dirichlet Multinomial Mixture model for short text clustering (GSDMM) is proposed [14]. GSDMM is capable of inferring the number of clusters automatically. The resulting clusters have good completeness and homogeneity, and are fast to converge. GSDMM can also be applied over sparse high dimensional short texts. It derives these remarkable features by identifying latent topics over the fed input. These topics could be used to obtain the representative features of a research article aka Document Topic Model (DTM) thereby modeling the article and further, the respective author (aka Author Topic Model) [9, 15] towards obtaining useful semantic authorial indices [16].

In this work, we attempt to alter the topic modeling parameters of GSDMM into Hierarchical Dirichlet Process (HDP). As we know that HDP could obtain enriched topics when compared to traditional topic models [16], use of HDP to pre-process the input text would result in rich topic collection upon which GSDMM is performed. In addition, the objective of the work is to cluster complete scientific articles unlike only abstracts (or short-texts) of scientific articles in earlier work [14]. Therefore, complete scientific article might result in rich pool of topics which are fed to the GSDMM. Therefore, the performance of clustering is expected to improve unlike clustering using short-texts [14].

GSDMM works on equal priors i.e. same topics for all clusters and same words for all topics. GSDMM does not give less emphasis for popular words. In other words, topic words that appear too often in the text are bound to escalate as top words with higher probability. As the aim is to not only use the *tf-idf* scores [14], GSHDMM aim at assigning higher probabilities for less important words as well. This is indirectly achieved by handling the popular words as such without removing duplicates. If the duplicate topics are normalized across their multiple occurrences and respective probabilities, unique topics would be obtained but, the underlying objective of clustering would be varying. Normalized words may have multiple occurrences under different topics with varied probabilities, indicating varied publications. With normalisation, it is like treating different documents which discuss the concept, the most and the least on par, which is not desirable. Additionally, unlike GSDMM, only the top-k words are fed as input to GSHDMM. By this approach, all the topics are given equal importance; therefore biasing the clustering towards popular topics is avoided.

3 Result and Discussion

For experimental validation of GSHDMM, we have attempted at four other clustering algorithms namely, (1) Traditional K-means (2) K-means over LDA topics (3) K-Means over HDP topics (4) Traditional GSDMM and (5) GSHDMM, and compared the results. The dataset we assumed encompasses 95 authors randomly picked from top Biomedical Journals like IEEE transactions on Biomedical Engineering, IEEE Journal of Biomedical & Health Informatics, Journal of Biomedical Informatics and Journal of Biomedical Semantics. 404 research articles published by the authors in Biomedicine domain are collected from DBLP [17]. GSHDMM automatically converges to 10 clusters. It should be recollected that GSDMM [14] is also automatic to convergence and hence is GSHDMM. Therefore we compared the performance with other variations of K-Means with $k = 10$. Table 1 lists the top 5 words under every cluster in GSHDMM.

The top-5 and least-5 words without removing duplicates reveal the information that the popular words are allowed to exist multiple times with their actual probabilities under various clusters. We have not normalised the multiple occurrences of topic-word probabilities for the reasons stated in Sect. 2. The comparison of cluster sizes across various algorithms is shown in Fig. 3. GSHDMM has greater affinity in the largest cluster when compared to other clustering algorithms. Further we evaluate the algorithms with positive cluster agreements (Figs. 4 and 5). The contingency matrix reveals the sharing of knowledge across other clusters for every cluster of GSDMM. The overall TP agreement is higher in GSHDMM than GSDMM (refer Fig. 6a). Absolute Entropy is also calculated and the results are presented in Fig. 6b. This is to be interpreted with respect to the information sharing of other algorithms thereby establishing that GSHDMM partitions the cluster in an efficient manner. Analysis of largest cluster of GSHDMM containing 174 articles reveals the sub research domain(s) which is concentrated by most of the 95 researchers.

The proposed algorithm is evaluated for purity (refer Table 2), Mutual Information and Normalised Mutual Information (refer Tables 3 and 4) [18]. Purity is an external

Table 1 Top 5 words in each cluster in GSHDMM

Cluster 1	Cluster 2	Cluster 3	Cluster 4	Cluster 5
Mappings Values Answers System Functions	Cost Ontologies Function **Specification** Descriptions	**Values** **Description** Example Satisfy Reduce	Engineering Publication Softgoals Assumptions **System**	Probabilistic Languages Defined Space User
Cluster 6	**Cluster 7**	**Cluster 8**	**Cluster 9**	**Cluster 10**
Data **System** **Set** **Specification** Requirement	Language Requirement Set Problem Current	Conversion **Functions** Quality Constraint Entity	Enterprise Instance Transactions Syntax Scheduler	Constraint Select Step Dimensions Note

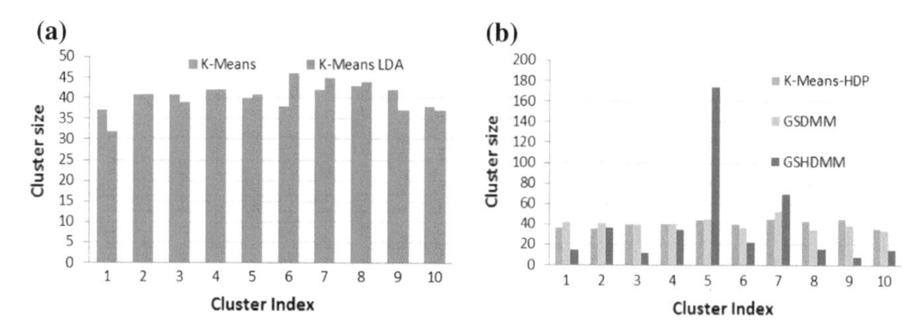

Fig. 3 Comparison of cluster size **a** K-Means versus K-Means LDA **b** K-Means HDP with GSDMM & GSHDMM

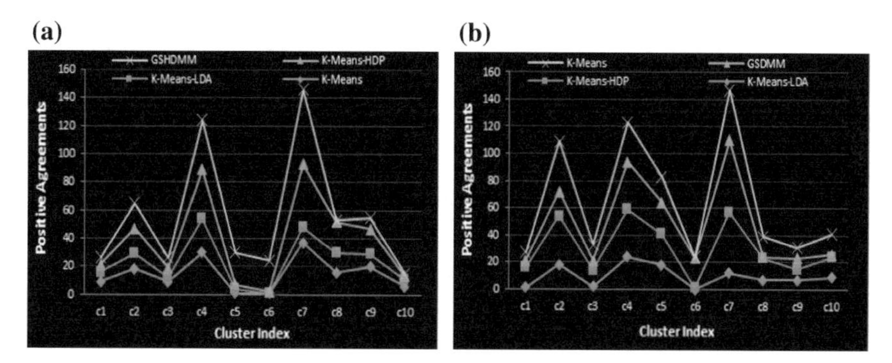

Fig. 4 Positive cluster agreement **a** GSDMM **b** GSHDMM

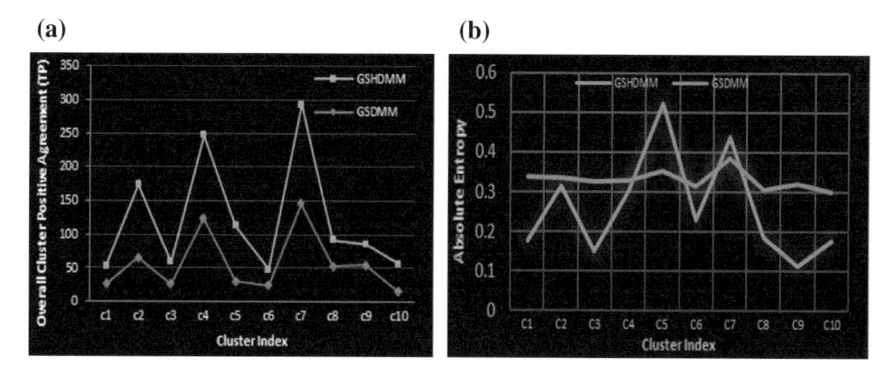

Fig. 5 **a** Overall TP **b** absolute entropy

Table 2 Purity of GSHDMM

Clusters	Class 1	Class 2	Max(C1, C2)	Cluster size	Purity
C1	12	3	12	15	0.800
C2	8	29	29	37	0.784
C3	10	2	10	12	0.833
C4	24	11	24	35	0.686
C5	89	85	89	174	0.511
C6	5	17	17	22	0.773
C7	38	32	38	70	0.543
C8	12	4	12	16	0.750
C9	3	5	5	8	0.625
C10	7	8	8	15	0.533
Total	208	196		404	0.684

Table 3 Entropy for GSHDMM cluster

Clusters	Cluster size	Entropy
C1	15	0.1764
C2	37	0.3159
C3	12	0.1507
C4	35	0.3057
C5	174	0.5234
C6	22	0.2286
C7	70	0.4382
C8	16	0.1845
C9	8	0.1120
C10	15	0.1764
Sum	404	2.6118

Table 4 Entropy for GSHDMM class

CLASS	Class size	Entropy
C1	208	0.4931
C2	196	0.5063
Sum	404	0.9994

evaluation criterion for clustering quality (Eq. 1). To compute purity, each cluster is assigned to the class which is most frequent in the cluster, and then the accuracy of this assignment is measured by counting the number of correctly assigned documents and dividing by N formally:

$$purity(\Omega, \mathbb{C}) = \frac{1}{N}\sum\nolimits_{k} \max_{j}\left|w_k \cap c_j\right| \tag{1}$$

where $\Omega = \{w_1, w_2, \ldots, w_k\}$ is the set of clusters and $\mathbb{C} = \left\{c_1, c_2, \ldots, c_j\right\}$ is the set of classes.

Mutual Information (Eq. 2) is an information theoretic measure which interprets how a cluster shares the information with other cluster. However, though the clustering algorithm is effective with small cluster size in assessing mutual information, as the clusters grow in size due to large data sets, overlap of information is quite common among clusters. This results in high mutual information for the same algorithm which performed better with a lower subset of the same dataset. Therefore, Normalised Mutual Information (NMI) (Eq. 3) is also assessed.

$$\begin{aligned} I(\Omega; \mathbb{C}) &= \sum_{k}\sum_{j} P(\omega_k \cap c_j) \log \frac{P(\omega_k \cap c_j)}{P(\omega_k)P(c_j)} \\ &= \sum_{k}\sum_{j} \frac{\left|\omega_k \cap c_j\right|}{N} \log \frac{N\left|\omega_k \cap c_j\right|}{|\omega_k|\left|c_j\right|} \end{aligned} \tag{2}$$

where $P(\omega_k)$, $P(c_j)$, and $P(\omega_k \cap c_j)$ are the probabilities of a document being in cluster, class and in the intersection respectively.

High purity is easy to achieve when the number of clusters is large—in particular, purity is 1 if each document gets its own cluster. Thus, we cannot use purity to trade off the quality of the clustering against the number of clusters. A measure that allows us to make this tradeoff is NMI

$$\begin{aligned} \mathrm{NMI}(\Omega; \mathbb{C}) &= \frac{I(\Omega; \mathbb{C})}{[H(\Omega) + H(\mathbb{C})]/2} \\ H(\Omega) &= -\sum_{k} P(\omega_k) \log P(\omega_k) \\ &= -\sum_{k} \frac{|\omega_k|}{N} \log \frac{|\omega_k|}{N} \end{aligned} \tag{3}$$

where "H" is entropy.

NMI evaluates the cluster quality via entropy measures. For all the above extrinsic evaluation measures, we assume the gold standard across which the clustering algorithm is compared is SVM classification. Bad clustering has purity close to zero and a perfect cluster has purity value 1. NMI also has to be interpreted in a similar manner. High purity could be well achieved when the number of clusters is large. The absolute entropy for all clusters across all the algorithms is presented in Table 5. The evaluation for GSDMM and the proposed GSHDMM are given in Table 6. Here, we calculate Average cluster purity to measure the overall cluster purity of the respective algorithms. Both the purity and NMI tend to increase for the proposed GSHDMM when compared to existing algorithms (Table 6).

Table 5 Clustering—absolute entropy across algorithms

Cluster No.	GSHDMM	GSDMM	KMEANS LDA	KMEANS-HDP
C1	0.17641	0.339524	0.289759325	0.31585161
C2	0.315852	0.334968	0.334967917	0.310837409
C3	0.150691	0.325593	0.325592973	0.325592973
C4	0.305724	0.330325	0.33952364	0.330325088
C5	0.523407	0.352688	0.334967917	0.348381965
C6	0.228646	0.315852	0.35691554	0.325592973
C7	0.438181	0.384419	0.352688434	0.352688434
C8	0.184484	0.305724	0.348381965	0.34399433
C9	0.112044	0.320769	0.31585161	0.352688434
C10	0.17641	0.300509	0.31585161	0.310837409

Table 6 Clustering extrinsic evaluation

Metric	GSDMM	GSHDMM
Mutual information	1.059	1.061
Average cluster purity	0.591	0.684
Normalised mutual information	0.492	0.587

Purity score around 1 depicts good cluster purity, and purity score towards 0 depicts that the obtained clusters are diluted. However, as like the standard GSDMM, the proposed algorithm also attains marginally better cluster purity. However, the proposed algorithm is not suitable for clustering big data. The reason is, obtaining an average of 10 clusters for big data is too abnormal and the natural distribution of words is not preserved when the collection is large. Therefore, improvements on split and merge of clusters based on various cluster parameters are required to make the proposed work more suitable to big data clustering.

4 Conclusion

This paper proposes a new technique of combining hierarchical Dirichlet process models with GSDMM clustering. The algorithm shows a competitive edge in performance when compared to other topic-based clustering algorithms. A better analysis on topic-word temporal occurrence hierarchies along with the topic probabilities and uniqueness would reveal more interesting insights about the bibliometric statistics of the derived clusters in the field of Biomedicine. More detailed analysis of cluster metrics is being carried out to enable GSHDMM to handle big data clustering.

References

1. Wagner, S., Wagner, D.: Comparing Clusterings: An Overview. Universität Karlsruhe, Fakultät fürInformatik, Karlsruhe (2007)
2. Miao, Y., Kešelj, V., Milios, E.: Document clustering using character N-grams: a comparative evaluation with term-based and word-based clustering. In: Proceedings of the 14th ACM International Conference on Information and Knowledge Management (2005)
3. Brants, T., Chen, F., Tsochantaridis, I.: Topic-based document segmentation with probabilistic latent semantic analysis. In: Proceedings of the Eleventh International Conference on Information and Knowledge Management. ACM (2002)
4. MuthuSelvi, G., Mahalakshmi, G.S., Sendhilkumar, S.: Author attribution using stylometry for multi-author scientific publications. Adv. Nat. Appl. Sci. **10**(8), 42–47 (2016)
5. Kanungo, T., et al.: An efficient k-means clustering algorithm: analysis and implementation. IEEE Trans. Pattern Anal. Mach. Intell. **24**(7), 881–892 (2002)
6. Mahalakshmi, G.S., MuthuSelvi, G., Sendhilkumar, S.: A bibliometric analysis of journal of informetrics—a decade study. In: International Conference on Recent Trends and Challenges in Computational Models (ICRTCCM'17), organized by Department of Computer Science and Engineering, University College of Engineering, Tindivanam (2017)
7. MuthuSelvi, G., Mahalakshmi, G.S., Sendhilkumar, S.: An investigation on collaboration behavior of highly cited authors in journal of informetrics (2007–2016). J. Comput. Theor. Nanosci. 3803
8. Mccallum, A., Nigam, K., Ungar, L.H.: Efficient clustering of high-dimensional data sets with application to reference matching. In: Proceedings of the 6th ACM SIGKDD International Conference on Knowledge Discovery and Data Mining, pp. 169–178. ACM Press, New York, NY (2000)
9. Mahalakshmi, G.S., MuthuSelvi, G., Sendhilkumar, S.: Generation of Author Topic Models Using LDA. Lecture Notes in Computational Vision and BioMechanics. Springer (2017)
10. Teh, Y.W., Jordan, M.I., Beal, M.J., Blei, D.M.: Hierarchical Dirichlet processes (PDF). J. Am. Stat. Assoc. **101**, 1566–1581 (2006)
11. Mahalakshmi, G.S., MuthuSelvi, G., Sendhilkumar, S.: Hierarchical modeling approaches for generating author blueprints. In: International Conference on Smart Innovations in Communications and Computational Sciences (ICSICCS-2017), organizing by North West Group of Institutions, Moga, Punjab, India during 23–24 June 2017, ID:145 (2017)
12. Sorg, P., Cimiano, P.: An Experimental Comparison of Explicit Semantic Analysis Implementations for Cross-Language Retrieval, Institute AIFB, University of Karlsruhe & Web Information Systems Group, Delft University of Technology (2009)
13. McLachlan, G., Basford, K.: Mixture Models: Inference and Applications to Clustering. Marcel Dekker, New York (1988)
14. Yin, J., Wang, J.: A Dirichlet multinomial mixture model-based approach for short text clustering. In: Proceedings of the 20th ACM SIGKDD International Conference on Knowledge Discovery and Data Mining (KDD '14), pp. 233–242. ACM, New York, NY, USA (2014)
15. Mahalakshmi, G.S., MuthuSelvi, G., Sendhilkumar, S.: Measuring author contributions via LDA. In: 2nd International Conference on Advanced Computing and Intelligent Engineering
16. Mahalakshmi, G.S., MuthuSelvi, G., Sendhilkumar, S.: Measuring authorial indices from the eye of co-author(s). In: International Conference on Smart Innovations in Communications and Computational Sciences (ICSICCS-2017), organizing by North West Group of Institutions, Moga, Punjab, India during 23–24 June 2017, ID:146 (2017)
17. Mahalakshmi, G.S., MuthuSelvi, G., Sendhilkumar, S.: Authorship analysis of JOI articles (2007–2016). Int. J. Control Theory Appl. **9**(10), 1–11 (2016), ISSN: 0974-5572
18. Amigó, E., Gonzalo, J., Artiles, J., Verdejo, F.: A comparison of extrinsic clustering evaluation metrics based on formal constraints. Inf. Retr. **12**(4), 461–486 (2009)

Experimental Evaluation of Dynamic Typing Mechanism: A Case Study

K. Indra Gandhi and K. Induja

Abstract In Model-Driven Engineering (MDE), the models are created and processed which conforms to a meta-model. In general, classes in meta-model are used as templates to create objects and as classifiers for them. These two things are tied together in most of the meta-modelling approaches but it lacks flexibility. In order to decouple the object creation from typing, a-posteriori typing is used. In a-posteriori typing, type-level and instance-level specifications are used to realize the retyping mechanism. The creation meta-model of the model is created with constructive types and the role meta-model of the model is created with a-posteriori types. The a-posteriori typing specifications are defined using transformation rules which maps the classes in the creation meta-model to the classes in the role meta-model. The correctness of the typing specification is evaluated using Object Constrained Language (OCL) constraints. The introduced typing mechanism is applied in three models such as Point Of Sale (POS), Digital Locker and E-Sign. The analysis of typing mechanism is performed on the typed models based on the executability, totality and surjectivity property. These properties represent the models that are typed validly using the a-posteriori typing mechanism.

Keywords Model-Driven Engineering · Typing · Model typing · Dynamic typing · Meta-model · Instance-level · Type-level

1 Introduction

Model-Driven Engineering (MDE) advocates the use of model as a principal component. In MDE, models are active components which are used to specify, simulate, test and generate code for the applications. MDE are generally a top-down approach, where classes in meta-models are used as templates to create objects in models,

K. Indra Gandhi · K. Induja (✉)
Department of Information Science and Technology, Anna University, Chennai, India
e-mail: indujakumar95@gmail.com

A. Elçi et al. (eds.), *Smart Computing Paradigms: New Progresses and Challenges*, Advances in Intelligent Systems and Computing 766,
https://doi.org/10.1007/978-981-13-9683-0_19

which then becomes classified by those classes. This kind of typing by meta-model element is called as constructive typing.

MDE is a top-down approach, where classes are used to create instances, which in turn are classified by those classes. This kind of typing is called constructive typing [1]. Models are instances of a meta-model and their logical and syntactical structures should obey the metamodel [2]. The main difference between the objects in classical object-oriented systems and the objects used within models is the presence of bidirectional relationship [3]. The major components of generic programming for meta-modeling are concepts, templates and mixin layers [4]. Steel et al. [5] proposed a simple strategy for typing models as a collection of interconnected objects, an extension to object-oriented typing. Lara and Guerra [6] proposed an approach where integrity constraints can be placed at any meta-level, but the meta-level below which they should be placed must be indicated.

The proposed work provides more flexible typing which then becomes multiple, partial and dynamic types. The a-posteriori typing [7] allows classification of objects by classes different from the ones that are used to create it. The creation meta-model is a set of classes which are of constructive types. It consists of classes, attributes and references to other classes without any roles assigned to them. The role meta-model is a set of classes which are of a-posteriori types. It consists of classes, attributes and references with roles assigned to them. The transformation rules are used to map the creation meta-model with the role meta-model. These rules are specified with the help of a-posteriori typing specifications. The a-posteriori typing can be specified in two ways: at the type level and at the instance level. The type-level specification provides a static relation between two meta-models, in order to show the instance of one class as an instance of another class.

The analysis is performed to check whether the retyping specification for mapping the creation meta-model with the role meta-model provides a valid instance of role meta-model. The analysis of dynamic typing checks the three properties: Executability, Totality and Surjectivity.

2 Problem Specification

The system consists of a source model for which the dynamic typing mechanism is applied. This source model is used to create the creation meta-model and role meta-model which is used to provide the typing specifications. The a-posteriori typing specification is provided in order to transform the instances of creation model to the instance of role model thereby performing dynamic typing. The typing specification can be at the type-level or at the instance-level. Both the type-level and instance-level specifications are analysed based on their well-formedness rules. Figure 1 shows the system architecture for the dynamic typing mechanism. Finally, the dynamic typing mechanism is analysed by using three properties namely executability, totality and surjectivity.

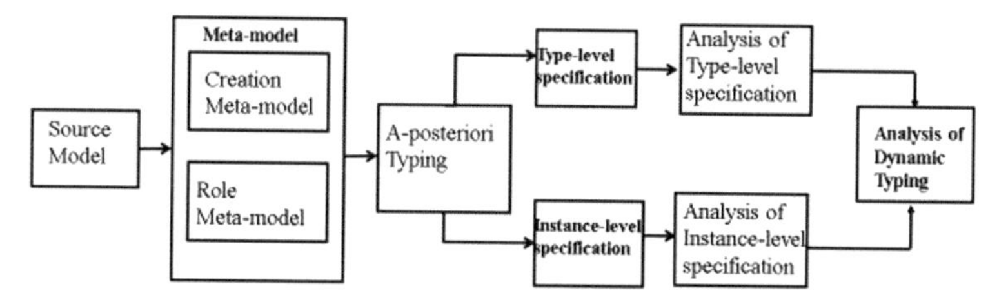

Fig. 1 Dynamic typing model

2.1 Type-Level Specifications

The typing specifications at the type-level is given by the static relation between the creation meta-model and the role meta-model. The creation meta-model consists of constructive types and role meta-model consists of a-posteriori types. The type-level specifications maps the classes, attributes and references from the creation meta-model to the role meta-model. A type-level specification is a collection of partial functions from elements of creation meta-model to elements of role meta-model [7].

2.2 Instance-Level Specifications

The specifications at the instance-level is given by a set of queries that are evaluated over the model which has to be typed and they are assigned types from the role meta-model. The specification requires construction of OCL invariants in order to type the model from creation meta-model. The instance-level typing specifications is a collection of partial functions from the instance of creation meta-model to the elements of role meta-model. These partial functions are specified by means of queries [7]. These functions return a set of objects of creation meta-model type in which each object is assigned a type from role meta-model.

3 Case Study

3.1 Point of Sale System

A Point Of Sale (POS) system [8] is a computer-based system used in stores to keep track of sales and to handle payments. It can be integrated with various applications like inventory control, tax calculator etc. The system should be more fault tolerant such that even if the remote services are unavailable it should be capable of handling sales so that the business is not stopped.

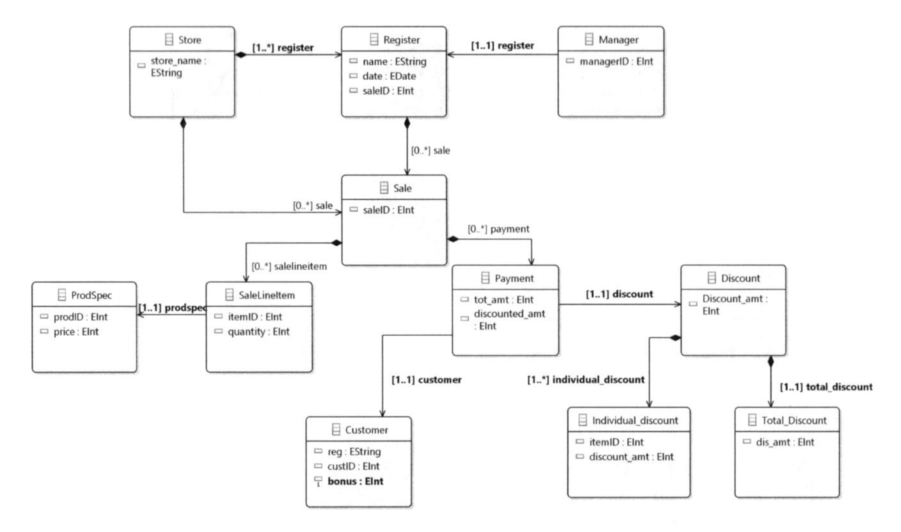

Fig. 2 Meta-model for POS system

The type-level specifications maps the static relation between the creation and role meta-model. In POS system, the type-level specification maps the static relation such as Store class, Manager class and Register class since these classes do not change the roles dynamically. Figure 2 shows the meta-model for POS system.

The instance-level specification maps the roles by evaluating queries over the creation meta-model. For instance, the special customer shall be provided with a discount. To assign special customer role, the specialID from the Customer class in the creation meta-model is retrieved and checked whether he/she is a regular customer and based on the result the roles from the role meta-model is assigned. After checking the specialID, the special customers will be provided with the discount. The transformation rules of instance-level specification should also obey the well-formedness rules of typing specification.

3.2 Digital Locker

Digital Locker [9] is a standardized mechanism to issue government documents to Aadhaar holders in electronic and printable formats, store them and make it shareable with various agencies. This allows government issued documents to be moved to electronic form and make it available for real-time access in a set of 'digital repositories'. In addition to supporting new documents to be made electronic and online accessible, this solution also offers a way to digitize older documents.

The type-level specifications maps the static relation between the creation and role meta-model. In Digital Locker, the type-level specification maps the static relation such as Issuer, DigitalRepository and Department class since these classes do not change the roles dynamically.

The instance-level specification maps the roles by evaluating queries over the creation meta-model. The instance-level specification make use of helper context to retrieve the value from the creation meta-model. The transformation rules of instance-level specification should also obey the well-formedness rules of typing specification.

3.3 E-Sign

E-Sign [10] facilitates digitally signing a document by an e-sign user using an Online Service. The authentication of the signer is carried out using e-KYC, the signature on the document is carried out on a backend server, which is the e-Sign provider. The service shall be offered only by Certifying Authorities(CA).

The type-level specifications maps the static relation between the creation and role meta-model. In E-Sign, the type-level specification maps the static relation such as E-Sign and ServiceProvider class since these classes do not change the roles dynamically.

The instance-level specification maps the roles by evaluating queries over the creation meta-model. Based on the result the roles from the role meta-model is assigned thereby typing it dynamically. The instance-level specification make use of helper context to retrieve the value from the creation meta-model. The transformation rules of instance-level specification should also obey the well-formedness rules of typing specification.

4 Analysis of Dynamic Typing

The analysis of dynamic typing is performed based on three properties, namely executability, totality and surjectivity. The typing rules for creation meta-model and role meta-model is defined. The instances of the classes in meta-model will obey certain rules and violate some of the rules. The executability property finds the instances that satisfy rules in both creation and role meta-model. The totality property finds the instances that satisfy rules in creation meta-model but violates some of them in role meta-model. The surjectivity property finds the instances that satisfy rules in role meta-model but violates some of them in creation meta-model.

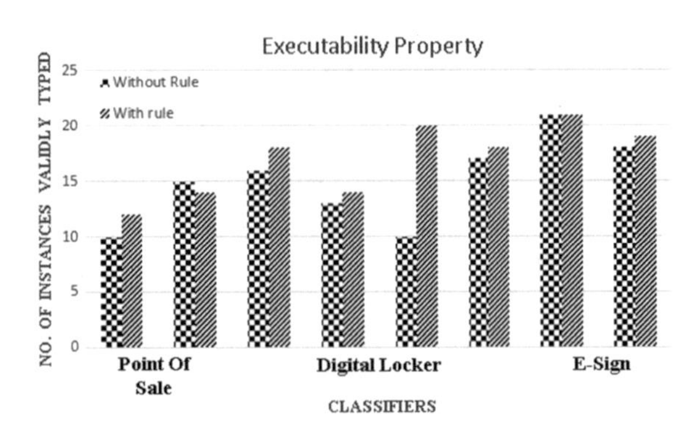

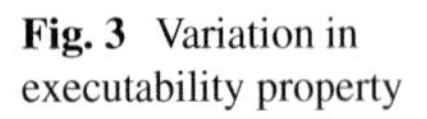
Fig. 3 Variation in executability property

4.1 *Executability Property*

The executability property finds the instance that are validly typed both in creation and role meta-model. The Fig. 3 shows the variations in the executability property with and without using typing rules. The number of instances that are validly typed with the help of typing rules is more when compared with instances that are typed without typing rules. Figure 3 shows the variation of the executability property with and without typing rules. The number of models that are validly typed is more when using typing rules.

4.2 *Totality Property*

The totality property finds the instances that are validly typed in creation but it violates some of the rules in role meta-model. The number of instances that are validly typed with the help of typing rules is more when compared with instances that are typed without typing rules. The number of models that are validly typed is more when using typing rules.

4.3 *Surjectivity Property*

The surjectivity property finds the instances that are validly typed in role meta-model but violates some of the rules in creation meta-model. The number of instances that are validly typed with the help of typing rules is more when compared with instances that are typed without typing rules. The number of models that are validly typed is more when using typing rules. Thus the number of models that are validly typed using typing rules is more when compared with the models that are typed without considering the typing rules.

5 Conclusion

The a-posteriori typing mechanism decouples the object creation from classification. The typing mechanism also improves flexibility and dynamicity. The correctness of type-level and instance-level specifications are analysed. The models that are classified with the help of typing specifications are more validly typed. It allows the existence of instances with multiple classifiers other than the constructive types. This case study provides the dynamic typing mechanism using two typing specifications with non-overlapping conditions in the role meta-model. In the future, the instance-level specifications can be provided with overlapping conditions in the role meta-model thereby increasing the efficiency of the typing mechanism.

References

1. Atkinson, C., Kennel, B., GoB, B.: Supporting constructive and exploratory modes of modeling in multi-level ontologies. In: Semantic Web Enabled Software Engineering, pp. 1–15 (2011)
2. Saeki, M., Kaiya, H.: On relationships among models, meta models and ontologies. In: Workshop on Domain-Specific Modeling (2007)
3. Steel, J., Jézéquel, J.M.: Model typing for improving reuse in model-driven engineering. Int. Conf. Model Driven Eng. Lang. Syst. **3713**, 84–96 (2005)
4. De Lara, J., Guerra, E.: Generic meta-modelling with concepts, templates and mixin layers. MODELS **6394**, 16–30 (2010)
5. Steel, J., Jézéquel, J.M.: Softw. Syst. Model **6**, 401 (2007). https://doi.org/10.1007/s10270-006-0036-6
6. De Lara, J., Guerra, E.: Automated analysis of integrity constraints in multi-level models. Data Knowl. Eng. **107**, 1–23 (2017)
7. De Lara, J., Guerra, E., Cuadrado, J.S.: A-posteriori typing for model-driven engineering. In: ACM/IEEE 18th International Conference on Model Driven Engineering Languages and Systems (MODELS), pp. 156–165 (2015)
8. Larman, C.: Applying UML and Patterns: An Introduction to Object-oriented Analysis and Design and the Unified Process. Prentice Hall Professional, Upper Saddle River (2002)
9. Digital Locker: Technology Specifications (DLTS). https://digilocker.gov.in/assets/img/technical-specifications-dlts-ver-2.3.pdf (2015)
10. eSign API Specifications. http://www.cca.gov.in/cca/sites/all/eSign-APIv1_0.pdf (2015)

Integrating Digital Forensics and Digital Discovery to Improve E-mail Communication Analysis in Organisations

Mithileysh Sathiyanarayanan and Odunayo Fadahunsi

Abstract In Digital Forensics and Digital Discovery, e-mail communication analysis has become an important part of the litigation process. Integrating these two can improve e-mail communication analysis in organisations and help both legal and technical professionals achieve goals of conducting analysis in a manner that is legally defensible and forensically sound. In this forensic discovery process, digital evidence plays an increasingly vital role in the court to prove or disprove an individual or a group of individual's actions in order to secure a conviction. However, e-mail investigations are becoming increasingly complex and time consuming due to the multifaceted large data involved, and investigators find themselves unable to explore and conduct analysis in an appropriately efficient and effective manner. This situation has prompted the need for improved e-mail communication analysis that can be capable of handling large and complex investigations to detect suspicious activities. So, our interactive visualisations aims to improve digital forensics discovery ability to search and analyse a vast amount of e-mail information quickly and efficiently.

Keywords Visualisation · Digital discovery · Digital forensics · E-mail communication

1 Introduction

Digital Forensics and Digital Discovery are an integral part of an organisation, as they are involved in collection, preservation, analysis and presentation of digital evidence for assisting/furthering other investigations of electronic data. Both investigations are conducted in isolation but the terms are muddled and misleading often. We aim to clarify few points in this paper and the contributions are listed as follows:

1. elucidate the terms 'Digital Forensics' and 'Digital Discovery',

M. Sathiyanarayanan (✉) · O. Fadahunsi
City, University of London, London, England, UK
e-mail: Mithileysh.Sathiyanarayanan@city.ac.uk

A. Elçi et al. (eds.), *Smart Computing Paradigms: New Progresses and Challenges*, Advances in Intelligent Systems and Computing 766,
https://doi.org/10.1007/978-981-13-9683-0_20

2. integrate both investigation methodologies for improving E-mail communication analysis,
3. develop interactive visualisations to improve digital forensics discovery ability to search and analyse a vast amount of E-mail information quickly and efficiently.

2 Integrating Digital Forensics and Digital Discovery

Electronic Discovery (E-discovery) is also called as Electronic Disclosure (E-disclosure) and/or Digital Discovery/Disclosure (sometimes called as computer discovery) which refers to 'a stage prior to a trial when a request is made by one party that the other hand over any and all archived electronic material that they hold in relation to the case (more often in civil or legislative case), which include E-mails, electronic documents, word processing documents, spreadsheets and other data' [1, 2]. In practice, E-discovery involves the process of sifting through large volume of 'raw' data to remove duplicates (called 'de-duping') and futile (called 'unpurposed') information, in order to search relevant (called 'pertinent') information easily and effectively by the investigators or lawyers representing the case. E-discovery is often used to investigate whether a company is compliant with the law in the way that it stores, handles and monitors electronic data and also include a comprehensive review of electronic materials which may involve the acquisitions and analysis of large volumes of e-materials. For example, in 2006, Morgan Stanley was fined 15 million USD on the grounds that it was found to have E-mail archiving that was not in accordance with that required by law.

On the other hand, Electronic Forensics (E-forensics) which is also called as Computer Forensics and/or Digital Forensics (sometimes called as cyber forensics) which refers to 'an analysis of computers and other electronic/digital devices in order to produce legal evidence of a crime or unauthorised action' [1, 2]. In practice, E-forensics involves the process of investigating the recovery of deliberately modified, deleted or hidden evidence and producing it as an evidence that they hold in relation to the case (more often in criminal case). As shown in the Fig. 1, the 'digital framework' in the E-forensics helps forensics specialists to follow certain guidelines in investigating and extracting the information, which is called as 'digital investigation'. Then the extracted information is presented in a digital form, called as 'digital evidence' which needs to be presented in a court (legal system). The legal proceedings include fraud, sexual harassment, theft, information leakage or misappropriation of trade secrets and other internal confidential informations, where the digital evidence can be used for law firms, corporations, government agencies, educational institutions and individuals.

Both E-forensics and E-discovery are important parts of an organisation [3]. The one thing that is common between E-forensics and E-discovery is to locate digital evidence for supporting investigations but their operations are quite different and they are used for different purposes [2]. In many cases, both E-discovery and digital forensics experts are needed to ensure a smooth and timely progression of a

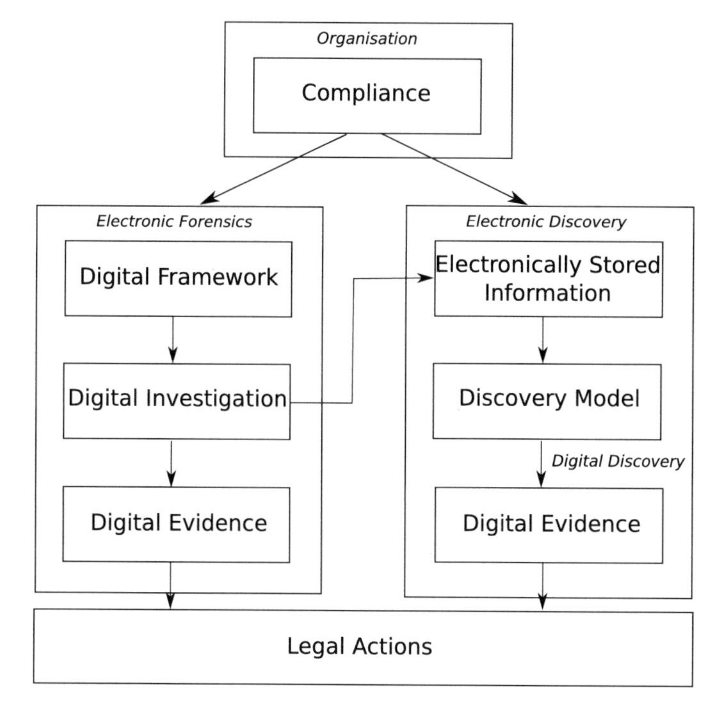

Fig. 1 Detailed structure of the Integration of Organisation Compliance with E-discovery and E-forensics

case. E-forensics requires a deep technical/practical knowledge of dealing with the electronic/digital technologies/systems/devices and knowledge of the investigation frameworks, which can vary from case to case. E-forensics employs strict protocols and procedures, that is completely in line with the guidelines of the investigation bureau for handling of computer based evidence and/or to gather information contained on a wide variety of devices to locate modified/deleted files and hidden information. If the guidelines are not properly adhered to, evidence will not be considered. Therefore, E-forensics experts are regularly called to the stand to affirm their findings and shield their methods and techniques under interrogation. E-discovery tasks usually begin after the files are captured and this needs only investigation (filtering and search methodologies) knowledge to search through a very large volume of archived electronic documents and e-mails. E-discovery does not employ any strict protocols and procedures to gather information but they have a set of instructions to carry out during investigation stages. To some extent, E-discovery experts are called to the stand to affirm their findings under interrogation [3].

In E-forensics, investigating modified and deleted data can have a place in the story about what happened on a device (locating and accessing deleted files). In E-discovery, only the existing data are investigated and such modified/deleted data may not be considered because they are not the focus of the case.

Example: An E-discovery [4] investigation process would identify a key e-mail or key information in the e-mail, where as a E-forensics investigation process would identify that e-mail and the device it resided on, how and when it arrived at the device, how often it was opened, if it was sent to any another location, if the e-mail was modified or deleted, and more. E-forensics can inform more technical information and the rest of the story.

Example: The Hillary Clinton case, involved both E-forensics and E-discovery teams. The former was involved in investigating Hillary Clinton's personal server and identified relevant/pertinent e-mails and also retrieved altered, deleted or otherwise rendered unreadable e-mails. The latter was involved in manually reading the relevant/pertinent e-mails to identify metadata such as time–date stamps, individuals' connections and context information to produce it as an digital evidence for the legal proceedings.

One of the traditional communication modes in organisations is e-mail system to exchange messages or documents. E-mail data are increasingly called upon for evidences in legal cases, either to protect organisations or even incriminate them. If organisations are unable to produce e-mail data or evidence when called upon by the courts or the authorities, they can face huge penalties. Hence, the need for analytic empowered solutions that covers organisations and leaves them better prepared when records/evidences are requested.

Figure 1 represents a bit more detailed structure of the interconnection between all the three investigation areas. The relation and distinction between the investigation areas are expressed below.

3 Interactive Visualisation Prototype

Many experts/analysts use E-discovery tools such as Brainspace Discovery5^{TM} [5], Jigsaw [6], Concordance by LexisNexis [7], IN-SPIRE [8], Radiance [9], Zovy Advanced E-discovery (AeD) [10], DocuBurst [11] to analyse unstructured data—there are many visualisations with limited success. Our functional prototype supports visual analysis for E-mail data exploration with the combination of matrices and bar charts that provide concurrent perspectives of multiple facets of the data. Heat matrices are to visualise relationship between two different granularities, shows the number of occurrences and help identify areas for further analysis, such as peak periods of activity (patterns/trends). Bar charts are to select components of interest, find changes in the matrices, and for comparing different subsets of data within the views. These two charts use crossfilter techniques along with D3 that help in search, navigation, drilling down and investigation, which aid in

(i) identifying 'pertinence' in data: to filter huge data and to fetch a relevant data using visual representations (from investigation point of view). Each bar charts will enable search towards finding relevance or subsets of data of interest based on the regular activities of individuals, for further investigation.

(ii) identifying 'key information' in data: the selection of components will aid in comparing two different subsets of data (for example, e-mail data of two different years) to find key information.
(iii) identifying 'points of interest (PoIs)' in data: the selections and filtering aid in discovering various PoIs in time, individuals and contents. Using the PoIs, further filtering can be done in the investigation process.

The multifaceted data is colour coded enabling the facets to be distinguished and compared quickly to find various tasks and information. The complete visualisation provides an informative and comprehensive overview of the entire dataset and exploration opportunities using the crossfilters for interaction (incremental filtering, reducing and comparing) (Fig. 2).

Applied context: an E-discovery investigation—we present an example for the applicability of our model in the investigation domains that contains multifaceted e-mail communication data in various forms: temporal, individuals/connections and context views (green, orange and blue, respectively, in Fig. 1). In the Enron scandal report [12], an US legal team had only the temporal information (October 2001) but they did not have any information about the individuals, connections and contents. The legal team had to manually comb through the E-mails to find pertinence and find individuals and keywords. Our interactive prototype helped in selecting the time frame of interest (based on the report) using bar charts. The selected bars in the temporal view visually represent frequency of e-mails sent by all the individuals in the particular year and month of interest, which further sampled (filtered) the data to give the days, days of the week, hours, individuals' connection and contents. Our tool identified 'j.kaminski@enron.com' self-emailed (cc'd) the most on October 2001 with the combination of contents 'plan, meeting, investigation & ferc', which is pertinence (relationship between the facets). Our prototype not only helped in 'finding' but also in 'discovering' pertinence, key information, PoIs, unexpected, unusual and interesting relationships in all the views, which are consistent with the legal report of the Enron case.

4 Future Work

We will work further to develop an effective analyst-friendly visualisation tool to explore and understand anomaly behaviours, pertinence, dynamic changes between two subsets and the underlying communication structures in E-mail communication which will help in improve E-discovery investigations [13–16]. The tool will be tested by the company partners, students and legal analysts on publicly available datasets in order to observe the efficacy of finding relevance, key information and PoIs in a selected set of e-mails and also to evaluate the visualisation design choices for some of the tasks, such as aggregation, comparison, etc. using Visual Data Reasoning (VDAR) [17]. The studies will help to determine the potential effectiveness of our techniques in actual ongoing e-mail investigations.

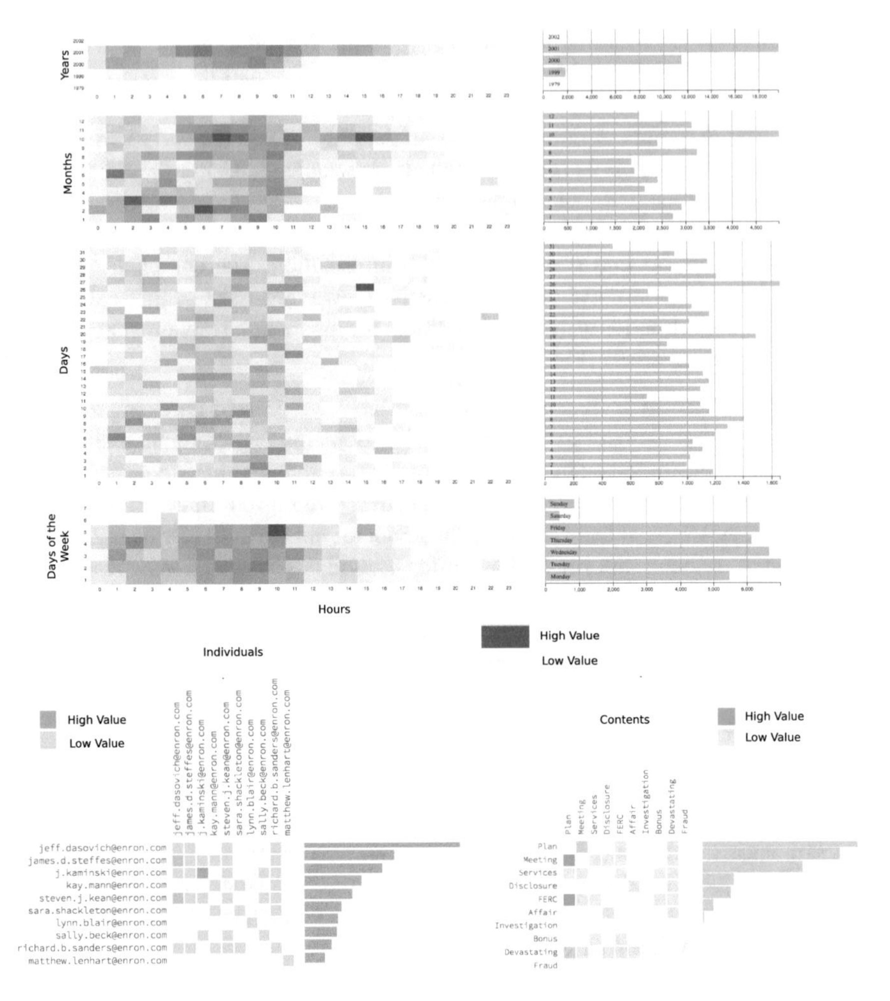

Fig. 2 D3 Prototype: a combination of matrices and bar charts for investigating multifaceted E-mail data

References

1. Casey, E.: Handbook of Digital Forensics and Investigation. Academic, Cambridge (2009)
2. Casey, E.: Digital Evidence and Computer Crime: Forensic Science, Computers, and the Internet. Academic, Cambridge (2011)
3. Attfield, S., Blandford, A.: Discovery-led refinement in e-discovery investigations: sensemaking, cognitive ergonomics and system design. Artif. Intell. Law **18**(4), 387–412 (2010)
4. Socha, G., Gelbmann, T.: The electronic discovery reference model (edrm). http://edrm.net/ (2009)
5. http://enterprise.brainspace.com/discovery
6. Stasko, J., Görg, C., Liu, Z.: Jigsaw: supporting investigative analysis through interactive visualization. Inf. Vis. **7**(2), 118–132 (2008)

7. http://www.lexisnexis.com/litigation/products/ediscovery/concordance
8. http://in-spire.pnnl.gov/
9. http://www.ftitechnology.com/radiance-visual-analytics-software
10. http://zovy.com/solutions/ediscovery/
11. Collins, C., Carpendale, S., Penn, G.: Docuburst: visualizing document content using language structure. In: Computer Graphics Forum, vol. 28, no. 3, pp. 1039–1046. Wiley Online Library (2009)
12. Klimt, B., Yang, Y.: The enron corpus: A new dataset for email classification research. In: Machine Learning: ECML 2004, pp. 217–226. Springer, Berlin (2004)
13. Sathiyanarayanan, M., Turkay, C.: Is multi-perspective visualisation recommended for e-discovery email investigations? (2016)
14. Sathiyanarayanan, M., Turkay, C.: Determining and visualising e-mail subsets to support e-discovery (2016)
15. Sathiyanarayanan, M., Turkay, C., Fadahunsi, O.: Design and implementation of small multiples matrix-based visualisation to monitor and compare email socio-organisational relationships (2018)
16. Sathiyanarayanan, M., Turkay, C.: Challenges and opportunities in using analytics combined with visualisation techniques for finding anomalies in digital communications (2017)
17. Lam, H., Bertini, E., Isenberg, P., Plaisant, C., Carpendale, S.: Empirical studies in information visualization: seven scenarios. IEEE Trans. Vis. Comput. Graph. **18**(9), 1520–1536 (2012)

Speaker Verification Systems: A Comprehensive Review

Ujwala Baruah, Rabul Hussain Laskar and Biswajit Purkayashtha

Abstract In this paper, we present a detailed review of the approaches for speaker verification. Speaker verification is a binary class problem of either accepting or rejecting a claimant speaker on the basis of the matching accuracy of the claimant's test utterance with his/her train utterance. It has been observed that text-independent speaker verification gives much more flexibility as compared to text-dependent speaker verification. However, it is easier to attain better verification accuracy in the text-dependent mode as compared to its counterpart. In addition, our focus has been on giving an overview of the most popular databases available for this purpose as well as on the various feature extraction methods and modeling tools used in the speaker verification systems.

Keywords Speaker verification system · Speech processing · Review · Text-dependent · Text-independent

1 Introduction

A speech signal carries information which includes the message that is being conveyed, the identity of the speaker, the language that is being used, the emotional state in which the speaker have uttered the speech, the background environment, the gender of the speaker, apart from other information. The major task of speaker verification is the ability to decide whether a claimant speaker is genuine or not [19, 45, 56]. In order to be able to do this, speaker-specific features are extracted from

U. Baruah (✉) · R. H. Laskar · B. Purkayashtha
National Institute of Technology Silchar, Silchar 788010, Assam, India
e-mail: ujwala@cse.nits.ac.in
URL: http://cse.nits.ac.in/ujwala/

R. H. Laskar
e-mail: rabul2u4u@gmail.com

B. Purkayashtha
e-mail: biswajit@cse.nits.ac.in

A. Elçi et al. (eds.), *Smart Computing Paradigms: New Progresses and Challenges*, Advances in Intelligent Systems and Computing 766,
https://doi.org/10.1007/978-981-13-9683-0_21

the given speech sample. The identity of a claimant speaker lies in the physiological properties of the vocal tract, the accent used while speaking, the rate of speaking and the emotional state of the speaker. The total task of a speaker verification can be divided into the following three stages:

- Feature extraction.
- Pattern classification.
- Decision making.

Feature extraction involves extracting the speaker-specific information from the speech signal. Features can be classified into short-term or long-term. Pattern classification, on the other hand, deals with classifying the concerned speaker as either belonging to a specific class or being alien to it. Last but not the least decision making involves a binary decision of either accepting or rejecting a claimant speaker. Experiments on speaker verification systems (SVS) have been rigorously carried out using diverse techniques under various assumptions, namely, heterogeneous channel SVS [65], speech forensic [1], anti-spoofing [13, 28, 34, 66], replay attack [58], uncontrolled noisy environment [39], and feature switching [40]. Speaker verification is a subtask of speaker recognition, in which the latter is broadly divided into speaker verification and speaker identification. Speaker identification is concerned with establishing the identity of the claimant with the set of enrolled speakers on the other hand speaker verification is concerned with verifying whether the claimant is who he/she claims to be. Both these tasks are carried out in two modes of operation: text-dependent and text-independent [9, 38, 44]. The text-dependent mode requires the test sample to be the same as the train sample, in contrast to that the text-independent mode does not employ this restriction on the speaker. Hence the length of speech recorded in the text-independent mode is more than its counterpart.

Speaker recognition research is traced to the advent of the sound spectrograph in the Bell telephone laboratories [30]. With the use of the sound spectrograph complex speech sounds are analyzed with the use of the spectrograms produced. The first person to suggest their comparison for speaker identification is Lawrence Kersta [26]. As the recognition rate achieved in this work is quite high and comparable to recognition by fingerprints the work received a lot of attention in the forensic setting [4], spectrogram-based techniques have received attention [7, 61] for a long period of time. During this time long-term averaged spectrum of a sentence-long utterance is also explored for speaker identification [16, 37, 53]. In addition, channel compensation issues also attracted a lot of attention as it is observed that recognition accuracy is greatly affected by channel mismatch. In recent years, text-independent NIST evaluations [43] have made major contribution in the development of new technologies. Hybrid speaker recognition systems consisting of several individual subsystems combining the benefits of the individual [35, 52, 60] have received a lot of attention. Although there are a variety of classification methods available Gaussian mixture models (GMMs) and support vector machines (SVMs) have attained a lot of popularity [14]. The performance of low-level spectral features is enhanced with the use of high-level features such as prosody, duration, and word frequencies. Among the

Table 1 Characteristics of biometric identification systems

Biometric type	Accuracy	Ease of use	User acceptance	Ease of implementation	Cost
Fingerprint	High	Medium	Low	High	Medium
Hand geometry	Medium	High	Medium	Medium	High
Voice	Medium	High	High	High	Low
Retina	High	Low	Low	Low	Medium
Iris	Medium	Medium	Medium	Medium	High
Signature	Medium	Medium	High	Low	Medium
Face	Low	High	High	Medium	Low

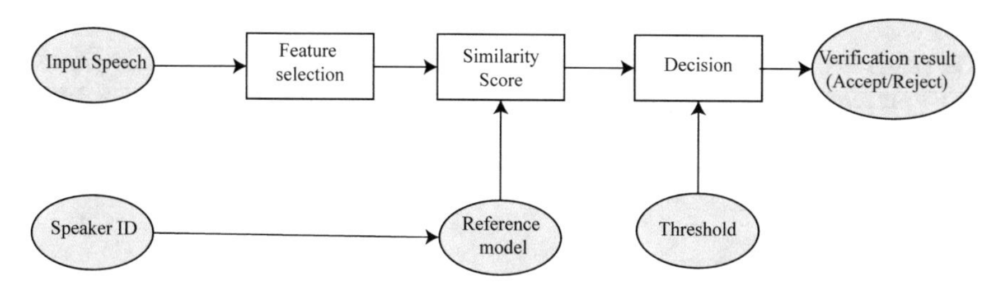

Fig. 1 Block diagram of speaker verification system

compensation techniques Factor analysis (FA) [25] and nuisance attribute projection (NAP) [59] have attracted a lot of attention for text-independent mode of research.

Moreover, Machine Learning is one of the most vibrant areas of research in the present day scenario [24, 55, 66]. The algorithms of this area are being deployed in diverse areas, namely, science, engineering, economics, politics, and government. Machine Learning is an integral part of speech processing as model-based speech systems rely heavily on it [69]. Speech processing has become indispensable for the recent IT industry for boosting up the revenue. In addition, it is evident from Table 1 that fingerprint technology has dominance over the other techniques in the global market despite having low user acceptance and medium cost, but it is clearly seen that voice technology fairs better than the rest of them in almost all the parameters examined. This is also the most natural and easily available source of biometric measure (Fig. 1).

1.1 Motivation

Some of the major research questions (RQ) which are partially answered till date are as follows:

RQ1 Who has spoken?
RQ2 Has the speaker been verified and validated?

The RQ1, and RQ2 call for a speaker verification system. Apart from this, when the speech sample collected is corrupted with background or channel noise finding answers to the above questions pose greater difficulties. A lot of research work is being done, but there is still a dearth of comprehensive information encompassing all the problems. Hence, we have given a comprehensive review of the trend from the 40's till date.

1.2 Organization

The paper is organized as follows: Sect. 2 exploits the speech databases. Sections 3 and 4 present the literature review of selected papers. Section 5 discusses the future scope of SVS. And finally, the paper is concluded by Sect. 6.

2 Database Review

There are numerous text-dependent as well as text-independent databases available [64]. The NIST database is treated to be the benchmark for text-independent speaker verification. But due to the lack of text-dependent database of a similar stature, many researchers have been developing their own databases according to the requirement of their application. An attempt has been made to give a comprehensive listing of the currently available speech databases. Table 2 includes the databases collected for a variety of different languages like English, Japanese, etc. Although it has been observed that YOHO is a widely accepted database for the evaluation of text-dependent speaker recognition systems but it lacks channel variations. While RSR2015 and the NITS speech database have incorporated channel variations as well as session variability. Apart from this, the NITS speech database is the only English database recorded by Indian speakers belonging to a significantly large number of different states, to study handset-channel variability issues for text-dependent, text-independent, and voice password-based tasks. The explosion of deep learning enhances learning system of a speaker verification system [63].

3 Literature Review

In [6], the authors compare and contrast the use of speech and fingerprints for the identification of a person. The accuracy of recognizing a person by listening to his/her voice is more than the accuracy attained by visual examination of the spectrograms.

Table 2 Available speech databases

Database	Language	Type of utterances	Telephone or microphone	Size of the database
$CNORM + DEMO_R VI$	English	Combination lock phrases	Telephone	35 speakers
Lamel and Gauvain [32]	French	Read (digit strings, sentences) + spontaneous speech	Telephone	100 target speakers 1000 imposters
Paper [5, 10]	French	17 words	Telephone	143 speakers
TUBTEL [20]	German	Sentences	Telephone	50 speakers
Ismail Shahin [57]	English	8 fixed sentences	Microphone	30 speakers
Isobe and Takahashi [22]	Japanese or English	Digit strings	Telephone	100 speakers
Charlet et al. [8]	French or English	5 short sentences	Telephone	55 target speakers 600 imposters
YOHO [21, 44]	English	Combination Lock phrases	High-quality telephone handset, no telephone channel	138 speakers
Rosenberg et al. [51]	English	A single phrase	Telephone	50 male target speakers 50 male imposters
Gupta et al. [18]	English	A phrase common to all speakers + speaker-specific phrases	Microphone	21 speakers
Nealand et al. [41]	English	Digit strings	Telephone	354 speakers
MSRI [12]	Indian and English	Digit strings and phrases	Microphone	344 speakers
RSR2015	English	Fixed short passphrases, speaker-loaded command, randomly prompted digit strings	Mobile devices	300 speakers
NITS speech database	English	Short prompts + 3 minutes of speech	Mobile devices	302 speakers

In [61], the authors use a speech database consisting of speech samples collected from 250 speakers who are all American students from Michigan State University. In [37], the author explores the linear prediction in both the time and frequency domain. An all-pole modeling is discussed because of its applicability and acceptance. In [23], the author proposes an algorithm for word recognition with minimum prediction error. Dynamic programming is employed to minimize the total log prediction error. In [7], the author presents a standalone algorithm for noise suppression. The noise magnitude spectrum is subtracted from the speech sample containing additive noise to estimate the magnitude of the frequency spectrum. In [50], the authors compare the performance of two of the noise compensation techniques for speaker identification in the text-independent mode. The database used for this study is prepared using four long distance calls made to ten male speakers. In [47], the author employs four different feature selection techniques, namely, MFCC, LPCC, LFCC, and PLPC. The baseline system has been developed using cepstral mean removal with the following specifications for each of the feature set: MFCC-24 (24 mel-frequency spaced filters, 23 cep coeff/vector), LFCC-40 (40 linear-frequency spaced filters, 39 cep coeff/vector), PLPC-I2 (17 Bark-frequency spaced filters, 12th-order LPC, 12 cep coeff/vector), LPCC-I2 (12th order LPC, 12 cep coeffvector).

In [71], the authors use the TIMIT database to carry out their experiments. A comparison of speaker recognition performance has been made using mixture-Gaussian VQ, ergodic CHMMs and phone-based left-to-right CHMMs. It is observed that ergodic CHMMs have given the best recognition accuracy. This is a key attribution to the little intra-speaker variance in the database used. In [30], the working of the Sound Spectrograph and its various component parts are discussed in great depth. The sound spectrograph gives a pictorial representation of the time variations of the short-term spectrum of the speech wave in the form of graphs. In [49], the authors carry out extensive experiments on text-independent speaker identification with total 630 speakers from the TIMIT and NTIMIT databases. Identification accuracies of 99.5 and 60.7% are reported on the TIMIT and NTIMIT databases, respectively. The authors further attempt to quantify the loss in performance due to the telephone degradations. In the paper [48], the authors advocate the use of Gaussian Mixture Models for speaker identification. The database used has conversational speech collected from 49 speakers. It is reported that GMMs are capable of representing general speaker-dependent spectral shapes and the mixtures are also capable of modeling arbitrary densities.

In [70], the authors present a robust speaker identification system with the use of CASA (Computational auditory scene analysis) in the front end in order to perform speech segregation. The authors propose a new set of speaker features which include gammatone feature (GF) and gammatone frequency cepstral coefficients ($GFCC$). A two module experimental setup is proposed wherein the first module uses the time-frequency binary mask generated by CASA to reconstruct the noise corrupted components. The second module carries out the bounded marginalization on the noise corrupted GF. In [62], the authors have carried out an experiment on speech samples extracted from the YOHO,TIMIT, TIDIGT, and UM databases. Comparison of the result thus obtained has been done with the commonly used feature-based methods, namely, MFCC, FDLP, and GFCC. In [33], the authors apply the RSR2015 database to verify the obtained results from the experiments. The authors develop a speaker verification system in the text-dependent mode which is robust in the presence of impostures of three types, namely, (a) when the test utterance is given by an imposture pronouncing the authenticated speech, (b) when the speech pronounced is not the authenticated one, and (c) when the test sample is played back. The modeling technique used is a combination of GMM and HMM. Combination of the score obtained in text-dependent speaker verification with the speaker-specific normalization score is seen to improve the accuracy.

In [17], the authors present an algorithm which is immune to the presence of background noise. The database used is created by recording the voice samples of 175 participants over the telephone channel in varied environments. The performance achieved is satisfactory although the work does not address the issues of signal modification and the effect of lexical information contained in the speech sample. In [11], the authors employ the NIST2012 database to evaluate their experimental results. MFCC feature vectors are extracted from the speech samples. A study has been carried out in order to find the effect of phonetic information present in a speech samples, in the accuracy of speaker recognition. For this, the modeling technique used has been built by hybridizing DNN with GMM. The results thus obtained show that addition of the phonetic information with the acoustic information increases recognition accuracy. In [67], the authors carry out their study using the RSR2015 database. The vulnerability of a speaker verification system to replay attacks is investigated and a solution to this problem is proposed. The proposed model is a hierarchical structure. The top layer of which is the UBM, the middle layer is GMM and the lowest layer is HMM. MFCC feature vectors are extracted and the silent zones are removed using VAD and RASTA filtering. Detection of the replayed speech sample is done by using spectrogram bitmaps. The results thus obtained reflect the danger that replay attacks pose to a speaker recognition system. The proposed method does elevate this problem to a great degree.

4 Literature Review on Recent Works

In [36], the authors have carried out their experiments on the Fisher database. The authors give a simple and efficient Deep Neural Network structure involving two each of convolutional layers and time delayed full-connection layers. These are connected by a bottleneck layer which consists of 512 hidden units. The features, thus learned are not fed to any backend models instead frame averaging of the speech segment is employed. The baseline of the experiments carried out is an i-vector system to which 19-dimensional MFCCs along with their log energy are fed. The features corresponding to particular speaker are extracted from the last hidden layer of the DNN. It has been observed that the accuracy improves with the length of the test utterances. As the residual noise within the d-vectors is not Gaussian the LDA approach improves the performance of the d-vector system while the same cannot be said about the

PLDA approach. The results thus obtained also suggest that speaker-specific traits is obtained from a relatively small number of frames, hence, this school of thought replaces the currently popular model-based approach to a large extent if not totally.

In [2], the authors use the King Saud University Arabic Speech Database to perform experiments. For the experiments conducted, the durations of the training and test sequences are kept different. The clean speech is corrupted with babble, and car noises. MFCC feature vectors are extracted, and these are modeled by using GMM-UBM which has been seen to be one of the most acceptable modeling techniques [3, 46]. It is investigated at four different levels: (a) by varying the length of training and testing data, (b) investigation on the noisy data set, (c) varying the time of recording of the train and test data sets, and (d) by investigating on test data collected through the mobile channel. The results are promising for all the levels except for when the test data are collected using the mobile channel as the experimental setup does not have a channel compensation features incorporated in it.

In [68], the authors perform an experiment on the English CTS (Switchboard and Fisher) corpora. Feature extraction involves extracting 40-dimensional log-filterbank features which are sampled every 10 ms. A 25 ms window is used for analysis. The hybrid system developed uses the advantages of the convolutional neural nets(CNN) and the long-short-term memory nets(LSTM). The CNN model has been reported to have given highest recognition accuracy. This is closely followed by the LSTM model. It is reported that increasing the architecture beyond 6-layers does not show any significant improvement in accuracy.

In [29], the authors present a new database for the study of spoofing on speech data. The database, thus presented, is prepared by exposing the existing RedDots database to replay attacks. The replay attacks are done under two conditions, namely, controlled and variable. In the former, the replay has been done in a quiet office environment, while for the latter the replay is done in the presence of real-world background noise. Thereafter, MFCC, LFCC, and constant q cepstral coefficients (CQCCs) are extracted. The MFCC feature vectors are normalized by using their mean and variance. Two flavors of modeling techniques are used to model these feature vectors, GMM-UBM, and i-vectors. The study reports a better performance by GMM-UBM as compared to i-vectors. This is attributed to the length of the speech samples used and the random selection of the speech files for testing.

In [42], the authors give a method to watermark a speech signal in order to secure a speech signal traveling over a network. For this work, the authors make use of the TIMIT database, MIT, and MOBIO. The baseline system is designed using two modeling techniques, GMM-UBM and i-vector probabilistic linear discriminant analysis (PLDA). MFCC feature vectors were extracted for modeling the speaker verification system. At the back-end, authentication is done by using a combined personal identification number and a one time password along with watermarked speech. For embedding the watermark in the speech signal Discrete Wavelet Packet Transform and Quantization Index Modulation are used. Watermarking is embedded in the wavelet sub-bands which are rich in speaker-related information. The testing for efficiency is done to detect the spoofing and attacks during communication. The results obtained show that detection of the watermark error decreases with the decrease in the number of bits in the watermark.

In [27], the authors use the RSR2015 and the IITG-MV databases to carry out an experiment and arrive at the results. In this work, the authors make use of speech enhancement techniques in the front end [31] in order to enhance the speech degraded by channel and background noise. This is seen to improve the accuracy of the speaker verification system. The study is carried out in the text-dependent mode. Thereafter, energy-based speech end-point detection is done. MFCC feature vectors, their delta, and double delta are then temporally aligned by using Dynamic Time Warping (DTW) [15, 54]. Enhancement of the signal is done both at the temporal as well as the spectral level. This has been seen to contribute significantly in the elimination of degradation of the speech sample as well as in improvement in the overall accuracy of the system. While temporal enhancement is done by emphasizing the regions with high SNR of the corrupted speech sample, spectral enhancement is done by applying the MMSE-LSA estimator to the spectrum of the corrupted speech. From the results thus obtained it is observed that employment of speech enhancement techniques before the detection of end-points gives better verification accuracy as compared to using the former after detecting the end-points.

5 Discussion

Starting from the acquisition of speech spectrograms with the use of sound spectrographs to acceptable performance of speaker verification systems, speech research has come a long way. But there are still lots more to be achieved. MFCC with its derivatives have been found to be the closest representation of the sound as perceived by human beings. GMM-UBM is the most widely used modeling technique. However, a study of the distribution of data in the feature set should govern the decision on the choice of distance measure while verifying. A hybrid system using MFCC features in conjunction with features from neural responses is expected to improve the accuracy. Although Dynamic Time Warping has been found to be effective in mapping signals with temporal differences in the presence of a large amount of noise its performance degrades. Use of a derivative of DTW is a better option in the presence of prominent channel degradations and background noise. Acquiring the clean signal is impossible in noisy public area. However, the clean speech signal eases processing and is one of the main factors contributing toward the increase in accuracy. Noise removal is a time consuming and computationally complex task, especially when the type of noise present is varied. Albeit this, there are numerous Machine Learning approaches to achieve high accuracy in the presence of noise, but Deep Learning takes a lion's share in the research of speech processing. In addition, deploying Deep Learning creates new possibilities of research direction in Speech processing.

6 Conclusion

Speaker verification is a promising field of research, especially in the presence of background noise and channel variations. Model-based approach is presently the most accepted approach, however, these require a relatively large number of frames. Hence an alternate school of thought with good promises is to look towards a feature-based approach. The performance of a feature-based speaker verification system is further improved by the use of features extracted from neural responses of the auditory periphery to the sample speech. This approach is expected to give better performance as it replicates the physical system to a large extent. The performance of the commonly used channel compensation techniques of CMS, RASTA filtering and Frequency warping is enhanced with the use of normalization techniques to wider the spectrum to linear noise filtering.

References

1. Al-Ali, A.K.H., Senadji, B., Naik, G.: Forensic speaker verification in noisy environmental by enhancing the speech signal using ica approach. World Acad. Sci. Eng. Technol. Int. J. Comput. Electr. Autom. Control. Inf. Eng. **11**(4), 420–423 (2017)
2. Algabri, M., Mathkour, H., Bencherif, M.A., Alsulaiman, M., Mekhtiche, M.A.: Automatic speaker recognition for mobile forensic applications. Mob. Inf. Syst. **2017** (2017)
3. Angkititrakul, P., Hansen, J.H.L.: Discriminative in-set/out-of-set speaker recognition. IEEE Trans. Audio Speech Lang. Process. **15**(2), 498–508 (2007)
4. Begault, D.R., Poza, F.: Voice identification and elimination using aural-spectrographic protocols. In: Audio Engineering Society Conference: 26th International Conference: Audio Forensics in the Digital Age (2005)
5. BenZeghiba, M.F., Bourlard, H.: User-customized password speaker verification using multiple reference and background models, vol. 48. Technical Report Idiap-RR-41-2004, IDIAP, 2004. Published in Speech Communication (2006)
6. Bolt, R.H., Cooper, F.S., David Jr, E.E., Denes, P.B., Pickett, J.M., Stevens, K.N.: Speaker identification by speech spectrograms: A scientists' view of its reliability for legal purposes. J. Acoust. Soc. Am. **47**(2B), 597–612 (1970)
7. Bolt, S.: Suppression of acoustic noise in speech using spectral subtraction. IEEE Trans. Acoust. Speech Signal Process. **27**(2), 113–120 (1979)
8. Charlet, D., Jouvet, D., Collin, O.: An alternative normalization scheme in hmm-based text-dependent speaker verification. Speech Commun. **31**(2), 113–120 (2000)
9. Chen, K., Xie, D., Chi, H.: Text-dependent speaker identification based on input/output hmms: an empirical study. Neural Process. Lett. **3**(2), 81–89 (1996)
10. Chollet, G., Cochard, J.-L., Constantinescu, A., Jaboulet, C., Langlais, P.: Swiss French Poly-Phone and PolyVar: telephone speech databases to model inter- and intra-speaker variability. Technical report (1996)
11. Cumani, S., Laface, P., Kulsoom, F.: Speaker recognition by means of acoustic and phonetically informed gmms. In: INTERSPEECH (2015)
12. Das, A., Chittaranjan, G., Anumanchipalli, G.: Usefulness of text-conditioning and a new database for text-dependent speaker recognition research (2008)
13. Dhanush, B.K., Suparna, S., Aarthy, R., Likhita, C., Shashank, D., Harish, H., Ganapathy, S.: Factor analysis methods for joint speaker verification and spoof detection. In: 2017 IEEE International Conference on Acoustics, Speech and Signal Processing (ICASSP), pp. 5385–5389. IEEE (2017)

14. Edwards, J.S., Ramachandran, R.P., Thayasivam, U.: Robust speaker verification with a two classifier format and feature enhancement. In: 2017 IEEE International Symposium on Circuits and Systems (ISCAS), pp. 1–4 (2017)
15. Furui, S.: Cepstral analysis technique for automatic speaker verification. IEEE Trans. Acoust. Speech Signal Process. **29**(2), 254–272 (1981)
16. Furui, S.: Cepstral analysis for automatic speaker verification. IEEE Trans. Acoust. Speech Signal Process. **27**, 01 (2003)
17. Gaka, J., Grzywacz, M., Samborski, R.: Playback attack detection for text-dependent speaker verification over telephone channels. Speech Commun. **67**(Supplement C), 143–153 (2015)
18. Gupta, H., Hautamäki, V., Kinnunen, T., Fränti, P.: Field evaluation of text-dependent speaker recognition in an access control application. Speech and Image Processing Unit. Department of Computer Science, University of Joensuu, Finland (2005)
19. Hansen, J.H.L., Hasan, T.: Speaker recognition by machines and humans: a tutorial review. IEEE Signal Process. Mag. **32**(6), 74–99 (2015)
20. Hardt, D., Fellbaum, K.: Spectral subtraction and rasta-filtering in text-dependent hmm-based speaker verification. In: Proceedings of the 1997 IEEE International Conference on Acoustics, Speech, and Signal Processing (ICASSP '97), ICASSP '97, vol. 2, pp. 867–870, Washington, DC, USA. IEEE Computer Society (1997)
21. Higgins, A., Bahler, L., Porter, J.: Speaker verification using randomized phrase prompting. Digit. Signal Process. **1**(2), 89–106 (1991)
22. Isobe, T., Takahashi, J.: A new cohort normalization using local acoustic information for speaker verification. In: Proceedings of the 1999 IEEE International Conference on Acoustics, Speech, and Signal Processing. ICASSP99 (Cat. No.99CH36258), vol. 2, pp. 841–844 (1999)
23. Itakura, F.: Minimum prediction residual principle applied to speech recognition. IEEE Trans. Acoust. Speech Signal Process. **23**(1), 67–72 (1975)
24. Jordan, M.I., Mitchell, T.M.: Machine learning: trends, perspectives, and prospects. Science **349**(6245), 255–260 (2015)
25. Kenny, P.: Joint factor analysis of speaker and session variability: theory and algorithms, 01 (2006)
26. Kersta, L.G.: Voiceprint identification. J. Acoust. Soc. Am. **34**(5), 725 (1962)
27. Khonglah, B.K., Bhukya, R.K., Prasanna, S.R.M.: Processing degraded speech for text dependent speaker verification. Int. J. Speech Technol. **20**(4), 839–850 (2017)
28. Kinnunen, T., Evans, N., Yamagishi, J., Lee, K.A., Sahidullah, M., Todisco, M., Delgado, H.: Asvspoof 2017: automatic speaker verification spoofing and countermeasures challenge evaluation plan. Training **10**(1508), 1508 (2017)
29. Kinnunen, T., Sahidullah, M., Falcone, M., Costantini, L., Hautamaki, R.G., Thomsen, D., Sarkar, A., Tan, Z.H., Delgado, H., Todisco, M., Evans, N.: Reddots replayed: a new replay spoofing attack corpus for text-dependent speaker verification research. In: 2017 IEEE International Conference on Acoustics, Speech and Signal Processing (ICASSP), pp. 5395–5399 (2017)
30. Koenig, W., Dunn, H.K., Lacy, L.Y.: The sound spectrograph. J. Acoust. Soc. Am. **18**(1), 19–49 (1946)
31. Krishnamoorthy, P., Prasanna, S.R.M.: Reverberant speech enhancement by temporal and spectral processing. IEEE Trans. Audio Speech Lang. Process. **17**(2), 253–266 (2009)
32. Lamel, L., Gauvain, J.: Speaker verification over the telephone. Speech Commun. **31**(2), 141–154 (2000)
33. Larcher, A., Lee, K.A., Ma, B., Li, H.: Imposture classification for text-dependent speaker verification. In: 2014 IEEE International Conference on Acoustics, Speech and Signal Processing (ICASSP), pp. 739–743 (2014)
34. Lavrentyeva, G., Novoselov, S., Simonchik, K.: Anti-spoofing Methods for Automatic Speaker Verification System, pp. 172–184. Springer International Publishing, Cham (2017)
35. Li, H., Ma, B., Lee, K.A., Sun, H., Zhu, D., Sim, K.C., You, C., Tong, R., Karkkainen, I., Huang, C.L., Pervouchine, V., Guo, W., Li, Y., Dai, L., Nosratighods, M., Tharmarajah, T., Ambikairajah, E., Siong E.C., Schultz, T., Jin, Q.: The i4u system in nist 2008 speaker recognition evaluation

36. Li, L., Chen, Y., Shi, Y., Tang, Z., Wang, D.: Deep speaker feature learning for text-independent speaker verification. In: INTERSPEECH (2017)
37. Makhoul, J.: Linear prediction: a tutorial review. **63**, 561–580 (1975)
38. Matsui, T., Furui, S.: Concatenated phoneme models for text-variable speaker recognition. In: 1993 IEEE International Conference on Acoustics, Speech, and Signal Processing, vol. 2, pp. 391–394 (1993)
39. Misra, S., Laskar, R.H., Baruah, U., Das, T.K., Saha, P., Choudhury, S.P.: Analysis and extraction of lp-residual for its application in speaker verification system under uncontrolled noisy environment. Multimed. Tools Appl. **76**(1), 757–784 (2017)
40. Saranya, M.S., Padmanabhan, R., Murthy, H.A.: Feature-switching: dynamic feature selection for an i-vector based speaker verification system. Speech Commun. **93**(Supplement C), 53–62 (2017)
41. Nealand, J.H., Pelecanos, J.W., Zilca, R.D., Ramaswamy, G.N.: A study of the relative importance of temporal characteristics in text-dependent and text-constrained speaker verification. In: Proceedings of the 2005 IEEE International Conference on Acoustics, Speech, and Signal Processing (ICASSP '05), vol. 1, pp. 653–656 (2005)
42. Nematollahi, M.A., Gamboa-Rosales, H., Martinez-Ruiz, F., De la Rosa, J., Al-Haddad, S.A.R., Esmaeilpour, M.: Multi-factor authentication model based on multipurpose speech watermarking and online speaker recognition **76**, 31 (2016)
43. NIST. Speaker recognition evaluation (2011)
44. Campbell Jr, J.P.: Testing with the yoho cd-rom voice verification corpus. In: Proceedings of the ICASSP, IEEE International Conference on Acoustics, Speech and Signal Processing, vol. 1, pp. 341–344 (1995)
45. Padmanabhan, J., Premkumar, M.J.J.: Machine learning in automatic speech recognition: a survey. IETE Tech. Rev. **32**(4), 240–251 (2015)
46. Prakash, V., Hansen, J.H.L.: In-set/out-of-set speaker recognition under sparse enrollment. IEEE Trans. Audio Speech Lang. Process. **15**(7), 2044–2052 (2007)
47. Reynolds, D.A.: Experimental evaluation of features for robust speaker identification. IEEE Trans. Speech Audio Process. **2**(4), 639–643 (1994)
48. Reynolds, D.A., Rose, R.C.: Robust text-independent speaker identification using gaussian mixture speaker models. IEEE Trans. Speech Audio Process. **3**(1), 72–83 (1995)
49. Reynolds, D.A., Zissman, M.A., Quatieri, T.F., O'Leary, G.C., Carlson, B.A.: The effects of telephone transmission degradations on speaker recognition performance. In: 1995 International Conference on Acoustics, Speech, and Signal Processing, vol. 1, pp. 329–332 (1995)
50. Rose, R.C., Fitzmaurice, J., Hofstetter, E.M., Reynolds, D.A.: Robust speaker identification in noisy environments using noise adaptive speaker models. In: Proceedings of the ICASSP 91: 1991 International Conference on Acoustics, Speech, and Signal Processing, vol. 1, pp. 401–404 (1991)
51. Rosenberg, A.E., Siohan, O., Parthasarathy, S.: Small group speaker identification with common password phrases. Speech Commun. **31**(2), 131–140 (2000)
52. Kajarekar, S.S., Scheffer, N., Graciarena, M., Shriberg, E., Stolcke, A., Ferrer, L., Bocklet, T.: The sri nist 2008 speaker recognition evaluation system, pp. 4205–4208 (2009)
53. Sadaoki, F.: Selected topics from 40 years of research on speech and speaker recognition. In: Proceedings of the Annual Conference of the International Speech Communication Association, INTERSPEECH, pp. 1–8 (2009)
54. Sakoe, H., Chiba, S.: Dynamic programming algorithm optimization for spoken word recognition. IEEE Trans. Acoust. Speech Signal Process. **26**(1), 43–49 (1978)
55. Schmidhuber, J.: Deep learning in neural networks: an overview. Neural Netw. **61**(Supplement C), 85–117 (2015)
56. Schuller, B., Steidl, S., Batliner, A., Nth, E., Vinciarelli, A., Burkhardt, F., van Son, R., Weninger, F., Eyben, F., Bocklet, T., Mohammadi, G., Weiss, B.: A survey on perceived speaker traits: personality, likability, pathology, and the first challenge. Comput. Speech Lang. **29**(1), 100–131 (2015)

57. Shahin, I.: Speaker identification in the shouted environment using suprasegmental hidden markov models. Signal Process. **88**(11), 2700–2708 (2008)
58. Singh, M., Mishra, J., Pati, D.: Replay attack: its effect on gmm-ubm based text-independent speaker verification system. In:2016 IEEE Uttar Pradesh Section International Conference on Electrical, Computer and Electronics Engineering (UPCON), pp. 619–623 (2016)
59. Solomonoff, A., Campbell, W., Boardman, I.: Advances in channel compensation for svm speaker recognition. vol. 1, pp. 629–632 (2005)
60. Sturim, D., Campbell, W.M., Karam, Z.N., Reynolds, D., Richardson, F.: The mit lincoln laboratory 2008 speaker recognition system. pp. 2359–2362 (2009)
61. Tosi, O., Oyer, H., Lashbrook, W., Pedrey, C., Nicol, J., Nash, E.: Experiment on voice identification. J. Acoust. Soc. Am. **51**(6B), 2030–2043 (1972)
62. van Vuuren, S.: Comparison of text-independent speaker recognition methods on telephone speech with acoustic mismatch. In: Proceedings of the ICSLP 96. Fourth International Conference on Spoken Language, 1996, vol. 3, pp. 1788–1791 (1996)
63. Variani, E., Lei, X., McDermott, E., Moreno, I.L., Gonzalez-Dominguez, J.: Deep neural networks for small footprint text-dependent speaker verification. In: 2014 IEEE International Conference on Acoustics, Speech and Signal Processing (ICASSP), pp. 4052–4056 (2014)
64. Wan, V., Renals, S.: Speaker verification using sequence discriminant support vector machines. IEEE Trans. Speech Audio Process. **13**(2), 203–210 (2005)
65. Wang, D.S., Zou, Y.X., Liu, J.H., Huang, Y.C.: A robust dbn-vector based speaker verification system under channel mismatch conditions. In: 2016 IEEE International Conference on Digital Signal Processing (DSP), pp. 94–98 (2016)
66. Wu, Z., Evans, N., Kinnunen, T., Yamagishi, J., Alegre, F., Li, H.: Spoofing and countermeasures for speaker verification: a survey. Speech Commun. **66**(Supplement C), 130–153 (2015)
67. Wu, Z., Gao, S., Cling, E.S., Li, H.: A study on replay attack and anti-spoofing for text-dependent speaker verification. In: 2014 Asia-Pacific Signal and Information Processing Association Annual Summit and Conference (APSIPA), pp. 1–5 (2014)
68. Xiong, W., Droppo, J., Huang, X., Seide, F., Seltzer, M., Stolcke, A., Yu, D., Zweig, G.: The microsoft 2016 conversational speech recognition system. In: 2017 IEEE International Conference on Acoustics, Speech and Signal Processing (ICASSP), pp. 5255–5259, (2017)
69. Young, S.: Statistical spoken dialogue systems and the challenges for machine learning. In: Proceedings of the Tenth ACM International Conference on Web Search and Data Mining, WSDM '17, p. 577. ACM, New York, NY, USA (2017)
70. Yu, K., Mason, J., Oglesby, J.: Speaker recognition using hidden markov models, dynamic time warping and vector quantisation. IEE Proc. Vis. Image Signal Process. **142**(5), 313–318 (1995)
71. Zhu, X., Millar, B., Macleod, J., Wagner, M., Chen, F., Ran, S.: A comparative study of mixture-gaussian vq, ergodic hmms and left-to-right hmms for speaker recognition. In: Proceedings of the ISSIPNN '94. International Symposium on Speech, Image Processing and Neural Networks, vol. 2, pp. 618–621 (1994)

A Fast Method for Segmenting ECG Waveforms

Dipjyoti Bisharad, Debakshi Dey and Brinda Bhowmick

Abstract Electrocardiography (ECG or EKG) is a medical test that is heavily used to assess human heart condition and investigate a large set of cardiac diseases. Automated ECG analysis has become a task of increased clinical importance since it can aid physicians in improved diagnosis. Most of the automated ECG analysis techniques require first identifying the onset and offset locations of its fiducial points and characteristic waves. Two of the important characteristic waves are P and T waves. They mark the beginning and end of an ECG cycle, respectively. In this paper, a fast technique is proposed that can segment ECG signals by accurately identifying the P and T waves. In this work, we evaluate the performance of our model on standard QT database (Laguna et al. Comput Cardiol 24:673–676, 1997 [1]). We achieved high accuracies above 99% and 97% while detecting P waves and T waves respectively.

Keywords Electrocardiogram · ECG features · ECG delineation · ECG segmentation

1 Introduction

ECG signal originates from the electrical activity of the heart that is synchronous with the contraction and relaxation of the atria and ventricles of the heart. Monitoring the electrical activities of the heart can help to identify various types of heart diseases. Nowadays, several methods are existing for ECG analysis and it has become a quite mature field. Some well-annotated datasets have been developed that have boosted the research in ECG data analysis [1–3]. Many works have been done till date for

D. Bisharad · D. Dey (✉) · B. Bhowmick
National Institute of Technology Silchar, Silchar 788010, Assam, India
e-mail: deydebakshi16@gmail.com

D. Bisharad
e-mail: dipjyotibisharad.nit@gmail.com

B. Bhowmick
e-mail: brindabhowmick@gmail.com

A. Elçi et al. (eds.), *Smart Computing Paradigms: New Progresses and Challenges*, Advances in Intelligent Systems and Computing 766,
https://doi.org/10.1007/978-981-13-9683-0_22

determining characteristic points in ECG signals. But most of them are computationally expensive because of using complex signal processing techniques. In [4], the QRS complexes are recognized using the information on the signal's slope, amplitude and width. Another proposed method uses the wavelet transform to detect all the P, QRS and T complexes but for high noise, the detection of P and T onsets and offsets is very difficult [5]. Hidden Markov Model is used to detect P wave along with QRS complex in [6]. In [7], P and T waves detection is based on length transformation technique. In [8] the QRS complexes have been clustered into several groups using self-organizing neural networks. The algorithm proposed in [9] is based on digital fractional-order differentiation for P and T waves detection and delineation. Though in this work, we built our model on single lead ECG system, the authors in [10] found that the detection of wave boundaries in multi-lead ECG signals gave better performance for measurements of T waves than the other characteristic waveforms.

In this paper, segmenting ECG cycles from the entire waveform is done by detecting the P and T complexes using local context window around R peaks. The proposed method shows very high detection accuracy and has linear computational complexity with respect to length of the ECG signal. All the ECG signals used in this work are obtained from modified limb lead II (MLII). This lead is placed on the chest.

The remainder of this paper is organized as follows. In Sect. 2, the composition of ECG signal is discussed. In Sect. 3, the methodologies and algorithms are discussed. The results of the proposed method are shown in Sect. 4. Finally, the paper concludes in Sect. 5.

2 Structure of ECG Signal

In this section, we provide a brief overview of the structure of ECG signal. Electrical signals generated during one heartbeat in humans undergo depolarization and repolarization. The magnitude and direction of these electrical events are what is captured by the ECG sensors. All the electrical events that take place are indicated by one of the multiple waveforms contained in the components of a normal ECG tracing during one cardiac cycle. The first short and upward wave on ECG tracing is P wave, which indicates atrial depolarization. Then, the QRS complex follows, which is due to ventricular repolarization. After this, the T wave is observed, which is usually a small upward waveform but it may be inverted in some cases [11]. These waves have a characteristic duration. The P wave exists for about 80 ms and one ST segment duration varies from 80 to 120 ms. One ST interval is about 320 ms [12] (Fig. 1).

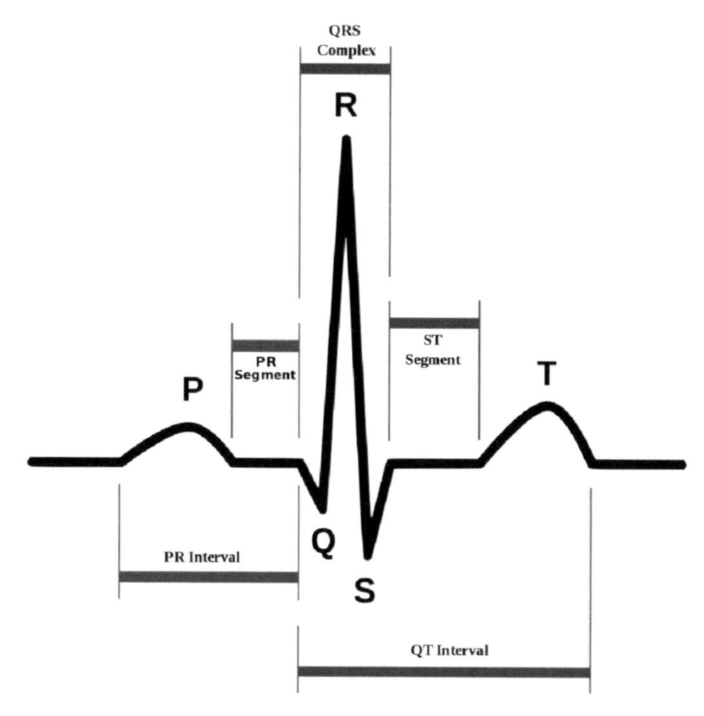

Fig. 1 Waveform of single ECG wave [16]

3 Methodology

From the review of ECG signal, it is clear that the P and T waves have distinct physical characteristics. Further, if R peak is known, then these two waves can be identified from its zone with fair accuracy. For instance, T peak can be taken as the local maxima between the R peak of the current wave and P peak of the subsequent wave. After detecting P and T waves, boundary of the ECG wave is determined.

3.1 Preprocessing the Signals

Electromyogram (EMG) signals that originate from muscles corrupt the raw ECG signals. High-frequency interferences, DC offset and baseline wandering occurring from electrical equipments can also corrupt it [13]. In order to reduce these noises, the signal is passed through a bandpass filter with a cutoff frequency of 3 and 45 Hz. To detect the R peaks, the Hamilton segmentation algorithm [14] is used on the filtered ECG signal.

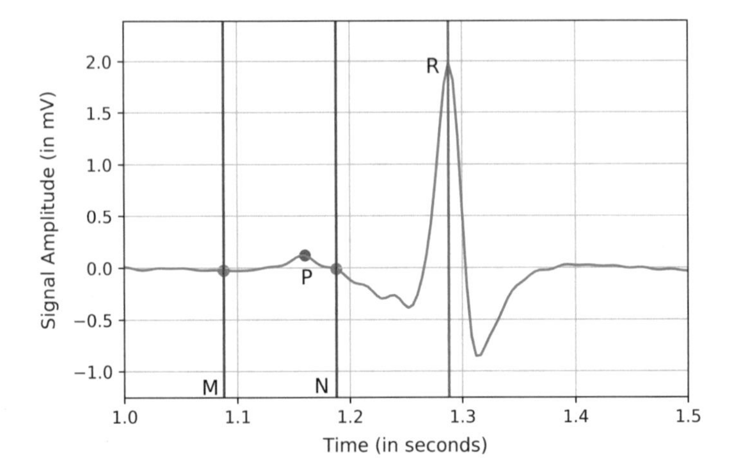

Fig. 2 *M* and *N* are the onset and offset of *P* wave, respectively; *M* and *N* are 100 ms apart; point *N* is 100 ms offset from *R* peak

3.2 Detection of the Peak of P Wave

After locating R peaks, we determine the location of P peaks. From the structure of ECG signal, we find that P peak can be approximated as the local maxima between the previous waveform T peak and the current waveform R peak. But the region between T and R peaks is quite wide, can be noisy and can have multiple peaks and troughs. So it can lead to increased false positives if the entire region is considered. Hence, a smaller context window of 100 ms duration is used which is offset from R peak by 100 ms on the left as shown in Fig. 2. The maximum of values in the context window is chosen as the peak of P wave.

3.3 Detection of the Peak of T Wave

It is more difficult to accurately detect T wave than detecting QRS complex. This is due to low signal-to-noise ratio (SNR), low amplitudes, variation in morphology and amplitude and probable overlapping of P and T waves [4]. Thus, the initially filtered signal is again passed through a second-order Butterworth filter having lower and upper cutoff frequency of 0.5 and 25 Hz. As already mentioned in Sect. 2, T wave may be inverted in certain cases. Thus, within the context window, T peak can be either the maxima or minima, depending on which one has the maximum absolute magnitude. To eliminate this uncertainty, all the values within the window are squared which makes the T peak to be located at the index of the value having maximum squared magnitude. But if the T peak is inverted, the voltage level at the peak might possibly lie between 0 and −1 mV. Since squaring a value between 0 and

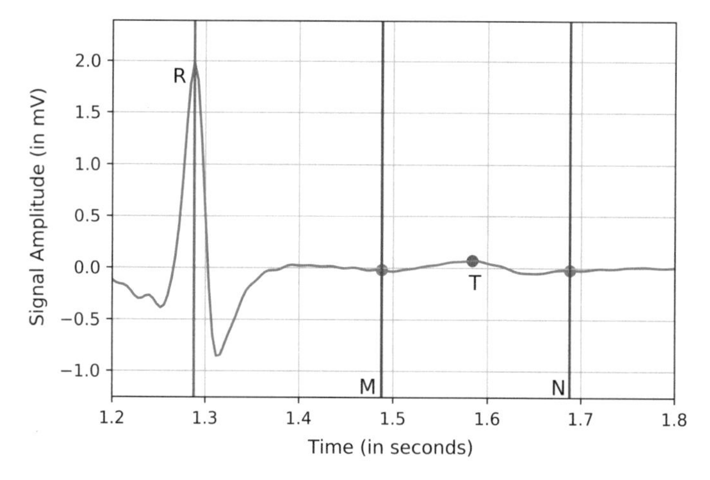

Fig. 3 *M* and *N* are the onset and offset of *T* wave, respectively; *M* and *N* are 200 ms apart; *N* is 200 ms ahead of *R* peak

1 will make the value even smaller, 1 mV is added to all the values before squaring. T waves have longer duration compared to P wave. Also, QT interval is longer than PR interval. Thus, the size of context window is set to 200 ms and its onset is kept at 200 ms ahead of R peak. Figure 3 shows the window boundaries M and N for locating T peak.

3.4 Estimating the Boundaries of ECG Wave

As stated earlier, P wave marks the beginning and T wave marks the end of one ECG cycle respectively. On having determined the locations of peak of P wave and peak of T wave, the onset and offset of ECG segments can be estimated. Since P wave is approximately 80 ms in duration, we set the onset of ECG wave as 50 ms off the peak of the P wave. Also, T wave is about 250 ms in duration. Thus, the end of ECG segment is set to 100 ms ahead of T peak. Figure 4 shows the boundary estimation of our proposed method on 2 ECG samples from the record *sel17453* of the QT database.

4 Results and Discussions

The results obtained after quantitative analysis of our model on all the 105 records in the database is discussed here. Though a total of nine sets of annotation files are present in the dataset, we only choose the annotation files *.pu0* since it has the waveform boundary measurements automatically determined for all beats.

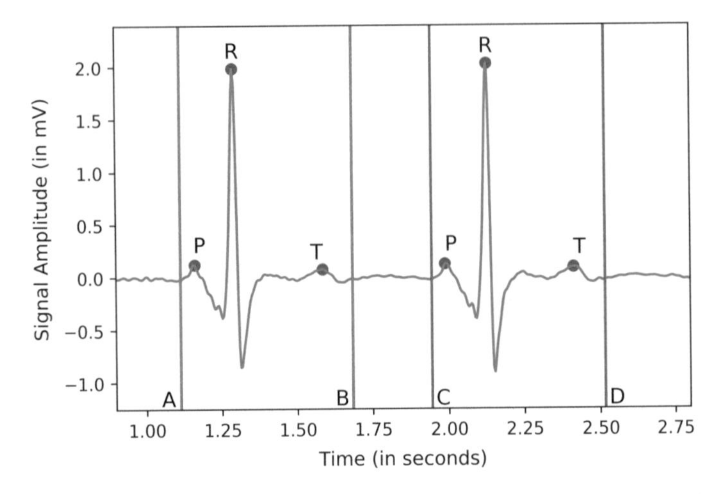

Fig. 4 The points *A* and *B* show the estimated boundary of 1st ECG cycle. The points *C* and *D* are the estimated boundary of second ECG cycle

The results are compared against reference annotations allowing for a 5% tolerance level; that is, a prediction is considered to be correct if its value lies within a range of ±5% of the reference value.

For each P and T, the total number of correct predictions are listed. Moreover, the total number of incorrect predictions, overall accuracy and median accuracy across all the 105 records are determined using our proposed method for the respective peaks and troughs. The results that we have obtained are exceptionally good. A median accuracy of 100% is achieved on 105 records for both the waves on both reference annotations.

Table 1, however, shows that the overall accuracy for T peak detection is lower than the P peak. It is to be mentioned that detecting T peaks is, in fact, an inherently difficult task due to its complex characteristics [15]. Certain algorithms proposed in some works give better accuracy but they have increased computational complexity. In this work, we trade-off accuracy for speed. However, the approximation used in this proposed method can detect peaks of T wave with high accuracy for most classes of ECG signal, but may provide inaccurate results for some abnormal ECG signals.

Table 1 Evaluation results

Wave	Annotation	Total number of correct predictions	Total number of incorrect predictions	Overall accuracy	Median accuracy
P	.pu0	88,652	88	0.9990	**1.0**
T	.pu0	73,133	2093	0.9721	**1.0**

5 Conclusion

In this work, we demonstrate an accurate and efficient method to segment ECG signals by detecting the peaks of P waves and T waves. The information of P and T waves gives an estimated boundary of an ECG cycle. The algorithm has O(n) complexity with respect to size of ECG input data. The proposed technique is shown to be highly robust for a wide class of ECG signals.

References

1. Laguna, P., Mark, R.G., Goldberger, A.L., Moody, G.B.: A database for evaluation of algorithms for measurement of QT and other waveform intervals in the ECG. Comput. Cardiol. **24**, 673–676 (1997)
2. Moody, G., Mark, R.: The impact of the MIT-BIH Arrhythmia database. IEEE Eng. Med. Biol. Mag. **20**, 45–50 (2001)
3. Taddei, A., Distante, G., Emdin, M., Pisani, P., Moody, G., Zeelenberg, C., Marchesi, C.: The European ST-T database: standard for evaluating systems for the analysis of ST-T changes in ambulatory electrocardiography. Eur. Heart J. **13**, 1164–1172 (1992)
4. Pan, J., Tompkins, W.: A real-time QRS detection algorithm. IEEE Trans. Biomed. Eng. **BME-32**, 230–236 (1985)
5. Cuiwei, L., Chongxun, Z., Changfeng, T.: Detection of ECG characteristic points using wavelet transforms. IEEE Trans. Biomed. Eng. **42**, 21–28 (1995)
6. Coast, D., Stern, R., Cano, G., Briller, S.: An approach to cardiac arrhythmia analysis using hidden Markov models. IEEE Trans. Biomed. Eng. **37**, 826–836 (1990)
7. Gritzali, F., Frangakis, G., Papakonstantinou, G.: Detection of the P and T waves in an ECG. Comput. Biomed. Res. **22**, 83–91 (1989)
8. Lagerholm, M., Peterson, C., Braccini, G., Edenbrandt, L., Sornmo, L.: Clustering ECG complexes using Hermite functions and self-organizing maps. IEEE Trans. Biomed. Eng. **47**, 838–848 (2000)
9. Goutas, A., Ferdi, Y., Herbeuval, J., Boudraa, M., Boucheham, B.: Digital fractional order differentiation-based algorithm for P and T-waves detection and delineation. ITBM-RBM **26**, 127–132 (2005)
10. Laguna, P., Jané, R., Caminal, P.: Automatic detection of wave boundaries in multilead ECG signals: validation with the CSE database. Comput. Biomed. Res. **27**, 45–60 (1994)
11. Hoffman, B.F., Cranefield, P.F.: Electrophysiology of the Heart. McGraw-Hill, Blakiston Division (1960)
12. Joshi, A., Tomar, M.: A review paper on analysis of electrocardiograph (ECG) signal for the detection of arrhythmia abnormalities. Int. J. Adv. Res. Electr. Electron. Instrum. Eng. **03**, 12466–12475 (2014)
13. Thakor, N., Zhu, Y.: Applications of adaptive filtering to ECG analysis: noise cancellation and arrhythmia detection. IEEE Trans. Biomed. Eng. **38**, 785–794 (1991)
14. Hamilton, P.: Open source ECG analysis. Comput. Cardiol. IEEE, 101–104 (2002)
15. Elgendi, M., Eskofier, B., Abbott, D.: Fast T wave detection calibrated by clinical knowledge with annotation of P and T waves. Sensors **15**, 17693–17714 (2015)
16. Badilini, F.F.: Method and apparatus for extracting optimum Holter ECG reading. U.S. Patent 8,560,054, issued 15 Oct 2013 (2013)

Removal of Eye-Blink Artifact from EEG Using LDA and Pre-trained RBF Neural Network

Rajdeep Ghosh, Nidul Sinha and Saroj Kumar Biswas

Abstract Electroencephalography (EEG) data are highly susceptible to noise and are frequently corrupted with eye-blink artifacts. Methods based on independent component analysis (ICA) and discrete wavelet transform (DWT) have been used as a standard for removal of such kinds of artifacts. However, these methods often require visual inspection and appropriate thresholding for identifying and removing artifactual components from the EEG signal. The proposed method presents a windowed method, where an LDA classifier is used for identification of artifacts and RBF neural network is used for correcting artifacts. In the present work, we propose a robust and automated method for identification and removal of artifacts from EEG signals, without the need for any visual inspection or threshold selection. Using test data contaminated with eye-blink artifacts, it is observed that our proposed method performs better in identifying and removing artifactual components from EEG data than the existing thresholding methods and does not require the application of ICA for identification of artifacts and can also be applied to any number of channels.

Keywords Artifact removal · Brain–computer interface (BCI) · Electroencephalogram (EEG) · Eye-blink · Linear discriminant analysis (LDA) · Radial basis function neural network (RBFNN)

1 Introduction

Electroencephalogram (EEG) records neural activity of the brain through electrodes placed non-invasively on the scalp. EEG is widely used by neurologists for diagnosing

R. Ghosh (✉) · S. K. Biswas
Computer Science and Engineering Department, NIT Silchar, Silchar 788010, Assam, India
e-mail: rajdeep.publication@gmail.com

S. K. Biswas
e-mail: bissarojkum@yahoo.com

N. Sinha
Electrical Engineering Department, NIT Silchar, Silchar 788010, Assam, India
e-mail: nidulsinha@gmail.com

A. Elçi et al. (eds.), *Smart Computing Paradigms: New Progresses and Challenges*, Advances in Intelligent Systems and Computing 766,
https://doi.org/10.1007/978-981-13-9683-0_23

various disorders of the brain, like epilepsy, Alzheimer's disease, and so on. Recently developed paradigm of brain–computer interface is also witnessing an increased use of EEG devices for the development of neuro-prosthetic devices, to facilitate paralyzed people in movement. However, EEG signals are often contaminated by artifacts of both biological and environmental origin [1–2]. Biological artifacts are the artifacts originating inside the human body; the examples include cardiac, ocular, and muscular activities which are not cerebral in origin. On the other hand, environmental artifacts originate externally and are not dependent on any biological phenomenon. Examples include artifacts arising due to movement of electrodes, power line interference, and so on. Artifacts degrade the quality of EEG signals and obscure information present in the EEG data. Hence artifact removal methods should remove artifacts and preserve as much information as possible. Moreover, conventional methods to remove EEG artifacts use linear filters or regressions, based on the time of occurrence or the frequency range of target artifacts [1]. However, filtering in either time or frequency domain can result in the loss of cerebral activity due to the overlap in the spectrum of the neurological activity and artifacts [3, 4]. Wavelet-based transforms have been used extensively in the literature to remove target artifacts while preserving the neural information of the EEG signal, both in time and frequency domains. On the other hand, ICA has become a popular method in removing artifacts from EEG data [5–8]. The drawback of ICA is that removing the components containing artifacts also removes some underlying EEG data. In recent years, artifact removal using a combination of wavelet and ICA has shown promising results in practical applications. Borna et al. used high-speed eye tracker to detect the eye movement activity, and used adaptive filters to correct the artifacts but required processing of the tracker data along with the selection of appropriate parameters for filtering the data [9]. Qinglin et al. used a combination of DWT and ANC to correct ocular artifacts online. They applied DWT to the contaminated EEG to derive a reference signal for subsequent ANC processing. The method achieved a good performance in cleaning the EEG but the execution time increased as the number of samples increased [10]. However, hybrid methods based on ICA-wavelet require complex computations [8, 11]. Adaptive filtering-based methods have also been used for real-time processing of the signals, but require an extra EOG channel for reference, and are generally not suitable for portable applications having a limited number of electrodes [12–14]. On the basis of earlier works, the objectives of the proposed work are as follows:

- Identification of artifact from EEG.
- Cleaning of the identified artifacts.

In view of the above objectives, the proposed method presents a novel combination of LDA and RBF neural network for identification and correction of EEG artifact, respectively. Section 2 describes the methods used in the proposed work. Section 3 describes the proposed work. Section 4 analyzes the results. Section 5 concludes the paper by describing the works that could be extended in future.

2 Materials and Methods

In the proposed work we have used LDA for identification of artifacts, and a pre-trained RBF has been used for correction of the artifacts. We describe them in the subsequent sections.

2.1 *Linear Discriminant Analysis*

Linear discriminant analysis is a generalization of Fisher's linear discriminant analysis and is used for classifying two or more objects by finding a linear combination of features. A d-dimensional vector x is transformed to a scalar z as:

$$Z = W^T X. \tag{1}$$

The LDA gives an optimal projection W so that the distribution of Z is easy to discriminate. The creation of LDA is given by:

$$\max J(w) = \frac{(m_1 - m_2)^2}{s_1^2 + s_2^2} \tag{2}$$

where m_1 and m_2 denote averages for $Z_n \in$ class1 and $Z_n \in$ class2, respectively. s_1^2 and s_2^2 denote the scatters for $Z_n \in$ class1 and $Z_n \in$ class2, respectively. In the proposed work, LDA has been used to classify whether an EEG segment contains eye-blink.

2.2 *Radial Basis Function Neural Network*

A Radial Basis Function Neural Network is a type of neural network consisting of three layers: an input layer, a hidden layer, and an output layer. It uses radial basis functions as the activation function. The activation function computes the distance of the input vector from the center of the hidden units. The RBFNN attempts to create a linear combination of the radial basis functions to approximate the nonlinear functions. The input can be modeled as a vector of real numbers $x \in R^n$. The output of the network is then a scalar function of the input vector, $\varphi: R^n \rightarrow R$. and is given by:

$$\varphi(x) = \sum_{i=1}^{N} a_i \exp\left[-\beta \|x - c_i\|^2\right]. \tag{3}$$

where N is the number of neurons in the hidden layer, c_i is the center vector for neuron i, and a_i is the weight of the hidden neuron i to the output neuron. There are two methods for selection of centers, one is the fixed center approach where the centers are selected at random, and the other is based on clustering where the data points are clustered to select the center of the cluster as the center of the RBF. In the proposed work we have used the fixed centering approach for selection of the centers. The RBF neural network has been used to correct eye-blink from EEG data. Parameters such as β, a_i, c_i are selected in a manner to optimize the approximation [15].

3 Proposed Methodology

The current section describes the overall methodology used to correct the eye-blink. The overall artifact removal task can be subdivided into two tasks, that is, identification of artifacts followed by correction of the artifact.

3.1 Data Description

In the proposed method we have taken a windowed approach for systematically correcting the eye-blinks. We select a sliding window size of 0.45 s because a typical eye-blink duration ranges between 0.3 and 0.4 s. The data were recorded with a standard medical-grade EEG equipment, with a sampling frequency of 512 Hz for 2 min from 10 subjects where the subjects were shown a video during recording. After recording, the EEG was segmented into a size of 0.45 s containing EEG corrupted with eye-blink and uncorrupted EEG. The selection was done based on a visual inspection by an expert. Thus 1000 data segments were formed for both of the classes, namely: contaminated EEG and uncontaminated EEG. These segments were used for training the RBFNN and LDA classifier. After training, we selected individual channels of the EEG data corrupted with eye-blink for correction. The procedure for correction is described in the next section.

3.2 Overview of the Overall Method

Figure 1 describes the overall system. EEG data corrupted with eye-blink are selected for correction. A sliding window of 0.45 s is applied on the data and each window is processed in accordance with the model as described below. The stages of the model are as follows:

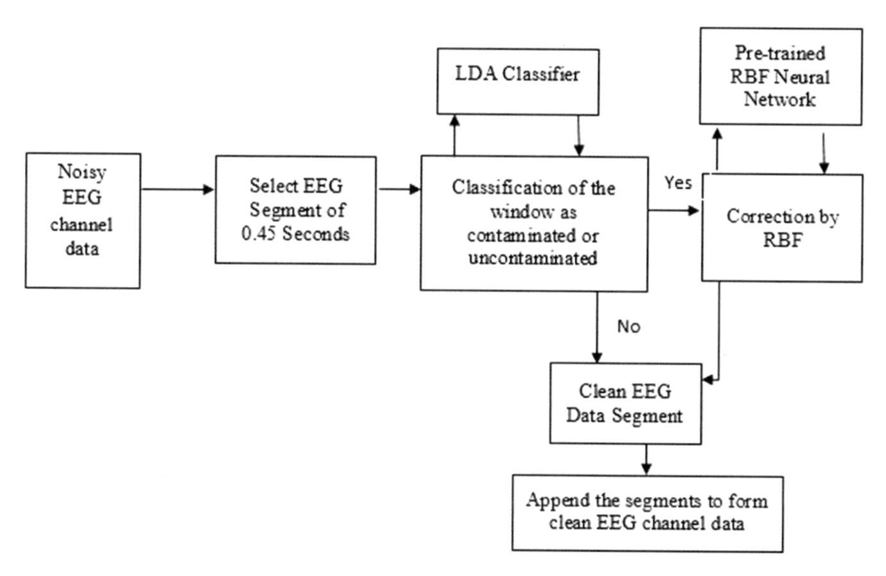

Fig. 1 Overall description of the proposed system

(a) The signal within the window is fed to the pre-trained LDA, and the LDA classifies whether the signal is an artifact or clean EEG. If it classifies the signal as artifact, it is fed to RBF for correction. On the other hand, if the signal is marked as non-artifact, then the window is slid forward.
(b) Signal window marked as artifact is then fed to the pre-trained RBF. The RBF corrects the specific window and gives the clean EEG signal.

After correction the window is slid forward. In the proposed work 50% overlap between the subsequent windows is found.

3.3 Identification of the Eye-Blink Artifact

In the proposed work, LDA has been used for identification of the eye-blink artifact. Figure 2a shows an uncontaminated EEG and an EEG contaminated with the eye-blink. From the characteristics of the EEG data, four features, namely variance, kurtosis, wavelet entropy, and peak-to-peak amplitude, have been used for classifying non-contaminated EEG data from contaminated EEG data. From Fig. 2b it is evident that the selected features discriminate contaminated EEG from non-contaminated EEG. The LDA classifier is trained with the features extracted from the clean EEG data.

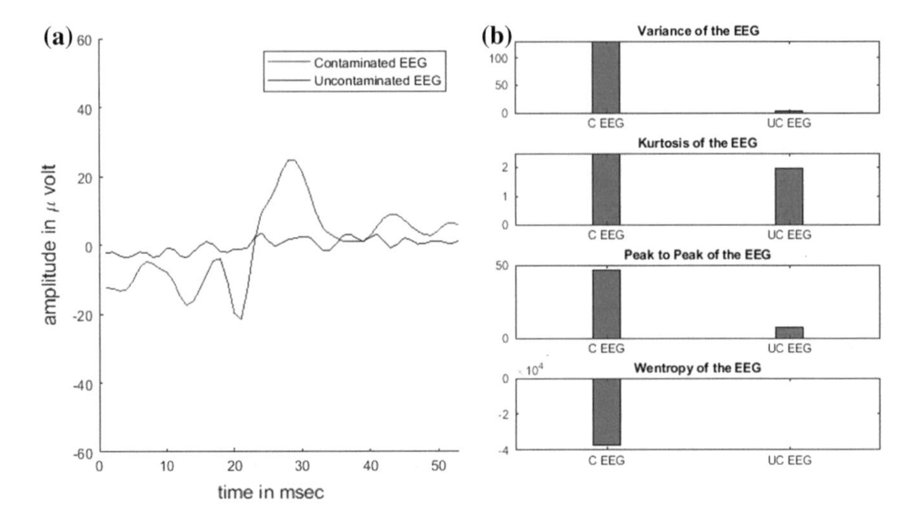

Fig. 2 **a** Sample EEG data segment (Red—EEG contaminated by eye-blink, blue—Uncontaminated EEG data segment). **b** Various feature values for contaminated and uncontaminated EEG (C EEG—Contaminated EEG, UC EEG—Uncontaminated EEG)

3.4 Correction of the Eye-Blink Artifact

In the proposed work, RBFNN has been used for correction of the eye-blink. RBFNN has been used as an eye-blink correction mechanism. The RBFNN has been used to remove noise which is the eye-blink in the present case. The RBFNN has been pre-trained with 1000 samples of artifact as input and 1000 samples of clean EEG as the target. The RBFNN is trained with 50 hidden units

4 Results

The overall method has been implemented using MATLAB on a PC with Intel i7 processor and 4 GB RAM. As mentioned earlier, an LDA classifier has been used to classify the contaminated EEG data from the non-contaminated EEG data. The performance of the classifier is shown in Table 1. We evaluate the classifier with respect to accuracy, sensitivity, and specificity. The classifier is trained with features

Table 1 Performance of the LDA classifier

Performance metrics	Percentage (%)
Accuracy	98.4
Sensitivity	99.1
Specificity	97.2

extracted from 1000 EEG segments from both classes. For testing, 100 EEG data segments from both the classes have been used. After identification of artifact by the classifier, the EEG data segment is fed to the pre-trained RBFNN. The RBFNN generates the clean EEG for the given window. It does not change any segment which has not been identified as artifact and hence preserves good correlation in frequency and time. Figure 3 shows a signal before and after correction of artifacts. Figure 4 shows how the proposed method corrects only the identified region and does not change the EEG anywhere else.

For evaluating the performance of the removal method, we have adopted mean absolute error (MAE) [10] as a measure of the signal quality after the application of the above model. MAE calculates the frequency difference between noisy data and corrected data. It is defined for each of the four bands of EEG data, that is, delta band ($0 < f < 4$ Hz), theta band ($4 \leq f < 8$ Hz), alpha band ($8 \leq f < 13$ Hz), and gamma band ($13 \leq f < 30$ Hz). MAE can be computed as:

$$MAE = \sum_{n=i}^{j} \frac{|P'(n) - P(n)|}{j - i}. \tag{4}$$

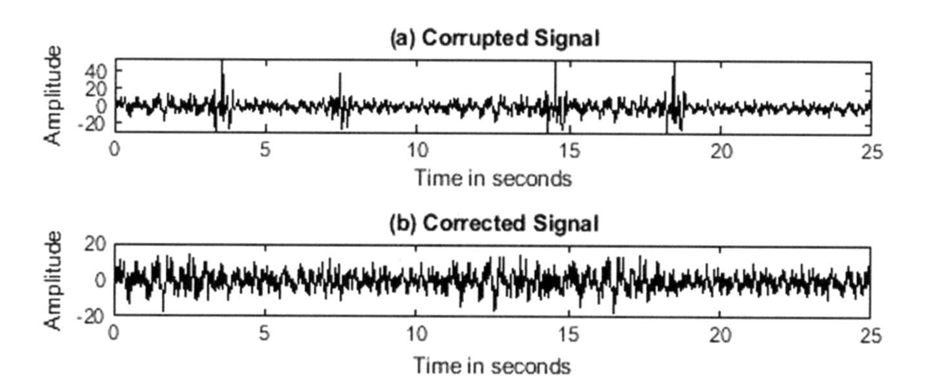

Fig. 3 **a** EEG signal corrupted with eye-blinks, **b** EEG signal corrected by the proposed method

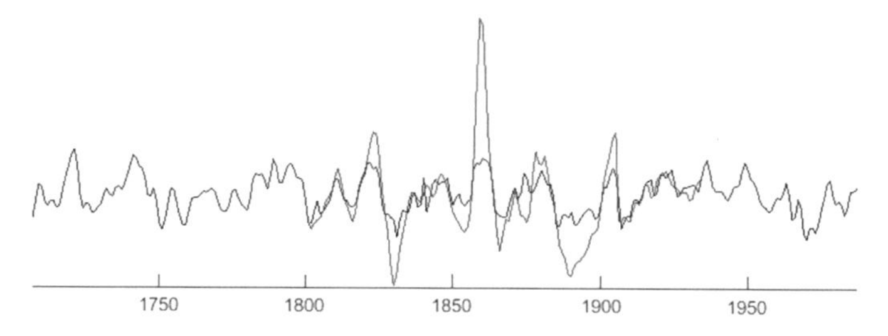

Fig. 4 Correction by the proposed method (red line denotes the corruption by artifact and black line denotes EEG data after artifact removal by the proposed method)

Table 2 MAE of the proposed system in comparison to other methods

MAE	Proposed method	Existing method [10]
Delta band	0.102	0.254
Theta band	0.094	0.201
Alpha band	0.073	0.094
Gamma band	0.012	0.014

where P′(n) represents the power spectral density after artifact removal and P(n) denotes the power spectral density before removal. i and j define the frequency range. Table 2 compares the MAE of the proposed method and an existing online method for artifact removal [10], and it is clear that the proposed system is better in removing artifacts than the existing methods.

5 Conclusion and Future Scope

In the present work we have implemented a hybrid system for systematic removal of eye-blink from EEG data. LDA was used to identify artifacts followed by RBFNN to remove artifact. It has been observed that LDA has a good accuracy in recognizing segments corrupted with eye-blinks from the EEG data. The proposed method only corrects the segment that has been marked as artifact but methods based on wavelets and adaptive filters correct the entire EEG data, resulting in loss of information from the data. The proposed system minimally alters the data and preserves as much information as possible, which was evident in Table 2, as the proposed system has minimal MAE in all the frequency bands. The proposed work can be extended to remove other kinds of EEG artifacts as well, such as cardiac artifacts, muscular artifacts, and so on for an overall artifact removal from EEG.

Acknowledgements This work was supported by the Ministry of Communications & Information Technology No. 13(13)/2014-CC&BT, Government of India.

Declaration Data were recorded under the project entitled “Analysis of Brain Waves and Development of Intelligent Model for Silent Speech Recognition” aimed at analyzing brainwaves to recognize silent speech from EEG at NIT Silchar. Adequate permission was obtained from the authority and informed consent was obtained from the subjects after explaining the details and possible outcomes of the study.

References

1. Selvan, S., Srinivasan, R.: Removal of ocular artifacts from EEG using an efficient neural network based adaptive filtering technique. IEEE Signal Process. Lett. **6**(12), 330–332 (1999)
2. Jervis, B.W., Garcia, M.I.B., Thomlinson, M., Lopez, J.M.L.: Online removal of ocular artefacts from the electroencephalogram. IEE Proc.-Sci. Meas. Technol. **151**(1), 47–51 (2004)
3. Vigon, L., Saatchi, M.R., Mayhew, J. E.W., and Fernandes, R.: Quantitative evaluation of techniques for ocular artefact filtering of EEG waveforms. EE Proc.-Sci. Meas. Technol. **147**(5), 219–228 (2000)
4. Mehrkanoon, S., Moghavvemi, M., Fariborzi, H.: Real time ocular and facial muscle artifacts removal from EEG signals using LMS adaptive algorithm. ICIAS 2007. In: IEEE International Conference on Intelligent and Advanced Systems, Kuala Lumpur, Malaysia, pp. 1245–1250 (2007)
5. Wu, X., Ye, Z.: Online Infomax algorithm and its application to remove EEG artifacts. In: Proceedings of the 2005 IEEE Engineering in Medicine and Biology 27th Annual Conference, Shanghai, China, pp. 4203–4207 (2005)
6. Guerrero-Mosquera, C., Vazquez, A.N.: Automatic removal of ocular artifacts from EEG data using adaptive filtering and independent component analysis. In: 17th European Signal Processing Conference, EUSIPCO 2009, Glasgow, Scotland, pp. 2317–2321 (2009)
7. Szibbo, D., Luo, A., Sullivan, T.J.: Removal of blink artifacts in single channel EEG. In: 34th Annual International Conference of the IEEE EMBS San Diego, California USA, pp. 3511–3514 (2012)
8. Liao, J.C., Fang, W.C.: An ICA-based automatic eye blink artifact eliminator for real-time multi-channel EEG applications. In: ICCE 2013, IEEE international conference on consumer electronics, Las Vegas, NV, USA, pp. 532–535 (2013)
9. Noureddin, B., Lawrence, P.D., Birch, G.E.: Online removal of eye movement and blink EEG artifacts using a high-speed eye tracker. IEEE Trans. Biomed. Eng. **59**(8), 2103–2110 (2012)
10. Zhao, Q., Hu, B., Shi, Y., Li, Y., Moore, P., Sun, M., Peng, H.: Automatic identification and removal of ocular artifacts in EEG—improved adaptive predictor filtering for portable applications. IEEE Trans. Nanobiosci. **13**(2), 109–117 (2014)
11. Sai, C.Y., Mokhtar, N., Arof, H., Cumming, P., Iwahashi, M.: Automated classification and removal of EEG artifacts with SVM and wavelet-ICA. IEEE J. Biomed. Health Inform. **22**(3), 664–670 (2018)
12. Kumar, P.S., Arumuganathan, R., Sivakumar, K., Vimal, C.: Removal of ocular artifacts in the EEG through wavelet transform without using an EOG reference channel. Int. J. Open Problems Compt. Math. **1**(3), 188–200 (2008)
13. Shahabi, H., Moghimi, S., Zamiri-Jafarian, H.: EEG eye blink artifact removal by EOG modeling and Kalman filter. In: BMEI 2012, 5th International Conference on Bio-Medical Engineering and Informatics, Chongqing, China, pp. 496–500 (2012)
14. Pereira, L.F., Patil, S.A., Mahadeshwar, C.D., Mishra, I., D'Souza, L.: Artifact removal from EEG using ANFIS-GA. In: IC-GET 2016, Online International Conference on Green Engineering and Technologies, Coimbatore, India, (2016)
15. Mateo, J., Torres, A.M., García, M.A.: Eye interference reduction in electro-encephalogram recordings using a radial basic function. IET Signal Process. **7**(7), 565–576 (2013)

Innovative Soft Computing Approaches for Pattern Recognition Applications

Skin Cancer Detection Using Advanced Imaging Techniques

Shivani Pal and M. Monica Subashini

Abstract Skin cancer exists in various forms like Melanoma, Basal and Squamous Cell Carcinoma among which Melanoma is the most hazardous and unpredictable. In this paper, we implement an image processing technique for the early detection of Melanoma Skin Cancer using MATLAB which is easy for use as well as detection of Melanoma skin cancer. There are two stages of detection, the first stage is a simple questionnaire which consists of all the common symptoms faced by melanoma affected person and if the report of the first stage comes out to be positive, the patient can go for the second stage, where the input to the system is the lesion skin image. Further, K-means segmentation is used to segment the images followed by feature extraction. The fetched parameter values are, therefore, used to determine whether the particular patient is suffering from skin cancer or not.

Keywords Skin image · Image processing · Segmentation · Feature extraction

1 Introduction

In skin health, diagnostics is the way towards perceiving a skin texture or surface by its signs, indications and the consequences of various conclusion methodologies. Skin development is an unsafe tumour that creates in the skin cells and records for more than 50% of all malignancies. More than 1 million Americans were diagnosed in 2007 to have non-melanoma skin development, and 59,940 were diagnosed to have melanoma, according to the American Growth Society. Fortunately, skin tumours (basal cell and squamous cell carcinoma, and debilitating melanoma) are remarkable in youths. Right when melanomas happen, they normally rise up out of pigmented nevi (moles) that are extensive (estimation more noticeable than 6 mm), asymmetric, with irregular outskirts and hue. Bleeding, itching and a mass under the skin are

S. Pal · M. Monica Subashini (✉)
School of Electrical Engineering, VIT University, Vellore, India
e-mail: monicasubashini.m@vit.ac.in

S. Pal
e-mail: shivani.pal2015@vit.ac.in

A. Elçi et al. (eds.), *Smart Computing Paradigms: New Progresses and Challenges*, Advances in Intelligent Systems and Computing 766,
https://doi.org/10.1007/978-981-13-9683-0_25

different indications of cancerous change. Skin Malignancy Detection Framework is the system to recognize and see skin malady signs and break down melanoma in starting circumstances. Skin Growth Detection Framework will help save heaps of specialist's chance and could break down more exact.

1.1 Literature Survey

Image processing techniques give a creative arm to classify cancer from the images. Lately, neural network strategies are also used to apprehend the most cancers to gather capable comes about. Some of them are discussed here. Azadeh et al. [1] completed a study in view of skin tumour recognition for the early discovery of skin growth. Sonali et al. [2] actualized a straightforward technique for the recognition of skin growth. Lau et al. [3] have portrayed skin disease grouping framework and the relationship of skin tumour picture utilizing distinctive sorts of the neural system. Abdul et al. [4] have exhorted a system for early discovery of Skin Cancer utilizing Artificial Neural Network. Jianli et al. [5], examined skin growth, utilizing the hereditary neural system and standard BP neural system. Sheha et al. [6, 7] has composed a computerized technique for melanoma determination, which is connected to an arrangement of dermoscopy pictures. Sigurdur et al. [8] have planned skin tumour arrangement instrument in light of vitro Raman spectroscopy, utilizing nonlinear neural system classifier. Ramteke et al. [9] have clarified the over a wide span of time advances for skin malignancy location alongside their effective instruments. Mahmoud et al. [10] depicted two cross-breed methods for the arrangement of the skin pictures. Amelard et al. [11] have depicted high-level intuitive features (HLIF), which measure fringe inconsistency of skin sore pictures got from standard cameras. Yuan et al. [12] built up a choice decision support network for skin disease determination. Ogorzałek et al. [13] have examined PC helped systems and picture preparing techniques, which can be utilized for picture filtering, feature extraction and pattern recognization in the chose skin pictures. Maglogiannis et al. [14] have evaluated the best in class in frameworks by first exhibiting the establishment, the visual highlights utilized for skin injury classification.

2 Materials and Methods

As stated above, there are two stages of detection as shown in Fig. 1, the first stage consists of a simple question–answer round in which all the basic physical symptoms are analysed and if the report of the first stage comes positive then the patient is recommended to go for second stage detection wherein the proposed approach for Skin Cancer Detection utilizing Image Processing is as shown in the block diagram. The input for the framework is the picture of the skin lesion [15, 16] which is suspected to be a melanoma sore. This picture is then preprocessed to improve the picture

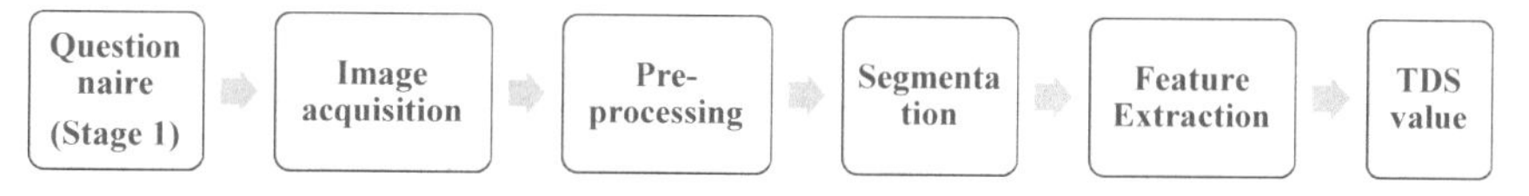

Fig. 1 Methodology

quality. The programmed thresholding procedure and edge detection are utilized for image segmentation. The segmented image is given to the feature extraction block, which comprises of lesion region analysis for its geometrical features and ABCD features. The geometrical features are proposed since they are the most unmistakable features of the skin cancer lesion. Lastly, the TDS esteem is calculated and the picture is delegated either carcinogenic or non-malignant.

2.1 Questionnaire

A set of questions were prepared based on the symptoms, which would reduce the analysis time in the diagnosis procedure. The questionnaire is based on queries regarding the existing mole in terms of size, colour, shape and borders. The tool allows the user to feel comfortable which is very much necessary in the medical environment. If, for more than one out of all questions, the patient replies with a yes then he/she is suspected to have skin cancer and hence MATLAB directs him/her to take second level test otherwise the stage one test is considered Negative.

2.2 Image Acquisition and Image Preprocessing

Image acquisition in image processing can be extensively characterized as the activity of recovering an image from some source. The input image [17] given to the system can be taken in any lighting condition or by using any camera such as a mobile camera. Hence, it needs to be preprocessed. Here, the preprocessing includes the image conversion (RGB to Grey), contrast–brightness adjustment and edge detection to compensate for the non-uniform illumination.

2.3 Image Segmentation and Feature Extraction

K-means is a least-squares partitioning technique which divides a collection of objects into K groups. The K-means is used to segment the image into three clusters—corresponding to two scripts and background, respectively. For each additional script, one more cluster is added. Here, each feature is assigned a different weight,

Table 1 Weight multiply factor

Criterion	Score	Multiply factor
Asymmetry	0–2	1.3
Border	0–8	0.1
Colour	1–6	0.5
Dermoscopic structure	1–5	0.5

which is calculated based on the feature importance. Then edge detection is applied to further segmentation portion. The main requirement for extracting the features is that the lesion must be separated from the surrounding normal skin area and also a type of dimensional reduction that efficiently represents interesting parts of an image as a compact feature vector.

2.4 TDS Calculation

The TDS Index is calculated using the equation below:

$$TDS = 1.3a + 0.1b + 0.5c + 0.5 \quad (1)$$

If the TDS Index is less than 4.75, it is benign (non-cancerous) skin lesion. If the TDS Index is greater than 4.75, and less than 5.45, it is a suspicious case of skin lesion. If the TDS Index is greater than 5.45, it is malignant melanoma (cancerous) skin lesion (Table 1).

3 Implementation

MATLAB R2015a is used for the implementation of the project.

3.1 STOLZ Algorithm

Step 1: Convert the segmented image into a binary image as shown in Fig. 2.

Step 2: Calculate asymmetry, border irregularity of the binary image.

Step 3: Calculate the colour variation and diameter of the lesion.

Step 4: Find the Total Dermatoscopic Value (TDS) by substituting all the values of calculated parameters.

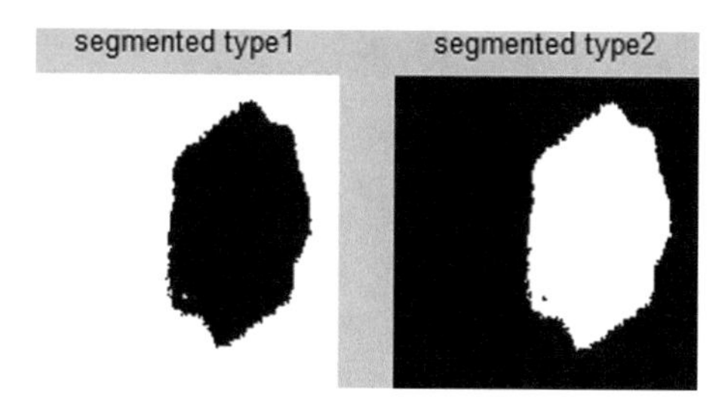

Fig. 2 Segmented binary images

3.1.1 Feature Extraction Techniques

(A) Asymmetry: Cancerous lesions are analysed for symmetry in 0, 1 or 2 axes. It is interpreted as 0—biaxial symmetry, 1—monoaxial symmetry, 2—biaxial asymmetry. To assess the degree of symmetry, Asymmetry Record is figured with the equation below:

$$AI = (\Delta A/A) \times 10 \tag{2}$$

where A = Area of the total Image. ΔA = Area difference between total image and lesion area (Table 2).

(B) Border irregularity: Most of the cancerous lesions are notched and blurred. Its value ranges between 0 and 8. Border irregularity can be determined using the formula below:

$$B = [(perimeter)^\wedge 2/4\pi A] \tag{3}$$

where A = lesion Area, P = Perimeter of lesion boundary.

(C) Colour: Cancerous skin lesion's pigmentation is never uniform. The presence of any 6 known colours must be detected—white, red, light brown, dark brown, slate blue and black. The value ranges in between 0 and 6 as shown in Fig. 3. This value is given to C.

(D) Diameter: The diameter value is said to be 5 if the diameter of the lesion area is more than 6 mm. For different values, the diameter is as much as its original

Table 2 Feature extraction values

Name	Value	Max	Min
Area	[47311;1;1;1;1;1]	47311	1
Perimeter	[1.0208e + 03;0;0;0;0;0]	1.0208e + 03	0
Major Axis	[328.2365;1.1547;1.1547]	328.23	1.1547
Minor Axis	[313.5781;1.1547;1.1547]	313.57	1.1547

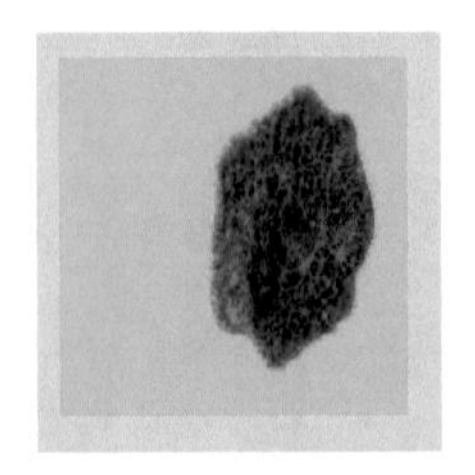

Fig. 3 Pigmented image. *Source* www.medicalnewstoday.com

rounded value. To calculate the diameter, we utilize the 'regionprops' function to get the minor axis length of the lesion area.

4 Discussion and Results

4.1 *Output of Stage 1*

The output of stage 1 gives us the information from the given set of questions, for each of which he has to answer with a Yes or No. When the person answers with a 'Yes' MATLAB displays that 'symptom matched, go for next question' and if he answers with 'No' then MATLAB displays 'symptom didn't match, go for next question'. And at the end, if answers for more than one question are Yes, only then he is advised to take stage 2 detection test otherwise his skin is declared non-cancerous at this stage itself.

4.2 *Output of Second Stage*

The outputs of stage 2 are shown in Figures. The original skin cancer image downloaded from Medical News Today [17], grey image, noisy image and median-filtered image is shown in Fig. 4. In Fig. 5, contrast adjusted image and its histogram displaying its even distributed black and white pixels. In Fig. 6, the binary mask which is segmented out for calculating its feature properties and in Fig. 7, the coloured segmented image for calculating its colour properties is shown. Hence, combining all we calculate the TDS value which help us in classifying the image as cancerous or non-cancerous as shown in Table 3. The TDS value for the above taken image is 7.6152. Hence it is malignant melanoma.

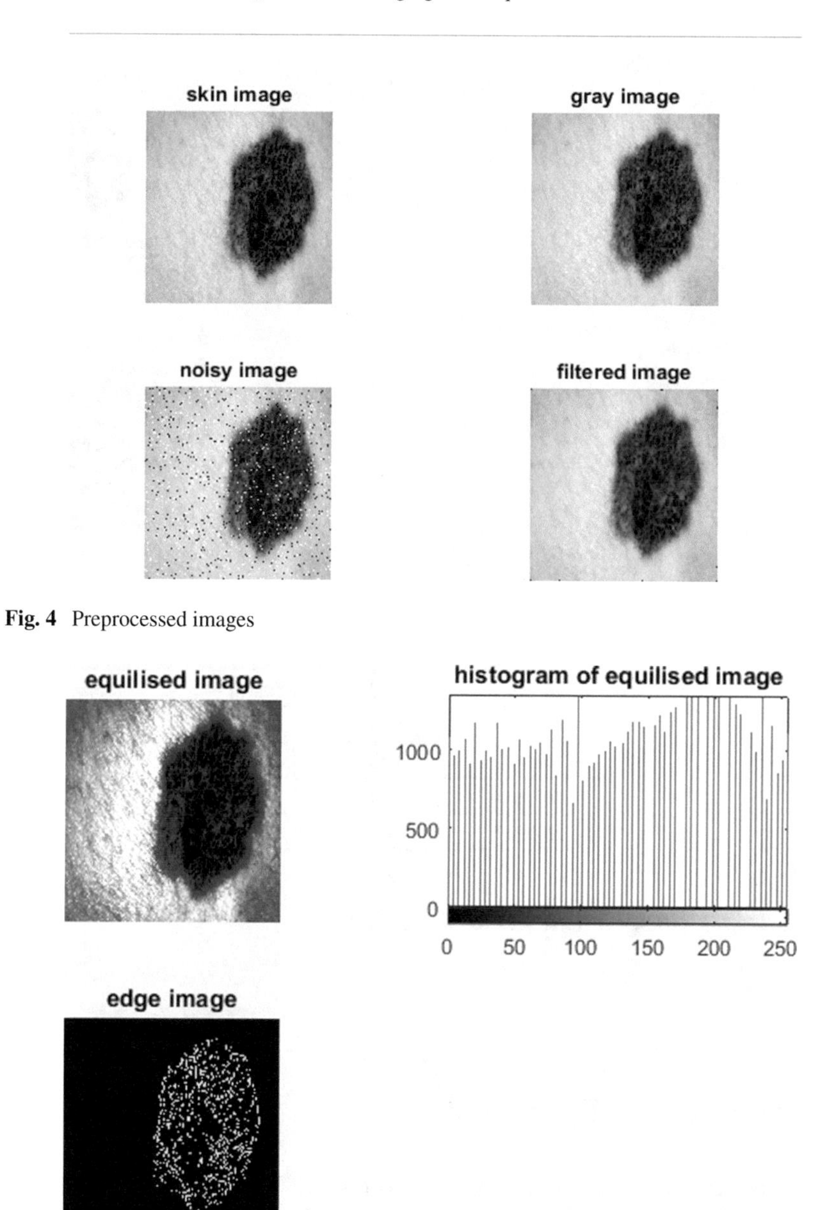

Fig. 4 Preprocessed images

Fig. 5 Histogram equalized and edge image

Fig. 6 Binary mask image

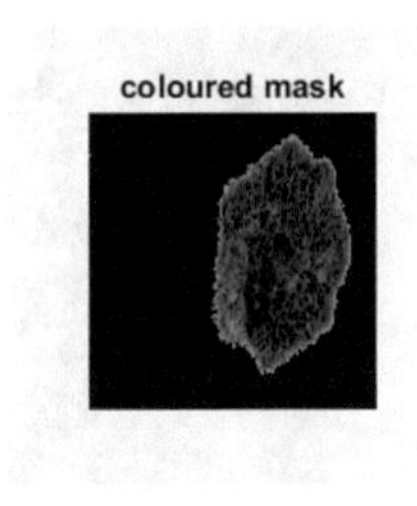

Fig. 7 Coloured mask image

Table 3 TDS values

Sample	TDS	Sample	TDS
Image 1	7.18	Image 6	6.13
Image 2	5.25	Image 7	3.11
Image 3	9.65	Image 8	8.19
Image 4	3.84	Image 9	13.92
Image 5	4.96	Image 10	6.24

4.3 Sample Images and Their TDS Values

See Table 3.

5 Conclusion and Future Scope

From most recent two decades, melanoma skin disease is on the increase. Along these lines, the early location of skin tumour is important. In the event that distinguished at a beginning period, skin tumour can be cured, and much of the time, the treatment is straightforward and includes extraction of the sore. Also, at a beginning period, skin disease is exceptionally efficient to treat, while at a late stage, melanoma skin disease gets extremely hard to cure and furthermore costs an exceptionally extensive sum for the treatment. Four examinations are done when skin sore is suspected as melanoma. In the event that the speculated skin sore experience just the three of these, it may be melanoma or not. For this reason, all four measures are considered to choose whether a skin sore is melanoma or not. The best approach to bring down the danger of melanoma is to restrict the introduction to solid daylight and other wellsprings of bright light.

Declaration We certify that no funding has been received for conducting this study and preparation of this manuscript. The database utilized in this study is downloaded from a verified source (Medical News Today) and no patients were involved in this work.

References

1. Hoshyar, A.N., Al-Jumaily, A., Sulaiman, R.: Review on automatic early skin cancer detection. In: 2011 International Conference on Computer Science and Service System (CSSS). IEEE (2011)
2. Jadhav, S.R., Kamat, D.K.: Segmentation based detection of skin cancer. In: IRF International Conference, 20-July-2014
3. Lau, H.T., Al-Jumaily, A.:. Automatically early detection of skin cancer: study based on neural network classification. In: International Conference of Soft Computing and Pattern Recognition, 2009. SOCPAR'09. IEEE (2009)
4. Dr. Jaleel, J.A., Salim, S., Aswin, R.B.: Artificial neural network based detection of skin cancer. Int. J. Advanc. Res. Electron. Instrument. Eng. **1.3** (2012)
5. Jianli, L., Zuo, B.: The segmentation of skin cancer image based on genetic neural network. In: 2009 WRI World Congress on Computer Science and Information Engineering, vol. 5. IEEE (2009)
6. Sheha, M.A., Mai, S.M., Sharawy, A.: Automatic detection of melanoma skin cancer using texture analysis. Int. J. Comput. Appl. **42** (2012)
7. Sadeghi, M., et al.: Detection and analysis of irregular streaks in dermoscopic images of skin lesions. IEEE Trans. Med. Imaging **32.5**:849–861 (2013)
8. Sigurdsson, S., et al.: Detection of skin cancer by classification of Raman spectra. IEEE Trans. Biomed. Eng. **51.10**:1784–1793 (2004)
9. Ramteke, N.S., Jain, S.V.: Analysis of skin cancer using fuzzy and wavelet technique–review & proposed new algorithm. Int. J. Eng. Trends Technol. (IJETT) **4.6** (2013)
10. Elgamal, M.: Automatic skin cancer images classification. Int. J. Advanc. Comput. Sci. Appl. **4.3** (2013)
11. Amelard, R., Wong, A., Clausi, D.A.: Extracting morphological high-level intuitive features (HLIF) for enhancing skin lesion classification. In: 2012 Annual International Conference of the IEEE Engineering in Medicine and Biology Society (EMBC). IEEE (2012)
12. Yuan, X., et al.: 28th Annual International Conference of the IEEE Engineering in Medicine and Biology Society, 2006. EMBS'06. IEEE (2006)
13. Ogorzałek, M.J., et al.: New approaches for computer-assisted skin cancer diagnosis. In: The Third International Symposium on Optimization and Systems Biology, Zhangjiajie, China, Sept. 2009
14. Maglogiannis, I., Doukas, C.: Overview of advanced computer vision systems for skin lesions characterization. IEEE Trans. Informat. Technol. Biomed. **13**(5), 721–733 (2009)
15. http://www.skincancer.org
16. http://www.medicinenet.com
17. Image Source. www.medicalnewstoday.com
18. Haykin, S.: Neural Networks: a Comprehensive Foundation. Prentice Hall PTR (1994)
19. Kopf, Alfred W.: Prevention and early detection of skin cancer/melanoma. Cancer **62**(S1), 1791–1795 (1988)

Auto-associative Neural Network Based Concrete Crack Detection

A. Diana Andrushia and N. Anand

Abstract Crack is an important sign to indicate the health of the concrete structures. It is mandatory to detect the cracks in the concrete structures. This paper presents a method for automatic detection of concrete cracks. Auto-associative neural network is used to detect the cracks. Initially, the necessary features are extracted from the input images which is given to the training algorithm to train the system. The experimental output produces reliable results in terms of training and testing accuracies.

Keywords Concrete crack · Crack detection · Classification · Auto-associative neural network

1 Introduction

Structural health analysis is the key parameter in all types of infrastructures. Concrete cracks are one of indicator to indicate the health condition of infrastructures. It is due to thermal conditions, human usage and damage, aging of structures, etc. It is mandatory to analyze the health condition of the structures because it directly affects the human life. In order to ensure the safety of the human, it is necessary to monitor the concrete cracks. It is very hard to check the cracks for large structures.

Nowadays, the automatic system which detects the concrete cracks and classify the behavior of the cracks. It is possible through the applications of image processing algorithms. The image processing methods which are involving in the detection and classification of concrete cracks are [1] threshold methods, morphological operators based methods, median filtering methods, neural network based methods, texture feature based methods, superpixel algorithm, Dijkstra algorithm, image correlation, transform-based methods, statistical approach, etc. The automatic detection of con-

A. Diana Andrushia (✉)
Department of ECE, Karunya Institute of Technology & Sciences, Coimbatore, India
e-mail: andrushia@gmail.com

N. Anand
Department of Civil Engineering, Karunya Institute of Technology & Sciences, Coimbatore, India
e-mail: davids1612@gmail.com

A. Elçi et al. (eds.), *Smart Computing Paradigms: New Progresses and Challenges*, Advances in Intelligent Systems and Computing 766,
https://doi.org/10.1007/978-981-13-9683-0_26

crete cracks are very useful among construction engineers because it identifies the cracks in a nondestructive way. There are different methods to perform nondestructive testing [2], which are laser testing, infrared testing, ultrasonic testing, etc. In recent years, the automated methods in crack detection are emerging among researchers. So in this paper also the automatic crack detection and classification is performed.

2 Backgrounds

2.1 Crack Detection

Talab et al. [3] proposed a method for concrete crack detection using multiple filtering processes. Three major steps are followed in this method. Initially, the Sobel filter is applied to find the edges of the input image. Second, threshold is applied to categorize the each pixel into fore ground and back ground. Finally, the noises are removed. This method effectively detects the cracks in cement based infrastructure.

Romulo Gonc et al. [4] presented a method for automatic crack detection. In this method, robot which is having single camera is used to find the crack dimensions. The method is validated with real structures. Vertical, horizontal, curve cracks are identified and checked. The crack length and crack width are calculated for each type of real-time cracks. The average error of 8.59% is occurs for all types of cracks.

Cubero Fernandez et al. [5] proposed a pavement crack detection and classification method using heuristic algorithm. Initially, the concrete cracks were identified by using logarithmic transformation, Canny filter, bilateral filter, and morphological filter. A decision tree based on heuristic algorithm is used to classify the cracks. The average of 88% accuracy has obtained in this method.

Zou et al. [6] developed CrackTree for pavement structures. Three stages are addressed in this method. Initially, shadow removal method is adopted to remove the shadows in the pavement. Probability map is constructed to check the crack continuity. The minimum spanning tree is developed from the probability map in order to identify the cracks in the pavements. Oliveira et al. [7] presented a method for automatic crack detection using unsupervised learning algorithms. Images are separated by nonoverlapping blocks. For each image blockwise the cracks are identified and the width of the cracks is also calculated. Real-time guidelines are taken from the Portuguese distress catalog. Many techniques are involved in crack detection and classification. In all methods, preprocessing stage and feature extraction are common. The detection and classification accuracies are varied with respect to the methods.

3 Proposed Method

The proposed methodology for crack detection and classification involves four major parts. Image collection, preprocessing, feature extraction and classification are the important blocks of the proposed method. Figure 1 shows the block diagram of the proposed method.

3.1 Preprocessing

Preprocessing stage is used to remove the noise and uneven illumination. The input images with uneven illumination influence the method to produce the wrong classification. So the preprocessing stage is necessary to denoise the uneven illumination. Spatial filtering and multi-image averaging is used to eliminate the noises and also smooth the input image [8]. It is used to strengthen the information of the input images without any loss. The output of the preprocessing stage is a uniform image with uniform illumination and which are not affecting the crack pixels.

3.2 Feature Extraction

Feature extraction is an important step to detect and classify cracks. Features are used to detect the cracks and non-crack region. The output of the preprocessing stage enhances the feature quality of the input image. It enhances the crack regions from background regions [9]. Many features are selected in the literatures to perform crack detection. Pixel-based feature selection and block based feature selection are

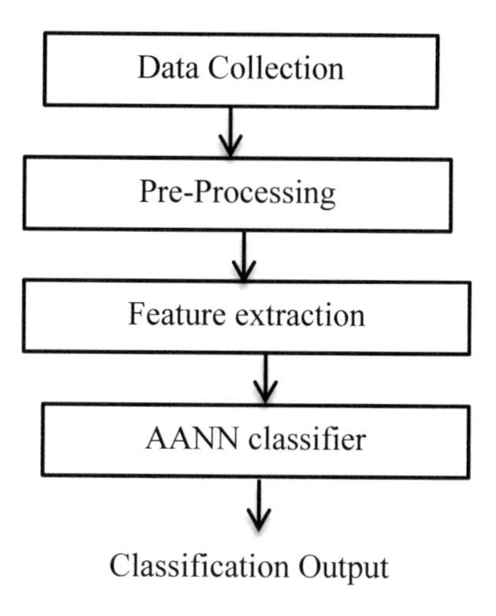

Fig. 1 Block diagram of the proposed methodology

two methods which are adopted in the literature of crack detection. The following features are extracted from the preprocessed image. Less intensity pixel, variance, mean intensity, higher order moments.

Less intensity pixel is identified within each image block. Usually, cracks are having lower intensity values compared to the others.

$$LIP^{(k)} = minC^{(k)}(Y) \tag{1}$$

$C^{(k)}(Y)$ represents the intensity at Y^{th} position in k block

The mean intensity is calculated by

$$\mu^{(k)} = \frac{1}{M^2}\sum C^{(k)}(Y) \tag{2}$$

where M is the size of the block.

The central moment is calculated by

$$\mu_j^{(k)} = \frac{1}{M^2}\sum\left(C^{(k)}(Y) - \mu^{(k)}\right)^j \tag{3}$$

All these features are extracted from the preprocessed input. The features are represented in terms of feature vector.

3.3 *Auto-associative Neural Network (AANN)*

Auto-associative Neural Network is the feed-forward neural network. The input and outputs of AANN are identical. It is used to find the distribution of feature vectors in the feature space. It is one kind of back propagation neural network. It is having linear input and output layers. The hidden layer nodes are nonlinear. The hidden layer nodes are lesser than the input and output nodes. AANN finds the application in the classification tasks in the area of voice recognition, network clustering, signal correction, bioinformatics, etc. In this proposed method also AANN is used to classify the concrete cracks from the feature vector. Initially, the network weights are taken randomly. The training method adopts the input from the given feature vector. The output is produced from the input feature vector. The training algorithm checks the input, output, and target. If the output matches with the target the learning process will stop. The weights are adopted through a recursive algorithm which begins from the output layer and ends with hidden nodes. The training inputs are given to the system. Once the training is finished, the testing images are given to the system for classification.

4 Results and Discussion

The performance of the system is experimented with the real-time crack images. The proposed method is experimented in personal computer with Intel i7 duo core, 8 GB RAM, Windows 10 and MATLAB 2013 version. In this proposed method, 80 images are taken. 60 images are given for training and 20 images are given for testing. All the input images are preprocessed and features are extracted. The feature vector is trained through AANN classifier. The results of AANN is given in Figs. 2 and 3.

The performance metric of the crack detection and classification of concrete structures are training and testing accuracies. The performance of the proposed method is compared to the two state-of-the-art methods (Table 1).

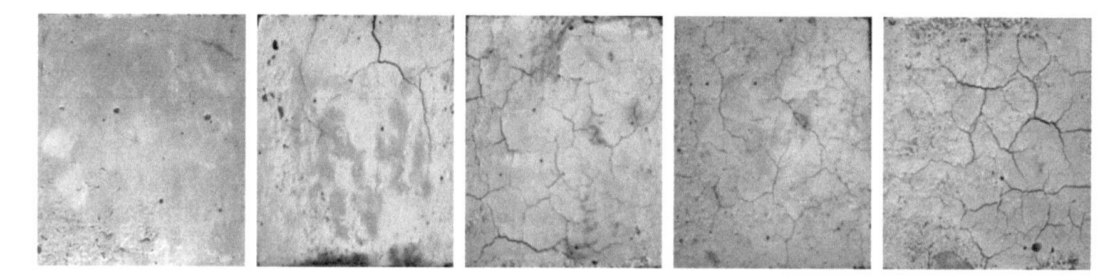

Fig. 2 Concrete crack images

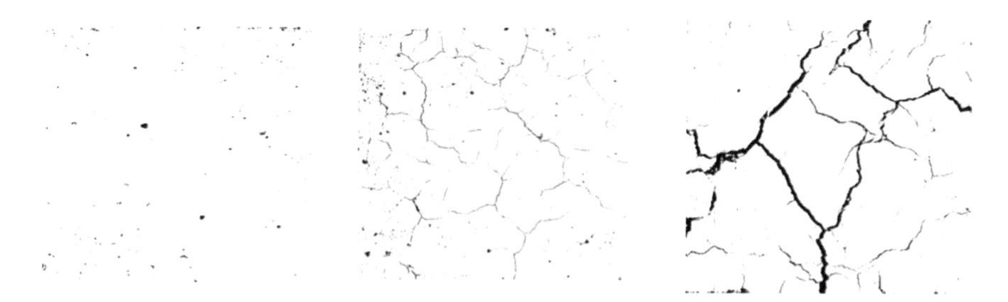

Fig. 3 Detected concrete cracks

Table 1 Classification accuracies of proposed and state-of-the-art methods

Method	Training accuracy	Testing accuracy
KNN	–	87.8
RBF	95.4	90.2
AANN	96.1	91.92

5 Conclusion

In this study, automatic crack detection and classification of concrete cracks using an image processing technique is proposed. AANN is a feed-forward neural network which is used to detect the cracks from the backgrounds. The experimental results of the proposed method are compared with two state-of-the-art methods. The proposed method can be extended by applying meta-heuristic algorithm for getting the optimal feature set.

Acknowledgements The authors wish to acknowledge the Science and Engineering Research Board, Department of Science and Technology of the Indian Government for the financial support (YSS/2015/001196) provided for carrying out this research.

References

1. Mohan, A., Poobal, S.: Crack detection using image processing: a critical review and analysis. Alexandria Eng. J. (2017). https://doi.org/10.1016/j.aej.2017.01.020
2. Fujita, Y., Hamamoto, Y.: A robust automatic crack detection method from noisy concrete surfaces. Mach. Vis. Appl. **22**(2):245–254 (2011)
3. Talab, A.M.A., Huang, Z., Xi, F., HaiMing, L.: Detection crack in image using Otsu method and multiple filtering in image processing techniques. Optik **127**:1030–1033 (2016)
4. Lins, R.G., Givigi, S.N.: Automatic crack detection and measurement based on image analysis. IEEE Trans. Instrum. Meas. **65**(3):583–590 (2016)
5. Cubero-Fernandez, A., Rodriguez-Lozano, F.J., Villatoro, R., Olivares, J., Palomares, J.M.: Efficient pavement crack detection and classification. EURASIP J. Image Video Process **39** (2017). https://doi.org/10.1186/s13640-017-0187-0
6. Zou, Q., Cao, Y., Li, Q., Mao, Q., Wang, S.: CrackTree: automatic crack detection from pavement images. Pattern Recogn. Lett. **33**:227–238 (2012)
7. Oliveira, H., Correia, P.L.: Automatic road crack detection and characterization, IEEE Trans. Intell. Transp. Syst. **14**(1):155–168 (2012)
8. Zhang, W., Zhang, Z., Qi, D., Liu, Y.: Automatic: crack detection and classification method for subway tunnel safety monitoring. Sensors **14**:19307–19328 (2014). https://doi.org/10.3390/s141019307 (2014)
9. Marques, A.S., Correia, P.L.: Automatic Road Pavement Crack Detection Using a Support Vector Machine. Proceedings of Conference on Telecommunications—ConfTele, Castelo Baranco, Portugal (2013)

Image Filter Selection, Denoising and Enhancement Based on Statistical Attributes of Pixel Array

Vihar Kurama and T. Sridevi

Abstract The choice of image filters in computer vision has a significant effect on the image reconstruction and feature extraction. Currently, the most filters are used to enhance images for human consumptions, programmed operations and to reduce the noise, frequency levels in the image. Though it is hard to select an optimal set of filters for a given series of images, in this work, we propose to choose the best assortment of different filters for a given image as the input. By generating the pixel array of the input image, we compute all the image attributes such as RGB colour mean, variance, mean squared error and signal-to-noise ratio values of the input image and then compare with the same, once the filter is applied. We verify the effectiveness of the filters by conducting an empirical evaluation with best-discovered traits.

Keywords Image filters · Convolution kernel · Mean filter · Median filter · SNR

1 Introduction

The main goal of image filter is to remove the unwanted features such as noise, pixel loss, higher frequencies. It is used to modify and reconstruct the image for enhancing an image. For example, you can filter an image to highlight significant features. Image processing operations implemented with filtering include smoothing, blurring, sharpening and edge enhancement. It is a neighbourhood operation in which the output is determined by applying the mathematical operations to the input image. A pixel's neighbourhood is some set of pixels, defined by their locations relative to that pixel.

V. Kurama (✉) · T. Sridevi
Chaitanya Bharathi Institute of Technology, Hyderabad 500075, Telangana, India
e-mail: vihar.kurama@gmail.com

T. Sridevi
e-mail: sridevi_t@cbit.ac.in

A. Elçi et al. (eds.), *Smart Computing Paradigms: New Progresses and Challenges*, Advances in Intelligent Systems and Computing 766,
https://doi.org/10.1007/978-981-13-9683-0_27

This image filtering is achieved by a mathematical operation named convolution. It is a general purpose filter effect for images. We take the input image in a matrix form consisting of all the pixel values, either RGB or grayscale. We apply a mathematical operation to this matrix which comprises the integers. It works by identifying the central or the median pixel by adding the weighted values of all its neighbours together. The output matrix is the new modified filtered image. This filtering is again categorised into two types: linear filtering and non-linear filtering. Linear filters blur sharp edges; destroy lines and other fine details present in the image. It is the filtering in which the value of an output pixel is a linear sequence of neighbourhood values, which can produce blur in the image. The median filter is one of the most popular non-linear filters.

In this work, we use several filters on different test images, to find the best filter for the image, which takes in an input a scalar matrix and outputs a scalar matrix converted to an image. To further validate the effectiveness of the filters, we compare them by visualising the mean squared error and variance graphically and choose the best fit filter. Our extensive experiments also include comparing the values of SNR after several filters are applied to the Input image. All the image features are tabulated in the below sections. The image features include (r, g, b) colour mean, variance, standard deviation, signal-to-noise ratio, size, mean squared error, extrema, count and median.

2 Linear Filters

Linear filters are also classified as convolution filters as they can be expressed using matrix multiplication. In linear filtering, the output will be a linear combination of the values of the pixels in the input pixel's neighbourhood. There are several types of linear filters out of which most commonly used are listed below.

2.1 Mean/Average Filter

The mean/average filter is one of the simplest linear filter, which is used to reduce noise in the image and then enhance. It mainly replaces each pixel value in an image with the mean ('average') value of its neighbours, including itself. Since the shot noise pixel values are often very different from the neighbouring values/pixels, they tend to vary the pixel average calculated by the mean filter significantly. Mathematically, this is achieved by convolving an image by a normalised box filter, with this operation, the noise and higher frequencies in the image will be reduced making the edges in the image blurred.

2.2 Gaussian Blurring

The Gaussian blurring operator is a two-dimensional convolution that is used to blur images to remove noise and detail. This operation is similar to mean filter, but it uses another kernel as a function. The Gaussian kernel gives a smaller weight to pixels further from the middle of the window. The Gaussian distribution in 1-D is represented by

$$G(x) = \frac{1}{\sqrt{2\pi\sigma}} e^{-\frac{x^2}{2\sigma^2}} \tag{1}$$

Gaussian functions have four main properties which are widely applied to computer vision applications. Gaussian functions are rotationally symmetric, which means the smoothness achieved will be the same in all the possible directions. The Gaussian function has a single lobe. This implies that a Gaussian filter smooths by replacing each image pixel with a weighted average of the neighbouring pixels such that the weight given to a neighbour decreases monotonically with distance from the central pixel. The value of, in the above formula which is the standard deviation of the distribution is directly proportional to the image smoothing and blurring. Gaussian filters are separable; this means we can apply these in large-scale efficiently.

2.3 Edge Enhancement Filter

It enhances the edge contrast of an image in an attempt to increase its sharpness. The filter operates by identifying the sharp edges in the image, such as the edge between a subject and a background of a different colour, and improving the image contrast in the area directly nearby the edge. This operation makes the image bright and leaving the edges more sharper. This filter is sometimes called as the unsharp filter, since it subtracts the blurred, unsharp and smooth versions of the picture. Edge enhance filter produces an edge image g(x,y) from an input image f(x, y).

$$g(x, y) = f(x, y) - f_{smooth}(x, y) \tag{2}$$

This formula is used on the image pixel array and can be performed as many times until the edges are sharpened.

2.4 Sobel Filter or Sobel Operator

Sobel filter or Sobel operator is extensively used in edge detection applications for computer vision and image processing. These filters create an image emphasising the edges. The kernel we use is convolved with the original image for the approximations of derivatives for horizontal pixels as well as vertical generating the output image. The Sobel operator is slower to the computer since it has a convolutional kernel. These filters can also compute the gradient for smoothing with normalisation. As the filters for image derivatives in different dimensions are applied, the edges enhance until one threshold point and then convert to the gradient.

2.5 Emboss Filter

Image embossing filter is a technique in which each pixel of an image is replaced either by a shadow or a highlight, depending on the background on the original image. The output image will represent the rate of colour change at each position of the original image. The embossed image has similar effects in both the greyscale image and the colour image. It is due to the claims of the embossing mask coefficients.

3 Non-linear Filters

The common idea in non-linear image filtering is that instead of using the spatial respirator in a convolution method, the mask is used to capture the neighbouring pixel values, and then adjusting devices produce the output pixel. That is, as the mask is changed about the image, the order of the pixels in the windowed section of the image is reconstructed, and the output pixel is produced from these rearranged input pixels.

3.1 The Median Filter

It is a non-linear image filtering technique. This filtering method preserves edges while separating noise. The median filter replaces each pixel entry with the median of neighbouring pixel entries. If the given image has an odd number of entries, then the median is simple to define: it is just the central value of all the entries in the image when arranged numerically. If the image has even number of pixel entries, then there is a chance of having more than one median. Median filters are broadly used for smoothing in computer vision and image processing, as well as in signal processing. The median filter is very efficient in eliminating salt and pepper or impulsive noise while preserving image detail. The output median filter at a moment t is determined as the median of the input values corresponding to the moments adjacent to t:

$$y(t) = median((x(t - T/2), x(t - T1 + 1), , x(t), , x(t + T/2)) \quad (3)$$

3.2 Bilateral Filter

A bilateral filter is a commonly used non-linear filter for images. The output image after the bilateral filter is applied will have fewer noise levels, enhanced edges with smoothness applied. It substitutes the intensity of the individual pixel with a weighted mean of intensity values from neighbouring pixels. To observe the image pixels attribute variation, we compare a few of these before and after the image filter operation is done.

4 Methodology

It is common to enhance images by changing the intensity values of pixels. Most software applications and programming languages for image processing have several alternatives for altering the image by transforming the pixels applying a single function that maps an input RGB or grey value into a new output value.

The first step includes transforming the image into the frequency domain which is an image matrix with the pixel colour values, multiplying it by the frequency filter function and re-transforming the result into the spatial area. The filter function is defined to attenuate some numbers and enhance others. Consider, a simple low pass function is 1 for frequencies smaller than the cut-off frequency and 0 for all others. The following process is to convolve the input image f(i,j) with the filter function h(i,j). Mathematically, this can be written as

$$g(i,j) = h(i,j) * f(i,j) \tag{4}$$

We perform the similar operation with mean, Gaussian, median, edge sharpening, emboss, Sobel and bilateral filters.

5 Searching for the Best Filter

We conduct all our searches for the best image filter concerning the change in the image attributes. Not all filters we apply are 'good'; few are even worse than the original image. At this point, we will need a criterion to choose the right filter. Here are the following factors that we consider for finding the best filter:

1. The image pixel array should not contain null values.
2. No solid nor near-solid coloured areas must occur in the filtered image.
3. Signal-to-noise ratio should decrease proportionately as the image filter radius increases.
4. Pixel array block details should increase based on the mean in at least some area of the image after the filter is applied.

5. Comparing the original variance of pixel array with its relative and absolute variance, if these values are lower than the original value, then we classify them as inferior filters.
6. Variance versus root mean square of pixel array should increase graphically.
7. Complicated image kernels take more time to operate, due to pixel reconstruction. Most used image kernels are of shape (3×3).
8. Denoising the image should change the median value based on the operation we do, which includes removing salt and pepper, Gaussian noise and speckle noise.

6 Image Transitive Comparison

To observe the image pixels attribute variation, we compare a few of these before and after the image filter operation is done.

As per the experimenting, the primary filter variations depend on the Signal-to-Noise Ratio (SNR) formula. It is defined as the ratio of average signal to the standard deviation.

$$SNR = 10\, log_{10} \frac{\sum_{i=1}^{k} S_i^z}{\sum_{i=1}^{k} (S_i^{'} - S_i)^z} \tag{5}$$

The ranges of SNR ratio is varying from −90dB to +15 dB. If the SNR value, for an applied filter, is less than the SNR value of the original image then it classified as a suitable filter. It is just the noise in the picture is decreased, but for few use cases like feature extraction, we need to find the edges by adding noise in the background which leads to increase in the SNR value. The image colour average is an essential factor for choosing the best filter. It is determined by taking the mean of every pixel intensity, for a picture of RGB type or a greyscale, then returning a single tuple. For a filter which is used for enhancement, the mean should increase. The next feature we compare is the standard deviation. The unbiased estimate of the standard deviation for an image is the brightnesses within the block of pixels and we define it as sample standard deviation, and mathematically, it is calculated as (formula). Extremes of the image define one maximum colour intensity of the pixel; with the increase in extrema, we can also predict the increase in the colour mean. If the extremes of an image with a filter are increased, then the filter can be classified as the good (Figs. 1 and 2).

The experimenting was made on a sample test image, and all the filters discussed above are applied, according to the numerical the best filter for image enhancement is edge enhancer with the least SNR value. The best filter to reduce the noise is a median filter as the SNR value is least when compared with other noise reduction filters. We conduct all our searches for the best image filter concerning the change in the image attributes. Not all filters we apply are 'good'; few are even worse than the original image. At this point, we will need a criterion to choose the right filter (Figs. 3 and 4).

Fig. 1 Image(4)—edge enhancer

Fig. 2 Image(5)—Sobel filter

Here are the following factors that we consider for finding the best filter: SNR, mean, extrema, RMS, median, standard deviation and variance. All the values are computed for the sample image using the pixel array with the respective formula for the factor. Few factors return a three index tuple as they are computed concerning the (r, g, b)—red, green and blue intensities in the image. Considering a greyscale or black-and-white image this factor will return a two index tuple (Tables 1 and 2).

6.1 *Standard Deviation Versus Mean Square*

The performance measure of the applied filter for the considered sample image is measured using the standard deviation and mean square error. These values are represented graphically for better interpretation. The graph is generated by computing the continuous change in values of RBG values throughout the image. By taking the RMS value of the tuple on X-axis and standard deviation on Y-axis, the graph is plotted (Figs. 5 and 6).

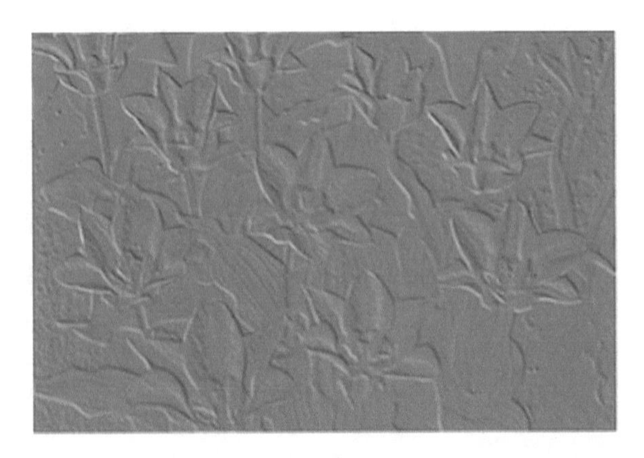

Fig. 3 Image(6)—emboss filter

Fig. 4 Image(7)—emboss filter

Table 1 SNR, mean, extrema and RMS computed by pixel array of several filters

Image	SNR	Mean	Extrema	RMS
1	1.453	[109.48, 104.01, 91.14]	[(0, 253), (3, 245), (0, 245)]	[138.14, 120.07, 110.02]
2	1.457	[109.37, 103.84, 90.95]	[(0, 245), (7, 235), (0, 236)]	[137.94, 119.72, 109.61]
3	1.476	[109.08, 103.58, 90.71]	[(2, 239), (6, 232), (0, 233)]	[137.23, 118.9, 108.63]
4	1.409	[109.8, 104.47, 91.71]	[(0, 255), (0, 255), (0, 255)]	[139.38, 121.98, 112.4]
5	0.47	[14.53, 16.32, 17.42]	[(0, 255), (0, 255), (0, 255)]	[35.33, 38.19, 39.99]
6	7.243	[127.6, 127.72, 127.71]	[(0, 255), (0, 255), (0, 255)]	[128.69, 128.91, 129.07]
7	1.468	[108.98, 103.51, 90.64]	[(1, 242), (7, 233), (0, 234)]	[137.29, 119.01, 108.8]

Table 2 Median, standard deviation and variance

Image	Median	Standard deviation	Variance
1	[69, 102, 76]	[84.23, 59.98, 61.63]	[7094.92, 3597.46, 3798.69]
2	[69, 102, 76]	[84.06, 59.58, 61.19]	[7066.31, 3549.69, 3743.73]
3	[69, 103, 77]	[83.28, 58.37, 59.78]	[6935.58, 3407.15, 3573.15]
4	[69, 100, 75]	[85.85, 62.98, 64.98]	[7370.71, 3966.85, 4222.26]
5	[0, 0, 0]	[32.2, 34.52, 36.0]	[1037.13, 1191.78, 1296.13]
6	[128, 128, 128]	[16.7, 17.49, 18.65]	[278.77, 305.89, 347.76]
7	[69, 103, 77]	[83.5, 58.72, 60.18]	[6971.49, 3448.02, 3622.23]

The performance measure of the applied filter for the considered sample image is measured using the Standard deviation and mean square error. These values are represented graphically for better interpretation. The graph is generated by computing the continuous change in values of RBG values throughout the image. By taking the RMS value of the tuple on X-axis and standard deviation on Y-axis, the graph is plotted. After the evaluation is made on the graph (Fig. 8), we found that edge enhancer filter is best for image enhancement. For noise reduction and smoothing the image, Gaussian filter performed better than bilateral and median filters. There is an enormous variation in the performance between Sobel and emboss filters concerning others as these are used for entirely different purposes. One of best-performed filter for feature extraction is Sobel, but then the performance reduces when we apply the same filter multiple times on the same image. The RMS values are taken for each filter to find the median for the pixel array using the box plots for comparative analysis. The median filter performed the best image reconstruction by visualising the result from (Fig. 9). The root means square of Sobel and emboss are entirely varied as there is the tremendous colour variation for the output values. The median is subjected to zero for this change as the higher and lower boundary pixels are almost nearer to each other (Fig. 7).

6.2 *RGB Colour Intensity*

As soon as the filter is applied, there will be complete pixel reconstruction leading to change in the colour intensities for every block. Based on the filter we choose these values to change, for an enhanced image filter the peak values in RGB tuple will be sharper when represented graphically, for a smooth filter the RBG tuple will be curvier when plotted concerning the width of an image. The below results are the graphs for the RGB colour means for the image before the edge enhancement filter is applied, the enhancement of the colour intensity is defined the peak value, as we compare the figures we find that the sharpness, high's and low's are varied (Figs. 10 and 11).

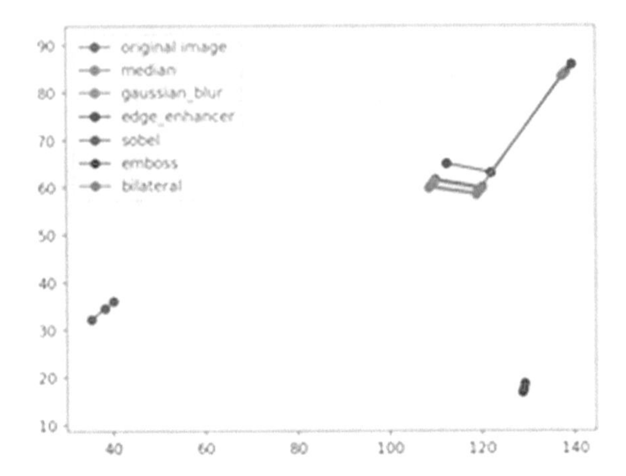

Fig. 5 Filter performance measure standard deviation versus mean square

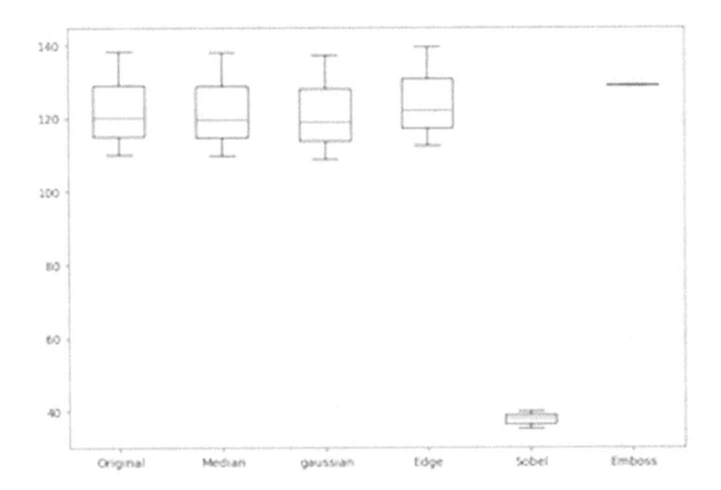

Fig. 6 RMS comparison on filters

6.3 Filters for Medical Fields

We also experimented these filters on medical images, which led to a better enhancement of images. For the initial detection of an anomaly, we sharpened the images by using the edge enhancer with a newer convolution kernel which gave us the following results. Furthermore when feature extraction is processed on the medical image, the Sobel filter is applied for better identification of edges. Though the noise in the

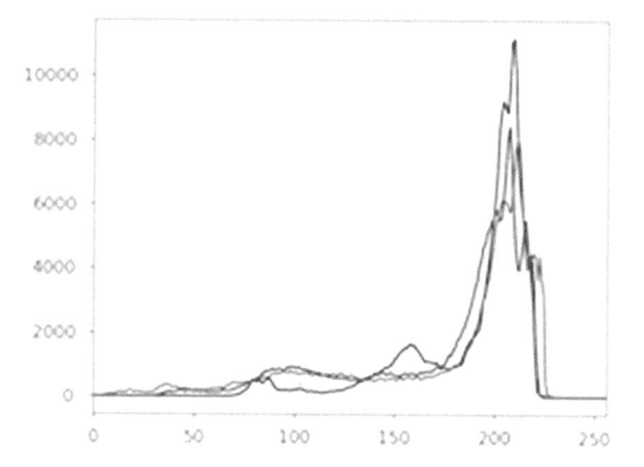

Fig. 7 Before edge enhance

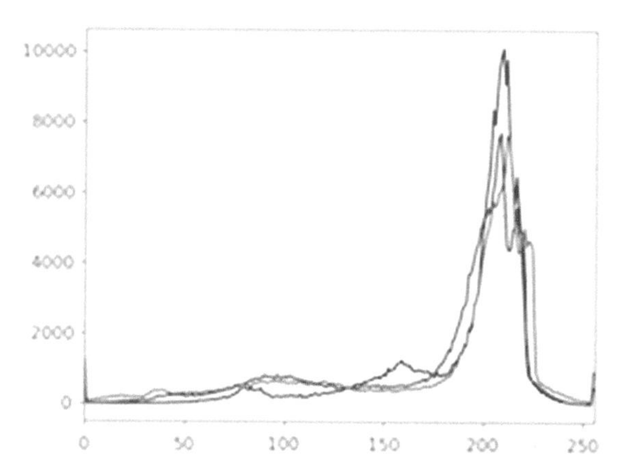

Fig. 8 After edge enhance

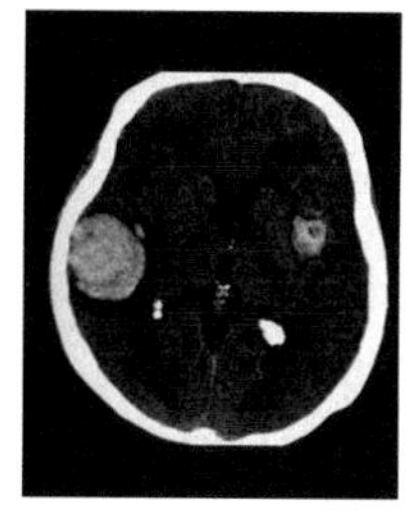

Fig. 9 Original image

image increased after the first iteration, the boundaries are identified. For the second derivative, the Sobel filter had more gradient which led to the threshold value. After the third iteration, the filter got pixelated resulting in losing the features (Figs. 12, 13 and 14).

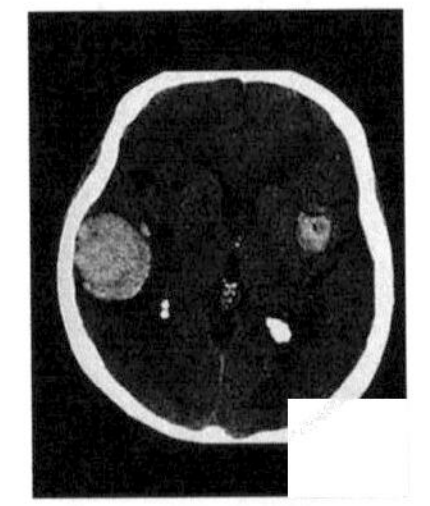

Fig. 10 First enhancement

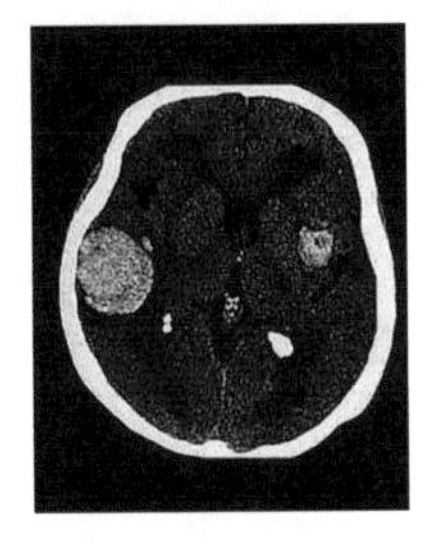

Fig. 11 Second enhancement

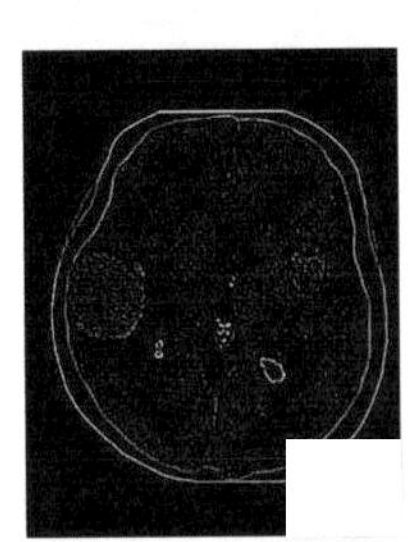

Fig. 12 Sobel filter

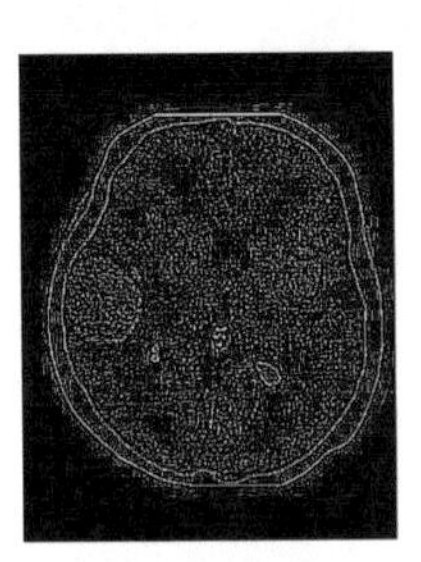

Fig. 13 Sobel gradient

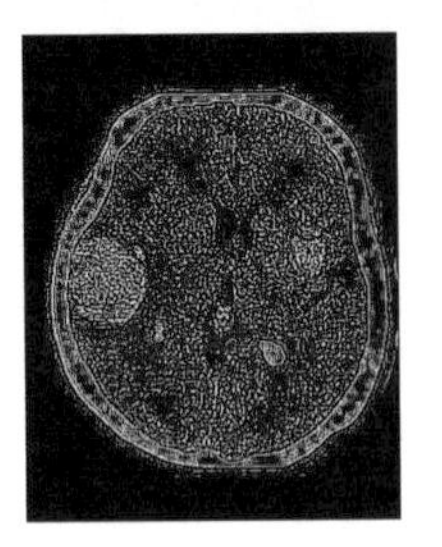

Fig. 14 Sobel enhanced

7 Conclusion

In this work, on finding the efficient image filters, we discovered the novel filters for different images depending on the application. Our experiments use models and hyperparameters that were defined for selection of best image filter like signal-to-noise ratio, root mean square error and variance. The performance for edge enhancer kernel is more, as the value of signal-to-noise reducing is reducing irrespective of the type of the image. We expect additional gains to be obtained when these models and hyperparameters are specifically created with new filters in mind. For other applications like image feature extraction and optimisation, we use Sobel and other edge enhancing filters which gave better noise reduction and highlighted features.

References

1. He, K., Sun, J., Tang, X.: Guided image filtering. IEEE Trans. Pattern Anal. Mach. Intell. **35**(6), 1397–1409 (2013). https://doi.org/10.1109/TPAMI.2012.213
2. Tierney, S., Gao, J., Guo, Y.: Affinity pansharpening and image fusion. In: 2014 International Conference on Digital Image Computing: Techniques and Applications (DICTA), pp. 1–8 (2014)
3. Bougleux, S., Elmoataz, A.: Image smoothing and segmentation by graph regularization. In: Bebis G., Boyle R., Koracin D., Parvin B. (eds.) Advances in Visual Computing. ISVC, Lecture Notes in Computer Science, vol. 3804. Springer, Berlin, Heidelberg (2005)
4. Dravida, S., Woods, J., Shen, W.: A comparison of image filtering algorithms. In: IEEE International Conference on Acoustics, Speech, and Signal Processing (ICASSP '84) (1984)
5. Thivakaran, T.K., Chandrasekaran, R.M.: Nonlinear filter based image denoising using AMF approach. Int. J. Comput. Sci. Inf. Secur. **7**(2) (2010)
6. Gonzalez, R.C., Woods, R.E.: Digital Image Processing. Prentice Hall, Upper Saddle River (2008)
7. Clementel, E., Vandenberghe, S., Karp, J.S., Surti, S.: Comparison of image signal-to-noise ratio and noise equivalent counts in time-of-flight PET. In: IEEE Nuclear Science Symposium and Medical Imaging Conference. Knoxville
8. Hardie, R.C., Barner, K.E., Sarhan, A.: Selection filters for signal restoration. In: Proceedings of the IEEE 1994 National Aerospace and Electronics Conference (NAECON 1994), vol. 2, pp. 827–834 (1994)
9. Eslahi, N., Mahdavinataj, H., Aghagolzadeh, A.: Mixed Gaussian-impulse noise removal from highly corrupted images via adaptive local and nonlocal statistical priors. In: 2015 9th Iranian Conference on Machine Vision and Image Processing (MVIP), pp. 70–75 (2015). ISSN 2166-6784
10. Song, D.-B., Zhang, J.-W., Zhou, J.: Case study for graph signal denoising by graph structure similarity. In: 2017 2nd International Conference on Image Vision and Computing (ICIVC), pp. 847–851 (2017)

Enhanced Security Credentials for Image Steganography Using QR Code

Prajith Kesava Prasad, R. Kalpana Sonika, R. Jenice Aroma and A. Balamurugan

Abstract The process of digital verification is a real needed factor in the current era of the technological world. Especially in the sector of online payment, verifying the authenticity of the user has always been a challenge and considered unsafe. A privileged process and current breakthrough of technology is the fingerprint recognition and the simplification of its usage in current smartphones and its simplicity. The aim of this work is to create a mode of digital verification that emphasis on verifying the user credentials, especially on secure payments. This is achieved by generation of a QR code by the given input fingerprint. The fingerprint link is converted to hexadecimal value and with that value, the QR code is generated. A signature of the user is saved as an image and is embedded with the QR using additive substitution and discrete wavelength transform (DWT) techniques in steganography. When a fingerprint is given as input to the fingerprint-generated and signature-embedded QR, the QR gets decoded and the given fingerprint is compared with the decoded fingerprint, now the similarity measure is compared among them, if the similarity measure is greater than the threshold set, the verification process will be successful else it asks for a digital doodle signature, now this signature is compared with the engrafted signature of the QR and again similarity of both the image is compared using the standard Image Quality Assessment (IQA) matrix.

Keywords Encryption · Secure authentication · Signature · Fingerprint · Biometrics · Steganography

1 Introduction

With the avid increase in online shopping, net banking has proved to become one of the most common methods of payment, with the increase in its popularity, the threats of the security has also become a major consideration, especially with the most common banks authenticate their users credentials by sending an OTP to their

P. K. Prasad (✉) · R. K. Sonika · R. J. Aroma · A. Balamurugan
Sri Krishna College of Technology, Coimbatore, India
e-mail: prajithprasad112@gmail.com

A. Elçi et al. (eds.), *Smart Computing Paradigms: New Progresses and Challenges*, Advances in Intelligent Systems and Computing 766,
https://doi.org/10.1007/978-981-13-9683-0_28

desired mobile number which is the least secure mode, as they can be easily accessed by a third party and could be used to access their bank accounts, online payments, etc.

QR-based steganography, in this approach relative to the subject, the user's fingerprint is converted to hexadecimal and then, with that series of numbers, a QR is generated. Thus when a user is required to input his fingerprint, the QR decodes, and the hexadecimal is converted to an image file and the verification of both of the fingers are compared,i.e., the similarity measurement is taken.

The fingerprint verification technique is implemented in lower complexity and in an easy approach. The verification process initially enhances the fingerprint image, and then after enhancement each line gets enhanced.

The next process of binarizing results in detection of ridges and valleys. The ridges tend to become dark whereas the valleys lighter. To simplify the distance between each line, the pixels are reduced to one pixel at the line, this process is called thining.

Minutiae help in the comparison process of one fingerprint with another fingerprint. Minutiae refer to the unique features or the primary key of a fingerprint using which the similarity could be compared.

2 Proposed Methodology

The proposed methodology uses the database FVC2002 [1] which was publicly available for Second International Competition for Fingerprint Verification Algorithms. It consists of four different databases with 110 fingerprints having eight different impressions. The methodology initially generates a secured QR with a hexadecimal value of users fingerprint link which is stored in the cloud and along with the generated QR the signature of the user is also embedded using steganography, thus creating a secured QR that contains the image and the hexadecimal value of the link which has the fingerprint image. The methodology includes two different phases: (i) Embedding (ii) Extraction.

2.1 Embedding: Generation of Hexadecimal Value for the Given Link

With the given image link, the link is converted to a hexadecimal value, the generated hexadecimal value will be saved as a text document. The generated link is converted to a hexadecimal value.

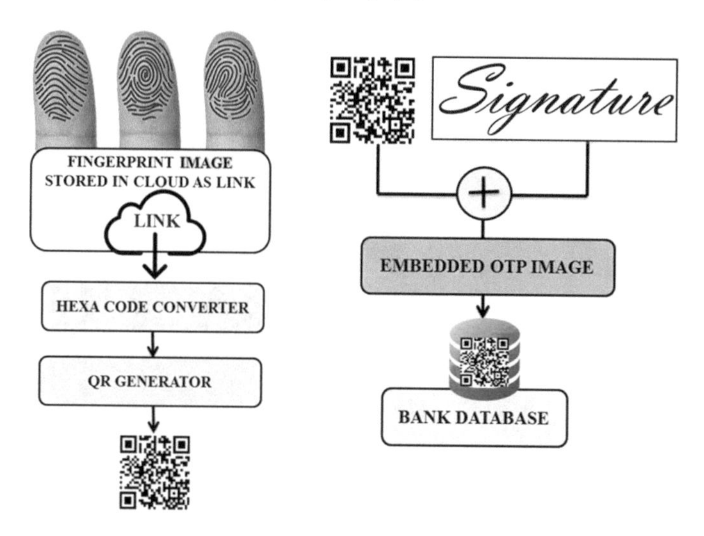

Fig. 1 Generation of QR and embedding of the signature

2.1.1 Generation of QR Code Based on the Given File

With the given text file as an input, the function generates a QR code [2, 3] which contains the hexadecimal value of the link of the image, the link is generated, based on the cloud storage of the particular user's fingerprint database as shown in Fig. 1. The link is converted to hexadecimal code and then with that code a QR is generated.

2.1.2 Embedding of the Signature in QR

The signature image is embedded into the QR using additive algorithm [4] and is now embedded [5–7] with the generated QR. The embedded image is sent by the bank when the user is in need of the verification of his identity during the online payment method.

2.2 Extraction: Decoding of the QR Code and Fingerprint is Generated

The QR code is decoded and the link is generated, from the given link the fingerprint image is downloaded and currently the image is retrieved from the PARENT database. The hexadecimal code of the link is converted back to its true form and, the image is downloaded from the cloud link as shown in Fig. 2.

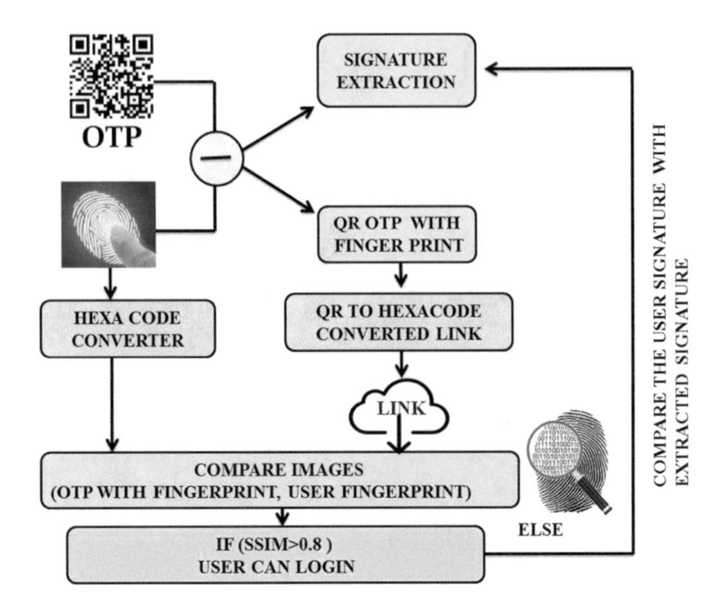

Fig. 2 Extraction of the QR and signature

2.2.1 Finger Print Verification

The verification of the user's fingerprint and the fingerprint retrieved from the QR code, both the fingerprints is now verified. The verification of the fingerprint is done by comparing the minutiae between two images thus resulting in the checking of similarities between two images. The verification process involves a few steps as shown in Fig. 3

- **Enhancement**: This process helps in enhancing the clarity of the fingerprint image, which helps in decreasing the noise of the image [8].
- **Binarize**: This process helps in increasing the boldness of the ridges and low lighting the rest.
- **Thinning**: This process helps in reducing the pixel size, thus generating one pencil line level of the fingerprint.
- **Minutiae**: Minutiae is the filter distribution that highlights the primary key, (unique) properties of the fingerprint.

2.2.2 Signature Verification

If the similarity measurement is lesser than 0.8, verification of the user's signature is also mandatory. Thus the signature will be extracted [9, 10] from the QR and the user will be required to, draw into the smartphone his signature in a generated "Doodle" screen, this signature is then compared [11] with the one extracted from

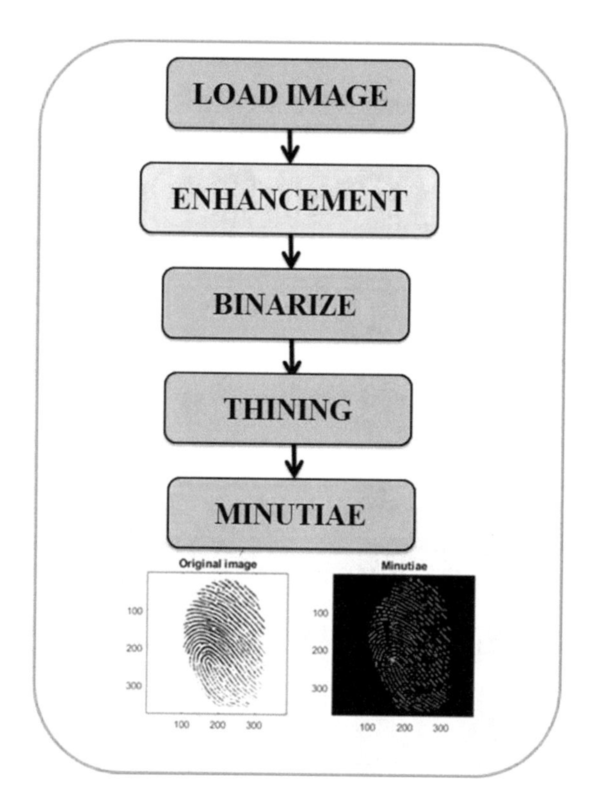

Fig. 3 Fingerprint verification

the QR and the similarity is again matched. The signature authentication would be highly efficient since it is executed quickly, and the threshold value is automatically set based on the similarity rate of the fingerprint. Thus it creates an efficient setup.

3 Experimental Results

The IQA metrics such as structural similarity (SSIM), peak signal-to-noise ratio (PSNR), correlation coefficient(CC) are used for comparing the images to identify the image variations, and the results are shown in Table 1.

The structural similarity (SSIM) index is a method for measuring the similarity between two images as shown in Eq. (1).

$$\mathrm{SSIM}(x, y) = \frac{(2\mu_x \mu_y + C_1)(2\sigma_{xy} + C_2)}{(\mu_x^2 + \mu_y^2 + C_1)(\sigma_x^2 + \sigma_y^2 + C_2)} \tag{1}$$

where μ_x is the average of x; μ_y is the average of y; σ_x^2 is the variance of x; σ_y^2 is the variance of y;σ_{xy} is the covariance of x, y;

Table 1 Quantitative assessment of fingerprint verification

Bank Fingerprint 1	User Fingerprint 2	PSNR	MSE	SSIM	CC	Normalized Absolute Error
		∞	0	1	1	0
		57.6732	0.1111	0.77067	0.2493	0.1918
		51.6526	0.4444	0.68476	0.2102	0.1570
		48.1308	1.0000	0.88396	0.1446	0.1594
		45.6320	1.7778	0.6742	0.2565	0.2256
		43.6938	2.7778	0.71263	0.3146	0.2217
		42.1102	4.0000	0.5766	0.1913	0.1691
		40.7713	5.4444	0.73983	0.2686	0.1798
		57.6732	0.1111	0.19739	0.2786	0.1864
		54.6629	0.2222	0.24535	0.2304	0.1838
		50.6835	0.5556	0.23262	0.2681	0.1850
		47.6732	1.1111	0.22975	0.1113	0.1785
		45.3687	1.8889	0.24618	0.2535	0.2479
		43.5235	2.8889	0.19978	0.2656	0.1613
		41.9912	4.1111	0.26673	0.2088	0.1631

$c1 = (0.01*L)^2$ $c2 = (0.03*L)^2$;

For the correlation coefficient where $i = 1$ to m is shown in Eq. (2).

$$R = \frac{\sum_i (x_i - x_m)(y_i - y_m)}{\sqrt{\sum_i (x_i - x_m)^2}\sqrt{\sum_i (y_i - y_m)^2}} \quad (2)$$

PSNR is most commonly used to measure the quality of reconstruction of lossy compression codes, as shown in Eq. (3).

$$\mathrm{PSNR} = 10.\log_{10}\left(\frac{\mathrm{MAX}_i^2}{\mathrm{MSE}}\right) \tag{3}$$

4 Conclusion and Future Work

The current reading of the SSIM, i.e., the similarity measure based on the minutiae is 1 if both are at the same level of percentage. But the possibility of the minutiae reading being wrongly calculated is high, and that could be improved by implementing the recognition by neural networks. The threshold value of SSIM is fixed has 0.8, with better test cases and reading capability even this could be improved to provide high authentication. Also, the security of the QR code is increased by the implementation of the RSA algorithm and thus encrypting the message in the QR by using the public and private key. The enhancement of this work could generally result in shorter time complexity and faster results. Using IoT-enabled fingerprint sensors and integrating it with the user's device, and creating a mobile application using Android Studio, thus integrating these sensors, along with this work could result in an advanced working prototype of this work (Fig. 4). Moreover, with proper analysis of the data, even the day-to-day activity of this work could be improved efficiently.

Declaration The authors have obtained all the images of the subjected fingerprints involved in this study from open available database FVC2002.

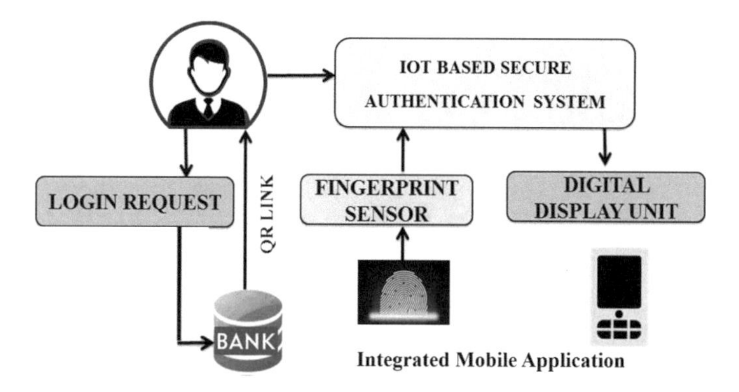

Fig. 4 Proposed secure authentication model for banking

References

1. Maltoni, D., Maio, D., Jain, A.K., Prabhakar, S.: Handbook of Fingerprint Recognition, 2nd edn. Springer, London (2009)
2. Maheswari, S.U., Jude, D.: Frequency domain QR code based image steganography using Fresnelet transform. AEU-Int. J. Electron. Commun. (2014)
3. Raheja, S., Kumar, D.: Personal authentication system using barcode generated by fingerprint
4. Tsai, M.-J.: Wavelet tree based digital image watermarking by adopting the chaotic system for security enhancement. Multimed. Tools Appl., 347–367 (2011)
5. Aroma, R.J., Raimond, K.: An Empirical study on the influence of image filters in effective closed contour extraction of lakes in satellite images. Indian J. Sci. Technol (2017)
6. Kalpanasonika, Agnes, A.: A resilient digital watermarking exploiting spatial and frequency domain against image hackers. Springer eBook Informatics and Communication Technologies for Societal Development, pp. 167–172 (2015)
7. Kalpanasonika, Agnes, A.: A scrutiny on digital watermarking technique. Int. J. Adv. Res. Comput. Sci. Softw. Eng., 175–179 (2013)
8. Lussona, F., Bailey, K., Leeney, M., Curran, K.: A novel approach to digital watermarking, exploiting colour spaces, pp. 1268–1294 (2013)
9. Hsieh, S.-L.: An image authentication scheme based on digital watermarking and image secret sharing. Multimed. Tools Appl., 597–619 (2011)
10. Leena, G.D., Dhayanithy, S.S.: Robust image watermarking in frequency domain. Int. J. Innov. Appl., 582–587 (2013)
11. Gunjal, B.L., Mali, S.N.: Strongly robust and highly secured DWT-SVD based color image watermarking. Int. J. Inf. Technol. Comput. Sci., 1–7 (2012)

Clustering-Based Melanoma Detection in Dermoscopy Images Using ABCD Parameters

J. Jacinth Poornima, J. Anitha and H. Asha Gnana Priya

Abstract Melanoma, dangerous among skin cancer, becomes fatal when not diagnosed and treated at the earliest. It can be correctly predicted only by the expert dermatologists. Owing to lack of experts, computer-aided diagnosis is preferred nowadays. Here we have proposed the image processing algorithm based on clustering to identify melanoma. A total of 170 images taken from the standard database are used to test the algorithm. Various filters and pre-processing techniques have been analyzed for better skin enhancement. The lesion portion is segmented using K-means clustering algorithm. Then the features are extracted from the segmented lesion, and total dermatoscopy score was calculated. This score was calculated for all images and are classified into melanoma and non-melanoma. Finally, the classification accuracy of the algorithm is computed.

Keywords Skin cancer · Dermatoscopic image · Melanoma · K-means clustering · ABCD features

1 Introduction

Melanoma is the collection of abnormal cells which develop mostly in white skin population. These cancerous cells multiple with the average human age range of 20–40 years [1]. The detection of cancerous development is complicated to justify by naked eye. Hence computer-aided diagnosis system is used for the classification of the melanomas. The K-means clustering method is not widely used; rather neural network ensemble model is used to classify the dermatoscopic images. The skin lesion segmentation is done by fully convolutional residual neural network and very

J. Jacinth Poornima (✉) · J. Anitha · H. Asha Gnana Priya
Karunya Institute of Technology and Sciences, Karunya Nagar, Coimbatore 641114, India
e-mail: jac.silent@gmail.com

J. Anitha
e-mail: anithaj@karunya.edu

H. Asha Gnana Priya
e-mail: ashapriya.h@gmail.com

A. Elçi et al. (eds.), *Smart Computing Paradigms: New Progresses and Challenges*, Advances in Intelligent Systems and Computing 766,
https://doi.org/10.1007/978-981-13-9683-0_29

deep residual neural network is done for classification [2, 3]. Dermoscopic images are segmented based on fixed-grid wavelet network (FGWN) which provides the comparison of methods, like AT, GVF, FBSM, and NN [4]. Various types of filter are used for pre-processing in which median filter is being used and segmentation is done by K-means clustering and fuzzy C-means algorithms [5]. The already-classified novel method has been segmented by self-generating neural network and classification is done by combining backpropagation (BP) neural network with fuzzy neural network [6, 7]. Then K-means clustering-based segmentation is done. To extract high-level details from MRI images, discrete wavelet transform is used and an SVM classifier is used for classification [8]. Fuzzy C-means (FCM) clustering method is used for segmentation, and the morphological operations are applied to get better segmentation results [9]. Thresholding-based segmentation is used and features are extracted by ABCD parameters and total dermatoscopy score (TDS) is used to classify the skin cancer [10]. Classification is studied by two variants: binary and multi-class, and the thickness is found from the dermoscopic images [11]. In this paper segmentation is done by K-means clustering and then classification is done based on ABCD rule.

2 Proposed Block Diagram

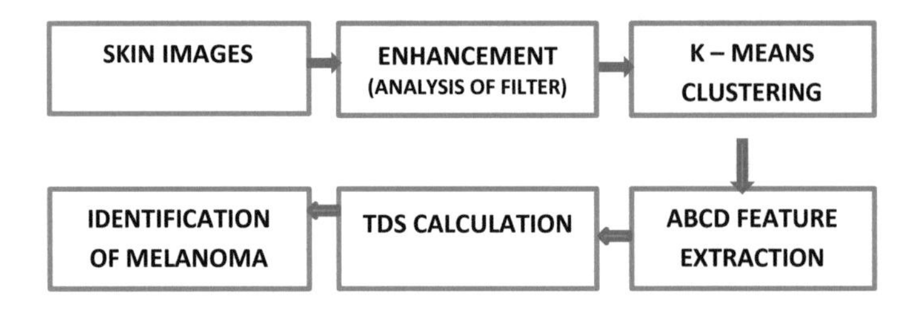

2.1 *Skin Images*

The skin images are obtained from the Med-Node database, which consists of 70 melanoma and 100 non-melanoma images. This Med-Node database is a digital image archive of the Department of Dermatology at the University Medical Centre Groningen (UMCG).

Table 1 Comparision of filters

S. no	Filters used	PSNR value
1	Bottom-hat	72.35
2	Gabor	51.41
3	Weiner	28.51
4	Median	26.94

2.2 *Enhancement (Analysis of Filter)*

The image quality has to be improved for further purpose by reducing or removing the unrelated parts in the background images [12]. The high frequency or noise can be removed by filters. Filters does not enlarge the content of images. It includes resizing, RGB to grayscale conversion, and hair removal. The collected database images are resized. The resized RGB images are converted to grayscale images. This grayscaling is done because color information in RGB images does not help us to identify important edges and many other features. This process in involved by using different types of filters, like Bottom-hat, Gabor, Weiner, Median, Canny, and so on.

The filtering process is done for all the 170 images and the average PSNR value is given in Table 1. On the basis of comparison, Bottom-hat filter is found to be efficient in removing the noise.

2.3 *K-means Clustering*

K-means clustering has been successfully used for brain tumor segmentation. This method is not widely used for skin cancer. In this paper, we have used K-means clustering for lesion segmentation which provides better results [13]. The algorithmic steps for K-means clustering are explained below.

Let data set points be $X = \{x_1, x_2, x_3, \ldots\ldots, x_n\}$ and cluster centers be $V = \{v_1, v_2\}$.

(1) Select "c" as random cluster centers.
(2) Compute the distance between each set of data points and cluster centers.
(3) Assign the data point to the cluster center whose distance from the cluster center is minimum of all the cluster centers.
(4) Recompute the new cluster center using:

$$Vi = (1/Ci)\sum_{j=1}^{Ci} x^i \tag{1}$$

where "c_i" represents the number of data set points in ith cluster.

(5) Recalculate the distance between each set of data point and the obtained next cluster centers.
(6) If no data point was reassigned, then stop; otherwise repeat from Step 3.

2.4 ABCD Feature Extraction

The features are extracted from the segmented image using ABCD parameters. The features extracted from the images are: area (A), perimeter (P), greatest diameter (GD), and shortest diameter (SD). Area is calculated as the total number of pixels with value (1) in the segmented binary image [14]. Perimeter is the total number of lesion boundary pixels. The greatest and shortest diameter is calculated from the length of line passing through the lesion centroid. The ABCD parameters are found using the following formulas:

1. Asymmetry/Symmetry—If the lesion value is approximately 0, then it is symmetry and non-melanoma. If the value is 1, then it is asymmetry and melanoma.

$$IrA = P/A \tag{2}$$

2. Border Irregularity—Cancerous edges are ragger, notched, or blurred. The value ranges from 0 to 8.

$$IrB = P/GD \tag{3}$$

3. Color—Cancerous skin lesion pigmentation is not uniform [15]. The value ranges from 0 to 6.

$$IrC = P * ((1/SD) - (1/GD)) \tag{4}$$

4. Diameter—Cancerous lesions are >6 mm.

$$IrD = GD - SD \tag{5}$$

2.5 Total Dermatoscopy Score

$$IrC = P * ((1/SD) - (1/GD))$$

Total dermatoscopy score (TDS) is used in the diagnosis of melanoma. Calculation of TDS is based on ABCD rule. TDS is given by:

$$\text{TDS} = 1.3 * \text{I}_\text{r}\text{A} + 0.1 * \text{I}_\text{r}\text{B} + 0.5 * \text{I}_\text{r}\text{C} + 0.5 * \text{I}_\text{r}\text{D} \tag{6}$$

If TDS is <4.75, then it is considered as non-melanoma; if it is >5.45, then it is considered as melanoma; and if it is between 4.75 and 5.45, then it is the suspicious case of skin lesion [16].

3 Results

The results are obtained for 170 images and the sample for one picture has been explained below. The results are classified using ABCD classification. The extracted features are shown in Fig. 1, and the TDS value for a sample of five images is shown in Table 2.

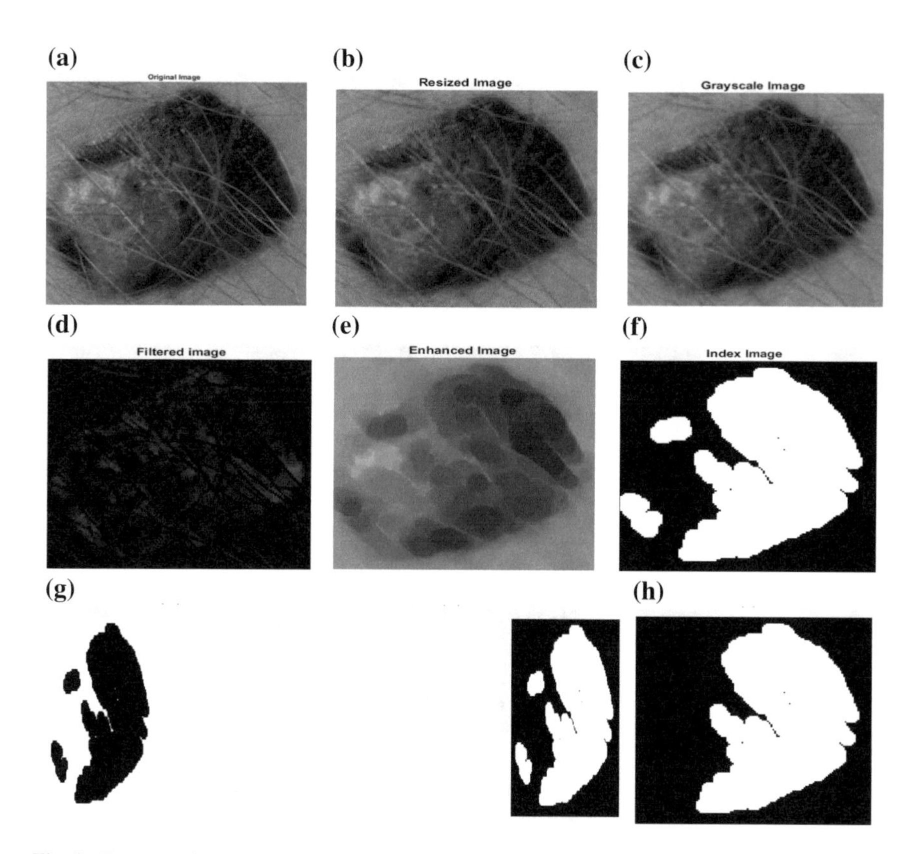

Fig. 1 K-means clustering output: **a** Original image, **b** resized image, **c** grayscale image, **d** filtered image, **e** enhanced image, **f** index image, **g** clustered image, **h** segmented image

Table 2 ABCD feature extraction-based classification

Image	A	P	D	IrA	IrB	IrC	IrD	TDS	M/NM	T/F
1	6792	331.892	96.11	0.048	3.06	0.88	24.16	12.89	M	T
2	2965	218.194	63.18	0.073	3.04	0.94	16.90	9.32	M	T
3	6196	497.476	190.5	0.080	2.59	0.02	2.118	1.43	NM	T
4	8543	497.476	179.3	0.058	2.74	0.06	4.472	2.62	NM	T
5	9974	497.476	172.9	0.049	2.82	0.09	5.645	3.21	NM	T

4 Conclusion and Future Work

In this paper, the efficiency of K-means clustering algorithm is tested for skin image analysis. The extracted ABCD features and total dermoscopy score calculated from segmented image is used for identification of melanoma. A total of 170 images is tested and achieved an accuracy of 88%. The accuracy can be further improved by using neural network approach.

References

1. Blank, C.U., Larkin, J., Arance, A.M., Hauschild, A., Queirolo, P., Del Vecchio, M., Garbe, C.: Open-label, multicentre safety study of vemurafenib in 3219 patients with BRAF V600 mutation-positive metastatic melanoma: 2-year follow-up data and long-term responders' analysis. Eur. J. Cancer **79**, 176–184 (2017)
2. Yu, L., Chen, H., Dou, Q., Qin, J., Heng, P.-A.: Automated melanoma recognition in dermoscopy images via very deep residual networks. IEEE Trans. Med. Imaging **36**(4), 994–1004 (2017)
3. Bi, L., Kim, J., Ahn, E., Kumar, A., Fulham, M., Feng, D.: Dermoscopic image segmentation via multi-stage fully convolutional networks. IEEE Trans. Biomed. Eng. **64**(9), 2065–2074 (2017)
4. Sadri, A.R., Zekri, M., Sadri, S., Gheissari, N., Mokhtari, M., Kolahdouzan, F.: Segmentation of dermoscopy images using wavelet networks. IEEE Trans. Biomed. Eng. **60**(4), 1134–1144 (2013)
5. Jose, A., Ravi, S., Sambath, M.: Brain tumor segmentation using k-means clustering and fuzzy c-means algorithms and its area calculation. Int. J. Innov. Res. Comput. Commun. Eng. **2**(3), 3496–3501 (2014)
6. Xie, F., Fan, H., Li, Y., Jiang, Z., Meng, R., Bovik, A.: Melanoma classification on dermascopy images using a neural network ensemble model. IEEE Trans. Med. Imaging **36**(3), 849–858 (2017)
7. Diniz J.B., Cordeiro F.R.: Automatic segmentation of melanoma in dermoscopy images using fuzzy numbers. In: Proceeding of IEEE 30th International Symposium on Computer-Based Medical Systems, pp. 150–155. (2017)
8. Labeed, K.A., George, A.: Detection of brain tumor using modified K means algorithm and SVM. Int. J. Comput. Appl. (2013)
9. Schmid, P.: Segmentation of digitized dermatoscopic images by two-dimensional color clustering. IEEE Trans. Med. Imaging **18**(2), 164–171 (1999)
10. Pauline, J., Abraham, S., Bethanney Janney, J.: Detection of skin cancer by image processing techniques. J. Chem. Pharm. Res. **7**(2), 148–153 (2015)
11. Abbas, A.H., Kareem, A.A., Kamil, M.Y.: Breast cancer image segmentation using morphological operations. Int. J. Electron. Commun. Eng. Technol. **6**, 08–14 (2015)
12. Ganster, H., Pinz, A., Rohrer, R., Wildling, E., Binder, M., Kittler, H.: Automated melanoma recognition. IEEE Trans. Med. Imaging **20**(3), 233–239 (2001)
13. Pezeshk, A., Petrick, N., Chen, W., Sahiner, B.: Seamless lesion insertion for data augmentation in CAD training. IEEE Trans. Med. Imaging **30**(4), 1005–1015 (2017)
14. Sáez, A., Sánchez-Monedero, J., Gutiérrez, P.A., Hervás-Martínez, C.: Machine learning methods for binary and multiclass classification of melanoma thickness from dermoscopic images. IEEE Trans. Med. Imaging **35**(4), 1036–1045 (2016)

15. Ma, Z., Tavares, J.M.R.: A novel approach to segment skin lesions in dermoscopic images based on a deformable model. IEEE J. Biomed. Health Inform. **20**(2), 615–623 (2016)
16. Satheesha, T.Y., Satyanarayana, D., Prasad, M.G., Dhruve, K.D.: Melanoma is Skin Deep: a 3D reconstruction technique for computerized dermoscopic skin lesion classification. IEEE J. Transl. Eng. Health Med. **5**, 1–17 (2017)

Automated Histogram-Based Seed Selection for the Segmentation of Natural Scene

R. Aarthi and S. Shanmuga Priya

Abstract Images have been widely used in today's life varying from personal usage with Flickr, Facebook to the analysis of hyperspectral images. Availability of such huge volume of images in digital form requires an automatic analysis on visual content. The major challenge in content labeling is in segmentation, which divides the image into regions. Our work focuses on the segmentation technique that adapts based on regions in the natural images. The proposed method used automatic seed selection by analyzing the dynamic color distribution of the image. The experimental results on datasets show the better performance of our method.

Keywords Histogram · Seeded region · Segmentation

1 Introduction

This paper proposes an improvised segmentation method to divide the natural images into regions using an automated histogram-based approach. Natural images in the digital form need to be stored with respect to its content. The manual cataloguing and indexing of the image content are difficult as millions of data get accumulated day by day. Hence, an idea of Content based Image Retrieval (CBIR) is brought into the field of image analytics to potentially derive content information automatically. The traditional approach of CBIR is to segment the image into distinct regions and classifies each region into their respective category. Segmentation plays a vital role in the content formation of image data.

The objective is to segment and classify a natural image containing regions like sea, sky, sand, rock, and vegetation into their respective classes. The major

R. Aarthi (✉) · S. Shanmuga Priya
Department of Computer Science and Engineering, Amrita School of Engineering, Coimbatore, India
e-mail: r_aarthi@cb.amrita.edu

S. Shanmuga Priya
e-mail: ss_priya@cb.amrita.edu

Amrita Vishwa Vidyapeetham, Coimbatore, India

A. Elçi et al. (eds.), *Smart Computing Paradigms: New Progresses and Challenges*,
Advances in Intelligent Systems and Computing 766,
https://doi.org/10.1007/978-981-13-9683-0_30

processes involved are segmentation, texture analysis, classification, and summarize the percentage of object present in the image with respect to the area covered.

2 Related Works

Segmentation is a preprocessing technique that facilitates the analysis of various regions and the connections in the image. The information is derived from local or global information of the images and grouped depending on the closeness of the image. The k-means clustering algorithm [1] starts with the random seed selection method followed by that it uses the pixel level information to group into appropriate regions. The good quality segmentation observed here has less cluster variance and being less sensitive to the initialization of center values are generated. Julie vogul et al. [2] applied both local and global information by dividing the image into grid in order to categorize the region as coastal, forest, and plain. Ashok Kumar et al. [3] has applied the k-means clustering to characterize the chemical properties of soil surface data. Eser Sert et al. [4] used color-based k-means segmentation to divide mushroom into various parts as cap width measurement.

Since in the above methods, seed selection places a vital role of segmentation, most of the time the process is done based on the unsupervised manner. Extending the method for unsupervised segmentation method, Rostom Kahouri [5] proposed a color similarity-based texture merging method for clustering. The method is applied on Corel1 [6] and Berkeley2 [6] datasets and it outperforms the state-of-the-art methods on visual accuracy comparison and Liu's factor measuring. Abdelali et al. [7] uses dynamically and automatically generating seed points and determines the threshold values for each region. The region-growing algorithm is used to divide mammogram into homogeneous regions, thereby resulting in segmenting different breast tissue regions. Patrícia Genovez et al. [8] applied image segmentation to detect oil slicks as dark spots, regions with low backscatter at sea surface. The segmentation process involves Region-fitting and Edge-based algorithms for segmenting at the pixel level, Region-growing and merging methods to a region level. The results proved that the region splitting of 3×3 yields good segmented results, when compared to window sizes of 5×5, 7×7 and necessitates the need for split up. The drawback of using grid-based segmentation is that the computation is time consuming. The standard k-means algorithm gives better results for smaller value of k, predetermination of k value is a difficult task. But the selection of choosing seeds randomly leads to wrong segmentation. In generalization, k-means algorithm and automatic seed generation using histogram are proposed.

3 Proposed Methodology

The studies show that grid-based segmentation gives good results for segmenting natural images by analyzing its local and global information. The disadvantage is that the processing time is linearly dependent on the size of the image and number of grids. It is highly affected by noise, variation in intensity level and results in holes or over segmentation.

First, we have investigated various natural images to find an optimal grid size for comparison. The given input image is divided into regions by varying grid sizes of 2×2, 5×5, and 10×10 in RGB color space [9]. Randomly, multiple grids are selected to be the seeds and color mean value for each grid is calculated across that color space. The seeds are then compared with all the other grids mean values in the image. The grids that lie within the particular threshold are merged together. This is done till the entire image is segmented. The sample results for three natural images are shown in the Fig. 1. As per the results, the grid size of 5×5 gives a better result of segmentation than the other grid sizes.

Most of the natural images are stored in a high-resolution format so that it attracts the human visual attention. The intricacies in the stored images make the segmentation process more challenging and it yields poor results. Since our objective of the segmentation process is to abstractly group the image into meaningful regions not retaining the finer details of the image. To handle the intensity variation, image smoothing is done by Gaussian filters. The process involves transforming the image into the frequency domain, multiplying it with the Gaussian filter function and re-transforming the result into the spatial domain. The Gaussian kernel (Eq. 3.1) size used for smoothing is 10×10 and $\sigma = 5$ so as to attenuate some frequencies and enhance others.

$$G(x, y) = \frac{1}{2\pi\sigma^2} e^{-\frac{x^2+y^2}{2\sigma^2}} \tag{3.1}$$

Fig. 1 Natural images with varying grid sizes

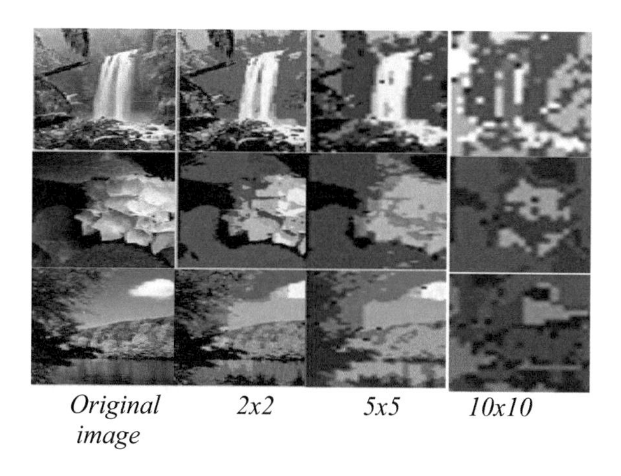

where

x,y—horizontal and vertical distance from the origin and σ—the standard deviation.

An analysis of various grid sizes applied for natural images is shown in Fig. 1. The visual observation shows a better result for grid size of 5×5. The above method uses random seed selection that may lead to false segmentation as well. So, there is a need of an algorithm that automates the random seed selection process to intelligently identify the seeds been distributed among the regions.

We propose a histogram-based automatic seed selection method to address the problem. In an image processing context, a color histogram is a representation of the distribution of colors in an image. The histogram is a graph showing the number of pixels in an image at each different intensity value found in that image. In a 2D plot, x-axis shows the intensity values, whereas y-axis shows the count. The peak and valley in a histogram represent most and least occurrences of the particular intensity value. The distribution of the intensity value between the peak and valley follows a Gaussian behavior, having a higher probability of belonging to the same region [10].

So, we apply the above inferences for selecting the seed by finding the peaks and/or valleys, which can then be used to determine a threshold value. This threshold value can then be used for image segmentation.

Algorithm for selecting seed pixel (RGB image)

Step 1 Generate a 3D histogram for the input image

Step 2 Divide each histogram into n sections of equal size as $x_0, x_1, \ldots x_{n-1}$, where x_i and x_{i+1} represent a section.

Step 3 Compute the peak ($P_{i,j}$) from the histograms where $1 \leq i \leq n$, $j = R,G,B$

Step 3.1 If $[P_{i,j}]$ is the peak in the ith region, compute the weighed mean $x_{i,j}$ for other j not in peak using Eq. 3.2

$$\frac{\sum_{x_{k_j}} \{\text{count}[x_{k_j}] \cdot [x_{k_j}]\}}{\sum_{x_{k_j}} \{\text{count}[x_{k_j}]} \quad x_{i-1} \leq k \leq x_i \tag{3.2}$$

Step 4 The seed is calculated as $([P_{i,j,}], G_{mean}, B_{mean})$

Figure 2 shows the result for 3D histogram with $n = 5$ sections and the peak for every section is marked in yellow. As an example, for region 1, Fig. 2a has the highest peak than Fig. 2b,c. So, the peak for that region is taken as $[P_{i,R}]$ and $x_{i,B}$ and $x_{i,G}$ is calculated for the region. Using the selected seed, the segmentation is done combined with the procedure discussed in the section.

Procedure for segmentation (RGB image, Seed Si)

Step 1 The image is divided into Red, Blue and Green Components.

Step 2 Each component is divided into grid of size 5×5. Mean value is calculated for each grid $M_i.(R_{mean}, G_{mean}, B_{mean})$

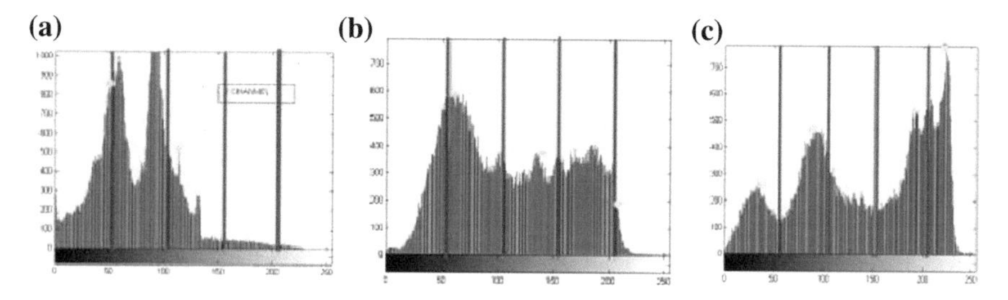

Fig. 2 Histogram of **a** Red **b** Green **c** Blue

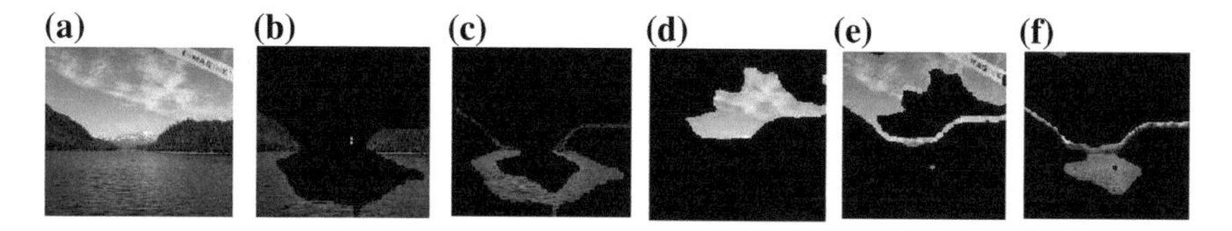

Fig. 3 **a** Original Image **b–f** five segments of the original image

Step 3 The distance between M_i and S_i is calculated. The grid is segmented to that particular region for which the seed and the grid mean Mi has minimum distance.

Step 3.1 Repeat Step 3 for all seed values

Figure 3 shows the original input image and the segmented image. The disadvantage of the above method is selection of the peak value, since single peak is restricted in a region leads to mixing of values, for example, light green and blue may be mixed as a single seed point as shown in Fig. 3b, so there may be several peak values in a region so while selecting a peak value we are going by the highest values and not by the regions, this algorithm is explained in the next section.

3.1 Improved Automated Seed Selection

The selection of seed point is same as the above method except for dividing the histogram into five regions and taking the peak values. Here it is done for the whole histogram without dividing into regions. Ten seed values are taken for improving the result and corresponding segments are obtained from the input image by the distance method [1] These ten segments may have similar mean value regions. So, these are then combined by the mean values of the RGB components. Figure 4 shows the segments after the segmentation and combining the similar mean value regions.

Improved automated histogram-based seed selection method for segmentation shows the better result for natural scenic images. Mixing of different colored regions is reduced in this algorithm because the seed selection is done for the entire histogram.

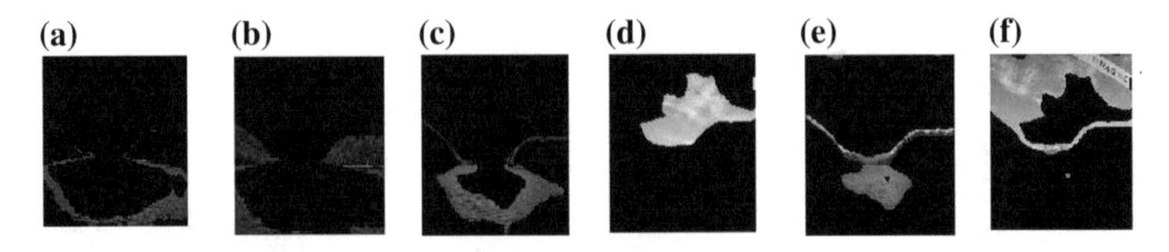

Fig. 4 **a–f** segments from the improved automated seed selection algorithm

The distance between M_i and S_i is calculated. The minimum distance for the grid with a seed is found. The grid is segmented to that particular region for which the seed and the grid mean M_i has minimum distance.

4 Conclusion

This paper mainly focused on the natural objects segmentation and classification in the scenic images. An extensive amount of testing has been done on various types of natural images and an accuracy of above 90% has been achieved. We found that histogram-based seed selection segmentation remains the best comparing with other descriptors in the experiment. Further areas, which need dwelling are segmentation of scenic image containing shadows, improper lighting. The application areas most likely to benefit from the adoption of "Natural object segmentation and classification" are those where indexing of images is necessary for faster and efficient usage of system resources, robotic navigation and image search on the web.

References

1. Raghesh Krishnan, K., Radhakrishnan, S.: Focal and diffused liver disease classification from ultrasound images based on isocontour segmentation. IET Image Process. **9**, (4, 1), 261–270 (2015)
2. Vogel, J., Schwaninger, A.: Categorization of natural scenes: local vs global information, University of British Columbia (2001)
3. Kumar, D.A., Kannathasan, N.: A study and characterization of chemical properties of soil surface data using K-means algorithm. In: Proceedings of the 2013 International Conference on Pattern Recognition, Informatics and Mobile Engineering (PRIME) 21–22 February, (2013). 978-1-4673-5845-3/13
4. Sert, E., Okumus, I.T.: Segmentation of mushroom and cap width measurement using modified K-means clustering algorithm. In: Digital Image Processing and Computer Graphics, Vol. 12, Issue 4 (2014)
5. Siddiqui, F.U., Isa, N.AM., Member, IEEE.: Enhanced moving K-Means (EMKM) algorithm for image segmentation. IEEE Trans. Consum. Electron. **57**(2) (2011)
6. Raju, P.D.R. et al.: Image segmentation by using histogram thresholding. IJCSET, **2**(1), 776–779 (2012)

7. Elmoufidi, A., Student Member IEEE, et al.: Automatically density based breast segmentation for mammograms by using dynamic k-means algorithm and seed based region growing. In: IEEE Instrumentation and Measurement Society, pp. 533–538. https://doi.org/10.1109/i2mtc.2015.7151324, 2015
8. Genovez, P. et al.: An image-segmentation-based framework to detect oil slicks from moving vessels in the Southern African oceans using SAR imagery. IEEE J. Sel. Top. Appl. Earth Obs. Remote. Sens. **99**, 1–9 (2017)
9. Machine learning data website available http://www.archive.ics.uci.edu/ml/datasets.html
10. Kachouri, R., Soua, M., Akil, M.: Unsupervised image segmentation based on local pixel clustering and Low-Level region merging. In: 2nd IEEE International Conference on Advanced Technologies for Signal and Image Processing ATSIP'. Monastir, Tunisia, vol. 16, (2016)
11. Aarthi, R., Padmavathi, S., Amudha J.: Vehicle detection in static images using color and corner map. In: International Conference on Recent Trends in Information, Telecommunication and Computing, kochi. pp. 12–13 (2010)

Application of Image Fusion Approaches for Image Differencing in Satellite Images

R. Jenice Aroma and Kumudha Raimond

The increasing thwart for the environment has lead to a revolutionary change in protective measures. The surveillance of the complete ecosystem was a laborious effort during the days of field sensors based assessment practices for spatial object monitoring. After the rise of satellite image based applications for observing the spatiotemporal changes in both coastal regions and urban ecosystems, environmental sustainability has got a balance. In order to implement such applications with more accurate assessment, the spatial image interpretation methods are in need to be improved. This paper proposes a significant approach for water body change visualization by applying feature-level image fusion methods for tracing the extent of change occurred within a water body over the chosen time period. The Image Quality Assessment (IQA) metrics have been applied for quantitative assessment. The results achieved on superimposing the various point descriptors, wavelet descriptors, and other reduced features could highly portray the effective change visualization of the chosen water body.

Water body · Spectral indices · Clustering · Image fusion · Landsat 8 · Change map

1 Introduction

During natural devastations like floods, earthquakes, and hurricanes, the post hazard situations of an environment can be instantly estimated through analyzing the multi-temporal satellite images, [1, 2]. With ground-based sensors, it is impossible to cover the entire affected region which includes huge manual intervention. Hence,

R. Jenice Aroma (✉)
Sri Krishna College of Technology, Coimbatore, India
e-mail: jenicearoma.r@skct.edu.in

K. Raimond
Karunya Institute of Technology and Sciences, Coimbatore, India

A. Elçi et al. (eds.), *Smart Computing Paradigms: New Progresses and Challenges*, Advances in Intelligent Systems and Computing 766,
https://doi.org/10.1007/978-981-13-9683-0_31

satellite image based change assessment has become the prime focus of research in environmental monitoring applications. The outburst of global warming alert, due to the increased use of carbon-emitting appliances has led to more serious threats. These seasonal changes attempt to thwart natural phenomena like rainfall, monsoons and plant health. It has been made evident from the recent study on El Nino-Southern Oscillation (ENSO) influence which is much responsible for climate variability, conducted in the Western United States. It brings the relationship between ENSO and terrestrial vegetation dynamics using the Moderate-resolution Imaging Spectro-radiometer (MODIS) and Advanced Very High-Resolution Radiometer (AVHRR) vegetation indices [3]. The Ocean monitoring, rainfall prediction, and flood monitoring are the predominant satellite image based models for monitoring the water level [4–7].

The change trend analyses over land cover regions require high-resolution data, where water body monitoring applications can suffice with moderate-resolution images. Generally, image fusion has been carried out in order to increase the spatial resolution of the data. A lower resolution image can be fused with a higher resolution image without any loss of information. It can be carried out with bands from the same sensor or can also be from different sensors susceptible to the limits on the need for appropriate geometrical corrections in case of different sensors [8, 9]. The image fusion can also be applied in change detection models where differences in spectral values of multi-temporal images can be well observed [10]. Though different fusion-based techniques are in use to observe the changes from multi-temporal images, the need for optimal change detection approaches is still an open research problem to exploit.

2 Image Fusion Based Change Mapping Model

In order to examine the spectral influence of spatial data, a water body has been chosen to construct a multi-temporal dataset. The Sambhar Lake in India has been chosen as the study area and Landsat 8 OLI imagery of the same has been acquired from USGS earth explorer. The different dates ranging from October and December from 2014 to 2016 have been considered. The Sambhar lake in Rajasthan (26°58′N 75°05′E) is the largest inland salt lake. It has been cropped from the collected Landsat 8 images using the shapefile acquired from the work of Ritesh Vijay et al. 2016 over the same Sambhar Lake region [11]. The collected dataset will then be atmospherically corrected for Surface Reflectance (SR) conversion and followed with water body segmentation using Normalized Difference Moisture Index (NDMI). These derived products are then subjected to image differencing methods to generate change difference and then fused for change maps generation.

Figure 1 shows the False Color Composite (FCC) of Sambhar Lake of size 500*1000, during 2014 and 2015, in which the land cover region and water bodies are in red and green color, respectively, where FCC is a combination of NIR,

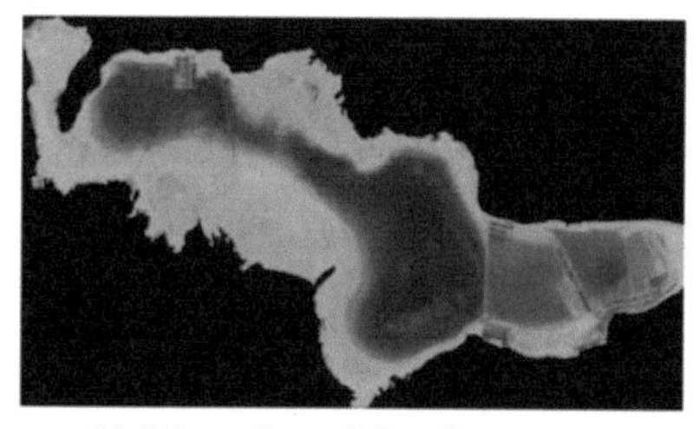
(a) SR product of October-2014

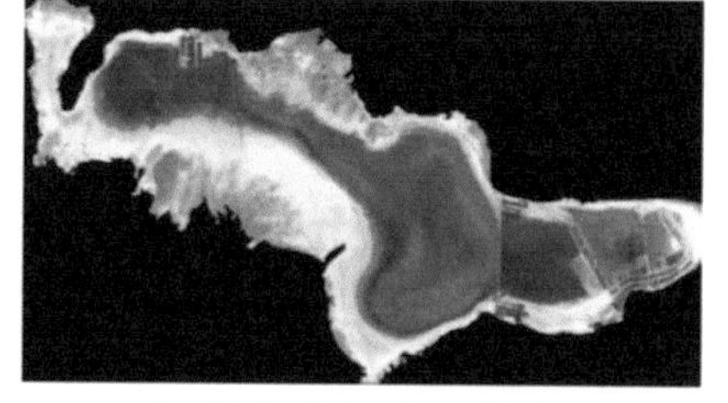
(b) FCC of October -2014

Fig. 1 FCC of Sambhar Lake during October and December of 2014–2016

green and blue bands. Here, the absence of water due to the presence of brackish water in salt pans is shown in brown color.

In order to achieve water body segmentation, the satellite images need to be scaled back to SR. The act of scaling back to SR through atmospheric correction methods may be tedious [12]. But using GIS software may reduce efforts [13]. Further, a new image fusion based approach has been proposed to monitor the changes in a water body from multi-temporal images which are merely a superimposition of images. A simple difference image of multi-temporal images can bring change observation.

However, in order to improve the discrimination power, and effective change observation with reduced uncertainty, a fusion of difference images computed using various change differencing methods is more advantageous. To identify the most appropriate approach, a comparison of different approaches such as image differencing, image ratio, wavelets, etc., including hybrid approaches has been carried out. In addition, different modes of widely used image fusion methods such as Discrete Wavelet Transform (DWT), Stationary Wavelet Transform (SWT), Principal Component Analysis (PCA), Scale Invariant Feature Transform (SIFT), and two other combined variants of DSIFT have been applied over the acquired change difference images for better change visualization through color composite maps generation. All the above-said phases of water segmentation and change difference image fusion are evaluated using standalone IQA metrics.

The proposed workflow for image fusion based water body change trend analysis is shown in Fig. 2. Generally, the change in a water body from multi-temporal images (I1 and I2) can be derived using simple image differencing and ratio methods. In this work, a color map based change evaluation has been adopted through superimposing the fused change difference image over the original image for better change trends.

Jing jing Ma et al. (2012) applied DWT on mean and log ratio images to generate effective fused satellite images [14]. Similarly, in this work, the widely used mean ratio and absolute differencing method have been chosen for observing the change difference from the chosen multi-temporal dataset. The mean ratio is defined to be the normalized ratio that considers the mean values computed at local neighborhood [3*3] of every pixel and the minimum value of the mean ratio between two images (I1 and I2) considered as given in the following Eq. (1):

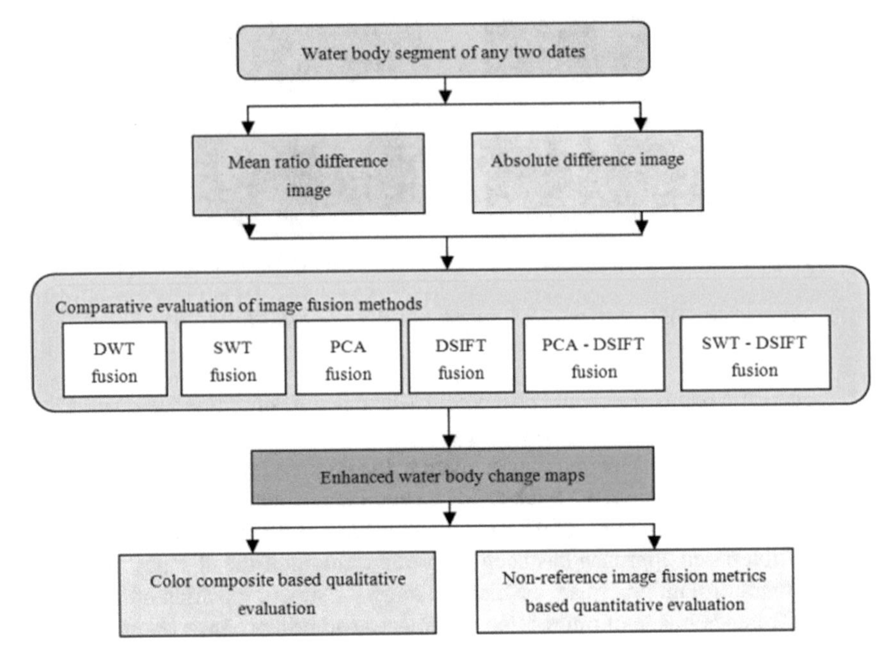

Fig. 2 Image fusion based change map generation

$$mean\,ratio = 1 - \min\left(\frac{\mu_1}{\mu_2}, \frac{\mu_2}{\mu_1}\right) \tag{1}$$

Similarly, the image absolute differencing has also been applied between these two images I1 and I2 to generate the difference image as in Eq. (2).

$$Absolute\ Difference = abs(I1 - I2) \tag{2}$$

Further, in order to improve the quality of change information acquired through change differencing, the image fusion methods can be adapted to reduce the loss of detail on ratio. The notable image fusion methods for change detection models using wavelet-based features, principal components, and point descriptors has been studied from the literature [15–17] and from the study, most prominent methods such as Discrete Wavelet Transform DWT fusion, PCA fusion, SWT fusion, and DSIFT features based fusion are chosen to be applied with change difference images.

In the case of DWT and SWT image fusion, the image detail preservation is high, due to multi-scale image transformation [18]. On applying DWT, it yields four coefficients such as CA, CH, CV, and CD which are the approximate, horizontal, vertical and diagonal details of an image, respectively [19]. On applying averaging and maximum rules for fusion, the average of two approximate coefficients from two images has been selected and maximum values of all the other three coefficients among two images have been selected [14]. These average and maximum values are used for generating the DWT-fused image.

The SWT is an extension of the standard DWT. It resolves the key limits of DWT upon scale invariability, i.e., translation invariance. In the case of SWT, the image has been reduced into four components in the frequency domain such as approximation, horizontal, vertical, and diagonal detail [20]. Here, fusion has been performed through scaling, the scaling factor for approximation coefficients be 0.5, whereas the other three coefficients are scaled with a positive constant which is the absolute difference of the respective coefficients from two images. Further, PCA which is a well known linear mapping technique has also been applied. Initially, on PCA-based fusion, the covariance among chosen bands has been computed. Then upon diagonalizing the covariance matrix, the corresponding high to low values have been chosen as eigenvectors [21]. Thus, PCA components from the two difference images are generated and the resultants are fused to generate the change maps.

As an extended step for image fusion, Yu Liu et al. in 2015 proposed DSIFT features and compared its performance over DWT, Dual-Tree Complex Wavelet Transform and other wavelet-based approaches [22]. Generally, the point descriptors are widely used in image registration applications for matching and enhancing the resolution of spatial data from different sensors or varied temporal resolution. Hence, the DSIFT features also have been chosen for fusion-based change detection in the proposed method and the results are found to be good.

In order to extract the DSIFT feature, a sliding window technique has been applied for processing the image as patches and an initial decision map has been obtained considering the entire scale invariance detail of the image. Followed by feature matching and focus measure comparison, a final decision map has been generated for DSIFT Fusion [23]. In addition, the combination of point descriptors with SWT and sparse PCA derived components has been attempted. For SWT-DSIFT, the DSIFT features have been extracted from the selected approximate coefficient of the image acquired from SWT and the image fusion has been performed. Likewise, DSIFT features have been extracted from the PCA derived components to perform the PCA-DSIFT fusion for change trend analyses. Through applying the abovementioned approaches for change trend analysis, the loss of detail in change maps can be prevented through image fusion.

3 Experimental Results of Proposed Change Trend Analysis Model

This spectrally segmented water body is then subjected to change differencing using mean and absolute differencing as shown in Fig. 3a, b, respectively.

These difference images are then fused for improving the image quality as shown in Fig. 4.

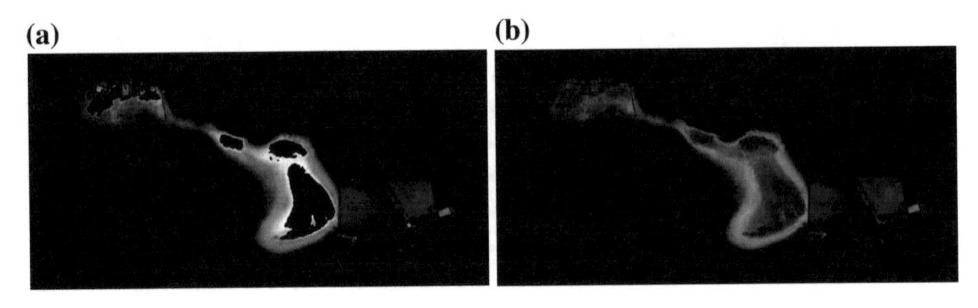

Fig. 3 **a** Mean ratio difference image. **b** Absolute difference image

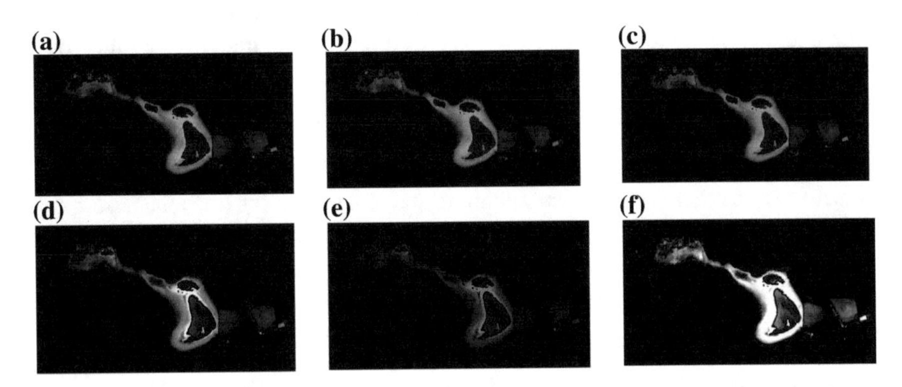

Fig. 4 OCT 2014 versus 2015: **a** PCA **b** DWT **c** SWT **d** DSIFT **e** PCA-DSIFT **f** SWT-DSIFT

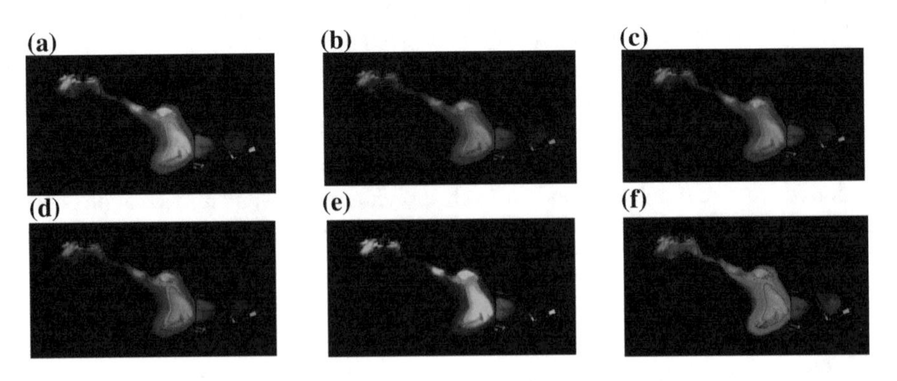

Fig. 5 Colormaps for October 2014 and 2015: **a** PCA **b** DWT **c** SWT **d** DSIFT **e** PCA-DSIFT **f** SWT-DSIFT

To produce the colormaps for change visualization these enhanced change difference images (Oct 2014 and 2015) are overlaid on the Oct-2015 image using color channels as in Fig. 5.

From the change maps generated, magenta color depicts the water body before water shrinkage as on Oct 2014 and the green color depicts the water body present in both Oct-2014 and 2015. Likewise, the water body not available in Oct 2014 but present in Oct 2015 is shown in gray color. It could symbolize the basic set theory combinations and it could be a simple strategy of change maps generation for better change observation of water bodies. On the qualitative assessment of the colormaps, both the hybrid variants of DSIFT with PCA and SWT does not achieve the exact mapping of gray color whereas all the other four methods are found to be good in bringing the change visualization. The reason for the misinterpretation of a water body may due to multiple scaling efforts on reduced data with PCA derived component and SWT approximate coefficients.

The entropy is widely used for measuring the randomness or quality of the image content. Here, the image gray levels have been dealt with for computing the randomness and p_i be the probability of respective gray levels and L be the overall gray scale levels as in Eq. (3).

$$E = -\sum_{i=0}^{L-1} p_i \log p_i \tag{3}$$

The Fusion Mutual Information is a non-reference image fusion metric is used for measuring the degree of dependency among the input change difference images (I_p and I_m) and the fused images (I_f) as in Eqs. (4) and (5).

$$FMI = MI_{i_p i_f} + MI_{i_m i_f} \tag{4}$$

where MI—Mutual information be

$$MI = \sum_{i+1}^{M} \sum_{j=1}^{N} h_{I_r I_f}(i, j) \log_2\left(\frac{h_{I_r I_f}(i, j)}{h_{I_r}(i, j) h_{I_f}(i, j)}\right) \tag{5}$$

where,

I_r be the corresponding change difference image (either I_p or I_m) with $h_{I_r I_f}(i, j)$ as the respective probability mass computed for the corresponding images of size M*N.

Similarly, the fusion quality index which is also a non-reference image quality index has been used for measuring the image similarity of fused images with the input images as in Eq. (6).

$$QI = \frac{\sum_{n=1}^{N} \sum_{m=1}^{M} Q^{AF}(n, m) w^A(n, m) + Q^{BF}(n, m) w^B(n, m)}{\sum_{i=1}^{N} \sum_{j=1}^{M} \left(w^A(i, j) + w^B(i, j)\right)} \tag{6}$$

where

$Q^{AF}(n, m)$ and $Q^{BF}(n, m)$ are the edge preservation value with w^A and w^B be the edge strength.

Table 1 Image quality evaluation for change difference image

Multi-temporal datasets	SSIM	PSNR
Oct-2014 versus Oct-2015	0.8851	19.5100
Dec-2014 versus Dec-2015	0.9725	28.8081

Further, the water body segmentation results that are applied with mean ratio and absolute differencing bring improved change difference images which are evaluated using the above-said image similarity and quality measures as shown in Table 1, both the difference images are found to hold certain detail in addition than the other.

The SSIM value shows the presence of certain dissimilarities and PSNR shows the good quality of image without any degradation after ratio. Hence image fusion methods can be applied for yielding the abstract features from both the difference images through fusion to attribute enhanced level of detail in change maps generation. The standard IQA metrics are chosen from the literature to evaluate the image fusion results quantitatively in case of unavailability of reference data.

Hence, the fusion-based change maps generated from the spectral-based water body segmentation results are evaluated using the image fusion metrics as shown in Table 2.

From the above quantitative assessment of the state-of-the-art image fusion algorithms, it is clearly found that point descriptors based fusion, i.e., DSIFT performs as good as PCA and SWT approaches. But the hybridization of DSIFT with PCA and SWT does not bring any promising improvement due to multiple scaling on reduced input space of PCA and SWT derived components.

4 Conclusion

The change maps generated using various Image fusion methods are found to be effective. The results of image differencing methods for change difference image generation is also found to be more promising. It can be further extended through the combination of instant template matching methods for faster change recognition and use of higher resolution images could bring more clarity in the change detail through color compositing.

References

. Jiang, H., Feng, M., Zhu, Y., Lu, N., Huang, J., Xiao, T.: An automated method for extracting rivers and lakes from landsat imagery. Remote Sens. **6**, 5067–5089 (2014)

Table 2 Quantitative assessment for image fusion methods

Image fusion methods	FMI	QI(ab/f)	Entropy
Yearly Change difference between Oct 2014 and Oct 2015			
DWT	0.9658	0.6581	1.8256
SWT	0.9680	0.7239	1.8164
PCA	0.9682	0.7146	1.8406
DSIFT	0.9640	0.7921	1.7551
SWT-DSIFT	0.9550	0.4121	1.7774
PCA-DSIFT	0.9606	0.3626	1.5637
Yearly Change difference between Dec 2014 and Dec 2015			
DWT	0.9800	0.7782	0.8466
SWT	0.9810	0.8102	0.8409
PCA	0.9811	0.8042	0.8568
DSIFT	0.9832	0.8328	0.8145
SWT-DSIFT	0.9740	0.4409	0.9084
PCA-DSIFT	0.9795	0.4360	0.7225

. Aggarwal, H.K., Minz, S.: Change detection using unsupervised learning algorithms for Delhi, India. Asian J. Geoinformatics **13**(4) (2013)
. Domínguez Gómez, J.A., Chuvieco Salinero, E., Sastre Merlín, A.: Monitoring transparency in inland water bodies using multispectral images. Int. J. Remote. Sens. **30**(6), 1567–1586 (2009)
. Klacka, J., Saniga, M.: Doppler effect and nature of light; earth, moon and planets (EM&P) **59**, 219–227 (1992)
. Jagalingam, P., Hegde, A.V.: A review of quality metrics for fused image. Aquat. Procedia **4**, 133–142 (2015)
. Aroma R.J., Raimond, K.: A review on availability of remote sensing data. In: Technological Innovation in ICT for Agriculture and Rural Development (TIAR). IEEE Xplore (2015)
. Yang, J., Li, Y., Chan, J.C.-W., Shen, Q.: Image fusion for spatial enhancement of hyperspectral image via pixel group based non-local sparse representation. Remote Sens. **9**, 53 (2017)
. Eslami, M., Faez, K.: Automatic traffic monitoring from satellite images using artificial immune system. In: SSPR & SPR 2010, LNCS 6218, pp. 170–179. Springer, Berlin Heidelberg (2010)
. Santos, M.D., Shiguemori, E.H., Mota, R.L., Ramos, A.C.B.: Change detection in satellite images using self organizing maps. In: 12th International Conference on Information Technology, pp. 662–667. IEEE (2015)
. Min, W., Zhang, W., Wang, X., Luo, D.: Application of MODIS satellite data in monitoring water quality parameters of Chaohu Lake in China. Environ. Monit. Assess. **148**, 255–264 (2009)
. Metwalli, M.R., Nasr, A.H., Allah, O.S.F., El-Rabaie, S.: Image fusion based on principal component analysis and high-pass filter, pp. 63–70. IEEE (2009)
. Haghighat, M.B.A., Aghagolzadeh, A., Seyedarabi, H.: A non-reference image fusion metric based on mutual information of image features. Comput. Electr. Eng. **37**, 744–756 (2011)

. Dhanachandra, N., Manglem, K., Chanu, Y.J.: Image segmentation using K-means clustering algorithm and subtractive clustering algorithm. Procedia Comput. Sci. **54**, 764–771 (2015)
. Radari, V., Amiri, F., Meleki, S.: Vegetation cover change monitoring applying satellite data during 1972 to 2007. Res. J. Environ. Earth Sci. **2**(3), 118–127 (2010)
. Venkatesh, H., Viswanath, K.: Fusion of satellite images in transform domain. In: International Conference on Communication and Signal Processing, pp. 1884–1888. IEEE (2016)
. Ruairuen, W., Jaroensutasinee, K., Jaroensutanee, M.: Flash flooding area prediction by GOES-9 satellite data. Walailak J. Sci. Technol. **2**(2), 135–148 (2005)
. Cui, W., Jia, Z., Qin, X., Yang, J., Yingjie, H.: Multi-temporal satellite images change detection algorithm based on NSCT. Procedia Eng. **24**, 252–256 (2011)
. Haibo, Y., Zongmin, W., Hongling, Z., Guo, Yu.: Water body extraction methods study based on RS and GIS. Procedia Environ. Sci. **10**, 2619–2624 (2011)
. Liu, Y., Liu, S., Wang, Z.: Multi-focus image fusion with dense SIFT. Inf. Fusion **23**, 139–155 (2015)
. Petrou, Z.I., Tarantino, C., Adamo, M., Blonda, P., Petrou, M.: Estimation of vegetation height through satellite image texture analysis. In: International Archives of the Photogrammetry, Remote Sensing and Spatial Information Sciences (ISPRS Archives), vol. XXXIX-B8, 2012 XXII ISPRS Congress (2012)
. Dong, Z., Wang, Z., Liu, D., Zhang, B., Zhao, P., Tang, X., Jia, M.: SPOT5 multi-spectral (MS) and panchromatic (PAN) image fusion using an improved wavelet method based on local algorithm. Comput. Geosci. **60**, 134–141 (2013)
. Mao, Z., Pan, D., He, X., Chen, J., Tao, B., Chen, P., Hao, Z., Bai, Y., Zhu, Q., Huang, H.: A unified algorithm for the atmospheric correction of satellite remote sensing data over land and ocean. Remote Sens. **8**, 536 (2016)
. Wang, Z., Ziou, D., Armenakis, C., Li, D., Li, Q.: A comparative analysis of image fusion methods. IEEE Trans. Geosci. Remote Sens. **43**(6), 1391–1402 (2005)

A Novel Method for Detecting Bone Contours in Hand Radiographic Images

D. Diana, J. Revathi, K. Uma, A. Ramya and J. Anitha

Abstract Rheumatoid arthritis (RA) is a disorder that causes pain and swelling in the hand and feet joints. From X-ray images, the exact eroded part of bone may not be visible. To overcome this, an Adaptive Snake Algorithm without edge detection is implemented to detect bone contours from hand radiographs. Without prior knowledge and locating initial contours for each bone specifically can be evaluated from both inside and outside regions. Morphological operations are done for binary images to get a better result. Techniques used for Active contour model are Neumann boundary condition and Gradient method (Curvature Center method). Subsequently, another algorithm named Multiple Kernel Fuzzy C-Means (MKFC) is also implemented to detect the eroded part of the bone.

Keywords Rheumatoid arthritis · Adaptive snake algorithm · Gradient method · Active contour

---

D. Diana (✉) · J. Revathi · K. Uma · A. Ramya
Department of Biomedical Instrumentation Engineering, Avinashilingam Institute for Home Science and Higher Education for Women, Coimbatore, India
e-mail: diana.dhevendaran@gmail.com

J. Revathi
e-mail: revathibmieau@gmail.com

K. Uma
e-mail: uma.088@gmail.com

A. Ramya
e-mail: ramyabmieau@gmail.com

J. Anitha
Department of Electronics and Communication Engineering, Karunya Institute for Technology and Sciences, Chennai, India
e-mail: anithaj@karunya.edu

A. Elçi et al. (eds.), *Smart Computing Paradigms: New Progresses and Challenges*, Advances in Intelligent Systems and Computing 766,
https://doi.org/10.1007/978-981-13-9683-0_32

1 Introduction

In common rheumatoid arthritis, wrist and finger joints were affected and cause swelling and pain [1]. Most of the women than men were affected by rheumatoid arthritis. X-rays and blood tests are used to diagnose rheumatoid arthritis. While X-rays alone are not usually sufficient to diagnose rheumatoid arthritis, they can provide doctors with good clues so they do blood tests to confirm the diagnosis. In order to detect the arthritis in early stage, automatic image segmentation algorithms are used in X-ray radiographs.

2 Methodology

The images of both diseased and non-diseased were collected from rheumatology image library [2] and used for further process. In order to detect the arthritis in early stages, image segmentation algorithms such as adaptive snake algorithm without using edge detection and Multiple Kernel Fuzzy C-Means algorithm were applied for processing the input radiographic images [3].

3 Bone Erosion in Rheumatoid Arthritis

A joint is protected by a capsule. The joint capsule is lined with synovium tissue. Synovial fluid is produced by this tissue and it lubricates and nourishes joint tissues. In rheumatoid arthritis, the synovium becomes inflamed and causes redness, swelling. and pain. As the disease progresses, the inflamed synovium damages the bone joints. It weakens the muscles, ligaments, and tendons. Rheumatoid arthritis causes more bone loss that leads to osteoporosis.

4 Segmentation Using Active Contours

Active contour models are being extensively used to solve the problem of image segmentation [4]. Segmentation means subdividing an image into its constituent regions. Segmentation of images is not an easy task because the nature of the underlying objects varies drastically which is crucial to detect. To add to the problem, images themselves may vary qualitatively, e.g., sampling artifacts affect the quality of medical images severely. Segmentation is the first step in image analysis and hence is an important part of image recognition and understanding process. The accuracy of segmentation determines the accuracy of automated analysis procedures used in computer vision.

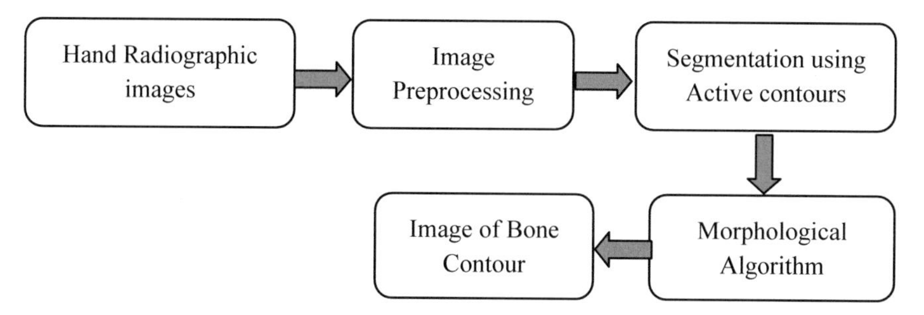

Fig. 1 Flow diagram—Segmentation using active contours

4.1 Active Contours

The active contour is a spline where the energy minimizes. The snake energy is based on its shape and location within the radiographic image. The snake minimizes its energy function and exhibits dynamic behavior. The energy function is minimized by the combination of both weighted internal and external forces [5] (Fig. 1).

The typical Neumann boundary condition is used so that the directional derivative normal to some boundary surface. The contour is the (x, y) plane of an image. It is a parametric curve, v(s) = (x(s), y(s)), where s is related to arc length.

Energy functional (E): the contour is specified as v(s), and the proposed model is defined as a total energy in the continuous spatial domain. The energy terms can be categorized as follows:

Internal Energy: It is a function of the contour v(s). It indicates the tension and smoothness of the curve. It is based on the internal properties of the snake.

External energy: It is the result of the image under inspection and it possesses local minima at the edges or the object boundaries.

Constraint energy: Constraint energy acts on the contour only if some sort of interpretation and response provided by a user. The mathematical model is

$$E_{snake} = E_{\text{int}} + E_{ext} + E_{con}$$

4.2 Algorithm

Step 1:	For a control point (x(i), y(i)) of the contour find the internal forces acting on it.
Step 2:	Assuming that, this control point moves under the influence of only its internal force E_{int}. Find the new position (x_{new}, y_{new}) of the control point. Apparently (x_{new}, y_{new}) will have lower internal energy than (x(i), y(i)).

(continued)

(continued)

Step 3:	Find the external energy $E_{image}\ (v_{new})$ and $E_{image}\ (v(i))$ and update $v(i)$ to v_{new} only if $E_{image}\ (v_{new})$ is lower than $E_{image}\ (v(i))$. If the energy of control point at the estimated new position is higher than current one, then v(i) is not updated because it reaches the local minima.
Step 4:	Repeat the above steps for all control points until the time derivative on RHS vanishes.

4.3 Morphological Algorithm

In the morphological algorithm, flood-fill operation was performed. It changes the background pixels (0 s) to foreground pixels (1 s), and it stops when it reaches the object boundaries. The intensity values of the dark areas in the grayscale images that are surrounded by lighter areas are very similar to the intensity level as in the neighboring pixels of the image. Then all connected components (objects) were removed from a binary image that has fewer than P pixels and produces another binary image BW2. The default connectivity is 8 for two dimensions and 26 for three dimensions.

4.4 Multiple Kernel Fuzzy C-Means Clustering Algorithm

Multiple Kernel Fuzzy C-Means (MKFC) clustering combines different information of image pixels in the kernel space by defining different kernel functions on specific information domains [6]. Kernel fuzzy c-means algorithm maps data with nonlinear relationships to obtain feature spaces accurately. Multiple kernels adjust the kernel weights automatically [7] (Fig. 2).

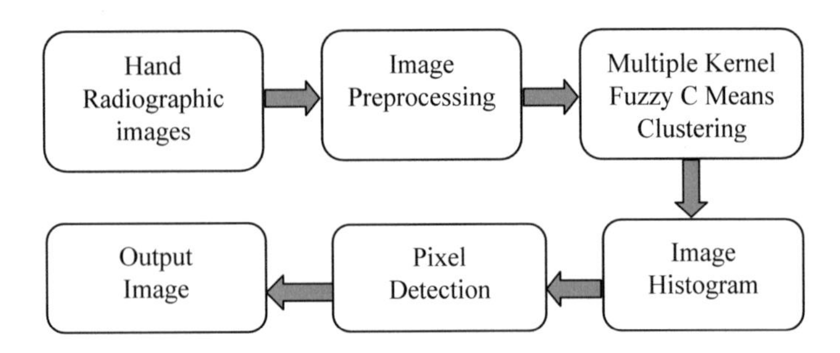

Fig. 2 Multiple kernel fuzzy C-means clustering algorithm

5 Results

Analysis of the proposed algorithm was performed using a data set of digitalized handwrist radiographs. Results from both the algorithms show a much better performance for the hand radiographic images. Adaptive threshold is applied and initial contours have been determined. Adaptive snake algorithm is then applied and 300 iterations have been done to extract the particular area of interest. Segmentation is done using active contour model, which is used to highlight the eroded part. Area of the particular eroded part is determined for all the eroded regions (Fig. 3).

Adaptive snake algorithm without edge detection is used to detect the eroded part of the bone externally and Multiple Kernel Fuzzy C-Means clustering is used to detect the eroded part of the bone internally through pixel intensity having high gradient values (Fig. 4).

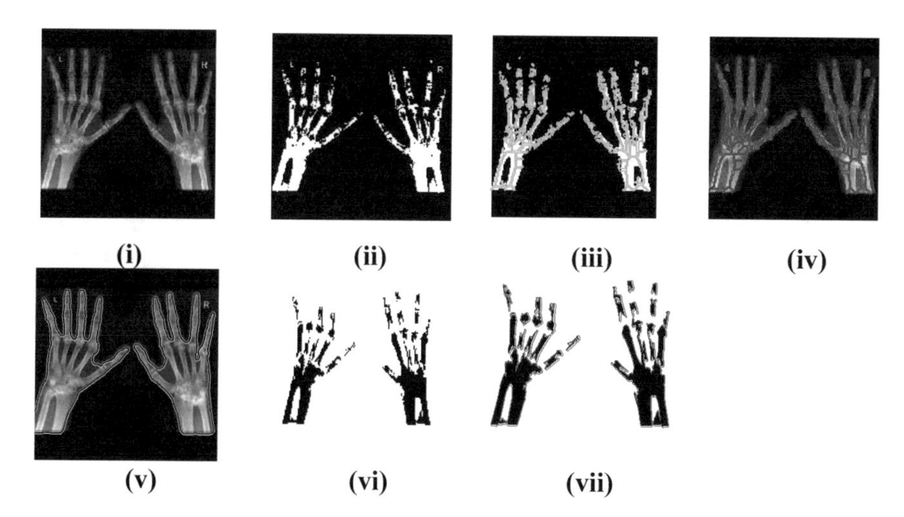

Fig. 3 (i) Input image. (ii) Adaptive thresholding. (iii) Skeleton of the hand. (iv) Skeleton with seed points. (v) Active contour model. (vi) Image complement. (vii) Image of bone contour

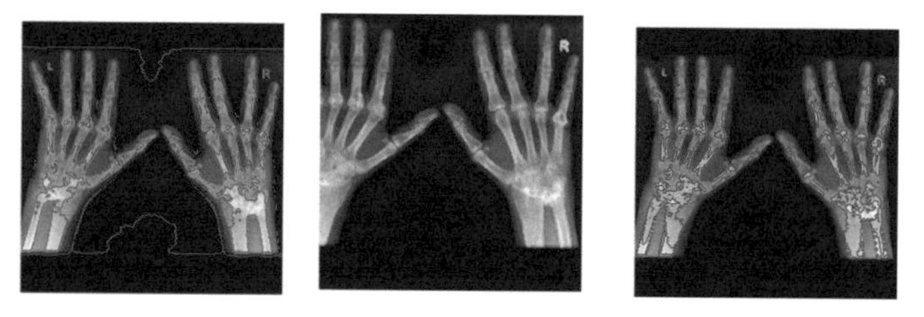

Fig. 4 Pixel detection of radiographic images

6 Discussion

The input image (2048 × 2048) is resized to (256 × 256) to get the better results. The adaptive thresholding (0.5) is done to identify the shadow of the image. The segmentation is done by taking the skeleton of the hand from the image which removes the pixels from the boundaries without any object loss. The branch points and endpoints were noted and merged with the threshold image as seed points to know the joints of the hand. The four parameters are calculated to distinguish the seed points—area, perimeter, centroid, and major axis length. After segmentation, active contour model is carried out by using various iteration without locating the initial contour. The bone contour of the image was highlighted externally by using Adaptive Snake Algorithm without using edge detection. To know the internal eroded part of the bone, Multiple Kernel Fuzzy C-Means Algorithm (MKFC) was performed.

7 Conclusion

The proposed adaptive snake algorithm makes the parameters of each bone are properly adjusted externally and prevents the external forces of the snake from extracting the contour outside the bone boundaries. Multiple Kernel Fuzzy C-Means is a new algorithm which provides a combination of different information of image pixels in the segmentation process. Results from both stages of the algorithm show a much better performance for images. Future work may be carried out by using the same algorithms for hip, spine, knee, and feet.

References

1. Böttcher, J., Pfeil, A.: Diagnosis of Periarticular Osteoporosis in Rheumatoid Arthritis using Digital X-ray Radiogrammetry (2008). https://doi.org/10.1186/ar2352
2. Huang, H.-C., Chuang, Y.-Y., Chen, C.-S.: Multiple kernel fuzzy clustering. IEEE Trans. Fuzzy Syst. (2012)
3. Manju, D., Pavani, K.: Brain tumor detection using multiple kernel fuzzy c-means on level set method. Int. J. Appl. Innov. Eng. Manag. (IJAIEM) **3**(4) (2014)
4. Xiang, Y., Chung, A.C.S., Ye, J.: An active contour model for image segmentation based on elastic interaction. J. Comput. Phys. **219**(1), 455–476 (2006)
5. De Luis-Garcia, R., Martin-Fernandez, M., Arribas, J.I. Alberola-Lopez, C.: A fully automatic algorithm for contour detection of bones in hand radiographs using active contours. In: International Conference on Image Processing, IEEE Proceedings on Image Processing, vol. 2, Issue 4, pp. 421–424 (2003)
6. Chen, L., Lu, M., Chen, C.L.P.: Multiple kernel fuzzy c-means based image segmentation. In: International Conference on Systems Man and Cybernetics (SMC), vol. 15, Issue 6, pp. 4123–4129. IEEE (2010)
7. Nascimento, J.C., Marques, J.S.: Adaptive snakes using the EM algorithm. IEEE Trans. Image Process. **14**(11), 1678–1686 (2005)

8. SyaifulAnam, E.U., Misawa, H., Suetake, N.: Texture analysis and modified level set method for automatic detection of bone boundaries in hand radiographs. Int. J. Adv. Comput. Sci. Appl. (IJACSA) **5**(10), 212–223 (2014)
9. Anam, S., Uchino, E., Suetake, N.: Image boundary detection using the modified level set method and a diffusion filter. In: International Conference in Knowledge Based and Intelligent Information and Engineering Systems, vol. 22, Issue 3, pp. 192 – 200. Elsevier (2013)
10. EI Hadji S.D., Burdin, V.: Bi-planar image segmentation based on variational geometrical active contours with shape priors, vol. 17, Issue 2, pp 165–181. Elsevier Science, B.V (2013)
11. Tripoliti, E.E., Fotiadis, D.I., Argyropoulou, M.: Automated Segmentation and Quantification of Inflammatory Tissue of the Hand in Rheumatoid Arthritis Patients using Magnetic Resonance Imaging Data (2006)
12. Behiels, Gert: Frederik Maes1, Dirk Vandermeulen, Paul Suetens,: Retrospective Correction of the Heel Effect in Hand Radiographs. Med. Image Comput. Comput-Assist Interv. **6**(3), 183–190 (2002)
13. Huang, H.-C., Chuang, Y.-Y., Chen, C.-S.: Multiple Kernel Fuzzy Clustering. IEEE Trans. Fuzzy Syst. **20**(1) (2012)
14. Zhang, Y., Matuszewski, B.J.; Shark, L., Moore, C.J.: Medical image segmentation using new hybrid level-set method, MEDIVIS'08. In: Fifth International Conference on BioMedical Visualization, vol. 12, Issue. 4, pp. 71–76 (2008)
15. Pietka, E.: Computer-assisted bone age assessment based on features automatically extracted from a hand radiograph. Comput. Med. Imaging Graph. **19**(3), 251–259 (1995)
16. de Luis-Garcia, R., Arribas, J.I., Aja-Fernandez, S., Alberola Lopez, C.: A neural architecture for bone age assessment. In: Proceedings of the IASTED International Conference on Signal Processing, Pattern Recognition & Applications, Creta, Greece, pp. 161–166 (1997)
17. Gross, G.W., Boon, J.M., Bishop, D.M.: Pediatric Skeletal Age: Determination with Neural Networks, IEEE Trans. Med. Imaging **195**(3), 689–695 (2002)
18. Kass, M., Witkin, A., Terzopoulos, D.: Snakes: active contour models. Int. J. Comput. Vis. 312–331 (2004)
19. https://www.rheumatology.org/Learning-Center/Educational-Activities/Rheumatology-Image-Library

Author Index

A. Elçi et al. (eds.), *Smart Computing Paradigms: New Progresses and Challenges*,
Advances in Intelligent Systems and Computing 766,
https://doi.org/10.1007/978-981-13-9683-0

Printed in the United States
By Bookmasters